Lecture Notes in Computer Science

Edited by G. Goos, J. Hartmanis, and J. van Lee

T0253659

Springer

Berlin
Heidelberg
New York
Barcelona
Hong Kong
London
Milan
Paris
Tokyo

Nataša Jonoska   Nadrian C. Seeman (Eds.)

DNA Computing

7th International Workshop on DNA-Based Computers, DNA7
Tampa, FL, USA, June 10-13, 2001
Revised Papers

 Springer

Series Editors

Gerhard Goos, Karlsruhe University, Germany
Juris Hartmanis, Cornell University, NY, USA
Jan van Leeuwen, Utrecht University, The Netherlands

Volume Editors

Nataša Jonoska
University of South Florida, Department of Mathematics
4202 E. Fowler A. PHY 114, Tampa FL 33620, USA
E-mail: jonoska@math.usf.edu

Nadrian C. Seeman
New York University, Department of Chemistry
New York, NY 10003, USA
E-mail: ned.seeman@nyu.edu

Cataloging-in-Publication Data applied for

Cataloging-in-Publication Data applied for

Die Deutsche Bibliothek - CIP-Einheitsaufnahme

DNA computing : revised papers / 7th International Workshop on DNA Based
Computers, DNA 7, Tampa, FL, USA, June 10 - 13, 2001. Nataša Jonoska ;
Nadrian C. Seeman (ed.). - Berlin ; Heidelberg ; New York ; Barcelona ;
Hong Kong ; London ; Milan ; Paris ; Tokyo : Springer, 2002
    (Lecture notes in computer science ; Vol. 2340)
    ISBN 3-540-43775-4

CR Subject Classification (1998): F.1, F.2.2, I.2.9, J.3

ISSN 0302-9743
ISBN 3-540-43775-4 Springer-Verlag Berlin Heidelberg New York

This work is subject to copyright. All rights are reserved, whether the whole or part of the material is
concerned, specifically the rights of translation, reprinting, re-use of illustrations, recitation, broadcasting,
reproduction on microfilms or in any other way, and storage in data banks. Duplication of this publication
or parts thereof is permitted only under the provisions of the German Copyright Law of September 9, 1965,
in its current version, and permission for use must always be obtained from Springer-Verlag. Violations are
liable for prosecution under the German Copyright Law.

Springer-Verlag Berlin Heidelberg New York
a member of BertelsmannSpringer Science+Business Media GmbH

http://www.springer.de

© Springer-Verlag Berlin Heidelberg 2002
Printed in Germany

Typesetting: Camera-ready by author, data conversion by Boller Mediendesign
Printed on acid-free paper   SPIN: 10869927   06/3142   5 4 3 2 1 0

Preface

Biomolecular computing is an interdisciplinary field that draws together molecular biology, chemistry, physics, computer science, and mathematics. DNA nanotechnology and molecular biology are key relevant experimental areas, where knowledge increases with each passing year. The annual international meeting on DNA-based computation has been an exciting forum where scientists of different backgrounds who share a common interest in biomolecular computing meet and discuss their latest results. The central goal of this conference is to bring together experimentalists and theoreticians whose insights can calibrate each other's approaches. DNA7, *The Seventh International Meeting on DNA Based Computers*, was held at The University of South Florida in Tampa, FL, USA, June 10–13, 2001. The organizers sought to attract the most significant recent research, with the highest impact on the development of the discipline. The meeting had 93 registered participants from 14 countries around the world. The program committee received 44 abstracts, from which 26 papers were presented at the meeting, and included in this volume. In addition to these papers, the Program Committee chose 9 additional papers from the poster presentations, and their revised versions have been added to this volume.

As is now a tradition, four tutorials were presented on the first day of the meeting. The morning started with general tutorials by Erik Winfree (Caltech) and Junghuei Chen (University of Delaware), designed to bridge between their respective areas of expertise, computer science and molecular biology. More specialized tutorials on encoding DNA sequences and on non-standard DNA motifs and interactions were given in the afternoon by Anne Condon (University of British Columbia) and Nadrian C. Seeman (New York University), respectively. Four plenary lectures were given during the conference by Nicholas Cozzarelli (University of California at Berkeley) on DNA topology, Richard Lipton (Georgia Technology Institute) on the state of DNA-based computation, John SantaLucia (Wayne State University) on DNA hybridization thermodynamics and Ronald Breaker (Yale University) on DNA catalysis. Those presentations are not included in this volume.

The research presented here contains a diverse spectrum of ideas and topics. The papers under *Experimental Tools* deal with issues such as optimization of biomolecular protocols or a computer program for designing DNA sequences that could be found useful in performing experiments. The papers in *Theoretical Tools* study theoretical properties of DNA sequences and structures that could be used in designing models and subsequently, experiments. Several papers deal with *Probabilistic Theoretical Models* which try to capture the inexact nature of the biomolecular protocols. As the experience of many has shown, sequence design and computer simulations can be very valuable before preparing an actual experiment and several researchers addressed these issues in *Computer Simulation and Sequence Design*. New algorithms for solving difficult problems such as the knapsack problem and SAT are introduced in *Algorithms*. Several researchers, in fact, reported on successful experimental solutions of instances of computational problems. Their results are included in *Experimental Solutions*. The papers in *Nano-tech Devices* report on the experimental design of DNA nano-mechanical devices. The section on *Biomimetic Tools* contains research on computational tools that primarily use processes found naturally in the cells of living organisms. Several papers deal with the theory of splicing systems and the formal language models of membrane computing. These papers are included in *Splicing and Membrane Systems*.

The editors would like to acknowledge the help of the conference's Program Committee in reviewing the submitted abstracts. In addition to the editors, the Program Committee consisted of Junghuei Chen, Anne Condon, Masami Hagiya, Tom Head, Lila Kari, George Paun, John Reif, Grzegorz Rozenberg, Erik Winfree, David Wood, and Bernard Yurke. The editors thank Denise L. Marks for helping us with her skillful typesetting abilities. The Organizing Committee (Anne Condon, Grzegorz Rozenberg, and the editors) is grateful for the generous support and sponsorship of the conference by the Center for Integrated Space Microsystems within the Jet Propulsion Laboratory, NASA, and the following branches of The University of South Florida: The College of Arts and Sciences, the Institute for Biomolecular Science, the Department of Mathematics, the Department of Biology, the Department of Chemistry; and the USF Research Foundation.

The meeting was held in cooperation with the ACM Special Interest Group on Algorithms and Computation Theory (ACM SIGACT) and the European Association for Theoretical Computer Science (EATCS).

We note with sadness the passing of Michael Conrad, who participated in several of the earlier conferences. His contributions will be missed by all.

Finally, the editors would like to thank all of the participants in the DNA7 conference for making it a scintillating and fruitful experience. This is a discipline that has not yet found its 'killer ap,' but the excitement that is generated when this group assembles is virtually palpable at the conference. The interactions, collaborations, and advances that result from each of the meetings on DNA Based Computers are the key products of the meeting. We hope that this volume has captured the spirit and exhilaration that we experienced in Tampa.

April 2002   Nataša Jonoska
                                                                                              Nadrian Seeman

Organization

Program Committee
    Junghuei Chen, University of Delaware, USA
    Anne Condon, University of British Columbia, Canada
    Masami Hagiya, University of Tokyo, Japan
    Tom Head, Binghamton University, USA
    Nataša Jonoska, University of South Florida, USA
    Lila Kari, University of western Ontario, Canada
    Gheorghe Paŭn, Romanian Academy, Romania
    John H. Reif, Duke University, USA
    Grzegorz Rozenberg, Leiden University, The Netherlands
    Nadrian C. Seeman, New York University, USA (Program Chair)
    Erik Winfree, California Institute of technology, USA
    David H. Wood, University of Delaware, USA
    Bernie Yurke, Bell Laboratories, Lucent Technologies, USA

Organizing Committee
    Anne Condon, University of British Columbia, Canada
    Nataša Jonoska, University of South Florida, USA
    (Organizing Committee Chair)
    Grzegorz Rozenberg, Leiden University, The Netherlands
    Nadrian C. Seeman, New York University, USA

Table of Contents

Experimental Tools

An Object Oriented Simulation of Real Occurring Molecular Biological
Processes for DNA Computing and Its Experimental Verification 1
    T. Hinze, U. Hatnik, M. Sturm

Towards Optimization of PCR Protocol in DNA Computing 14
    S. Kashiwamura, M. Nakatsugawa, M. Yamamoto, T. Shiba, A. Ohuchi

DNASequenceGenerator: A Program for the Construction of
DNA Sequences 23
    U. Feldkamp, S. Saghafi, W. Banzhaf, H. Rauhe

DNA Computing in Microreactors 33
    D. van Noort, F.-U. Gast, J.S. McCaskill

Cascadable Hybridisation Transfer of Specific DNA between Microreactor
Selection Modules 46
    R. Penchovsky, J.S. McCaskill

Theoretical Tools

Coding Properties of DNA Languages 57
    S. Hussini, L. Kari, S. Konstantinidis

Boundary Components of Thickened Graphs 70
    N. Jonoska, M. Saito

Probabilistic Computational Models

Population Computation and Majority Inference in Test Tube 82
    Y. Sakakibara

DNA Starts to Learn Poker 92
    D.H. Wood, H. Bi, S.O. Kimbrough, D.-J. Wu, J. Chen

PNA-mediated Whiplash PCR 104
    J.A. Rose, R.J. Deaton, M. Hagiya, A. Suyama

Computer Simulation and Sequence Design

Biomolecular Computation in Virtual Test Tubes 117
    M.H. Garzon, C. Oehmen

Developing Support System for Sequence Design in DNA Computing 129
    F. Tanaka, M. Nakatsugawa, M. Yamamoto, T. Shiba, A. Ohuchi

The Fidelity of the Tag-Antitag System 138
    J.A. Rose, R.J. Deaton, M. Hagiya, A. Suyama

PUNCH: An Evolutionary Algorithm for Optimizing Bit Set Selection 150
    A.J. Ruben, S.J. Freeland, L.F. Landweber

Algorithms

Solving Knapsack Problems in a Sticker Based Model 161
    M.J. Pérez-Jiménez, F. Sancho-Caparrini

A Clause String DNA Algorithm for SAT 172
    V. Manca, C. Zandron

A Proposal of DNA Computing on Beads with Application to SAT
Problems 182
    T. Yoichi, H. Akihiro

Experimental Solutions

Aqueous Solutions of Algorithmic Problems: Emphasizing Knights on a
3×3 191
    T. Head, X. Chen, M.J. Nichols, M. Yamamura, S. Gal

Solutions of Shortest Path Problems by Concentration Control 203
    M. Yamamoto, N. Matsuura, T. Shiba, Y. Kawazoe, A. Ohuchi

Another Realization of Aqueous Computing with Peptide Nucleic Acid . . . 213
    M. Yamamura, Y. Hiroto, T. Matoba

Experimental Conformation of the Basic Principles of Length-only
Discrimination 223
    Y. Khodor, J. Khodor, T.F. Knight, Jr.

Experimental Construction of Very Large Scale DNA Databases with
Associative Search Capability 231
    J.H. Reif, T.H. LaBean, M. Pirrung, V.S. Rana, B. Guo, C. Kingsford,
    G.S. Wickham

Nano-tech Devices

Operation of a Purified DNA Nanoactuator 248
    F.C. Simmel, B. Yurke

DNA Scissors 258
    J.C. Mitchell, B. Yurke

Biomimetic Tools

A Realization of Information Gate by Using *Enterococcus faecalis*
Pheromone System 269
    K. Wakabayashi, M. Yamamura

Patterns of Micronuclear Genes in Ciliates 279
    A. Ehrenfeucht, T. Harju, I. Petre, G. Rozenberg

Peptide Computing - Universality and Complexity 290
    M.S. Balan, K. Krithivasan, Y. Sivasubramanyam

Programmed Mutagenesis Is a Universal Model of Computation 300
    J. Khodor, D.K. Gifford

New Computing Models

Horn Clause Computation by Self-assembly of DNA Molecules 308
    H. Uejima, M. Hagiya, S. Kobayashi

DNA-based Parallel Computation of Simple Arithmetic 321
    H. Hug, R. Schuler

Splicing Systems and Membranes

On P Systems with Global Rules 329
    A. Păun

Computing with Membranes: Variants with an Enhanced Membrane
Handling 340
    M. Margenstern, C. Martín-Vide, G. Păun

Towards an Electronic Implementation of Membrane Computing:
A Formal Description of Non-deterministic Evolution in Transition
P Systems 350
    A.V. Baranda, F. Arroyo, J. Castellanos, R. Gonzalo

Insertion-Deletion P Systems 360
    S.N. Krishna, R. Rama

A Universal Time-Varying Distributed H System of Degree 1 371
    M. Margenstern, Y. Rogozhin

A Note on Graph Splicing Languages 381
    N.G. David, K.G. Subramanian, D.G. Thomas

Author Index

Author Index 391

An Object Oriented Simulation of Real Occurring Molecular Biological Processes for DNA Computing and Its Experimental Verification

Thomas Hinze[1], Uwe Hatnik[2], and Monika Sturm[1]

[1] Institute of Theoretical Computer Science, Dresden University of Technology,
Mommsenstr. 13, i Dresden, D-01062, Germany
{hinze, sturm}@tcs.inf.tu-dresden.de
http://www.tcs.inf.tu-dresden.de/dnacomp
[2] Institute of Computer Engineering, Dresden University of Technology,
Mommsenstr. 13, i Dresden, D-01062, Germany
hatnik@ite.inf.tu-dresden.de

Abstract. We present a simulation tool for frequently used DNA operations on the molecular level including side effects based on a probabilistic approach. The specification of the considered operations is directly adapted from detailed observations of molecular biological processes in laboratory studies. Bridging the gap between formal models of DNA computing, we use process description methods from biochemistry and show the closeness of the simulation to the reality.

1 Introduction

It is well-known that DNA operations can cause side effects in a way that the results of algorithms do not fit to the expectation. Any molecular biological operation used for DNA computing seems to be closely connected with certain unwanted effects on the molecular level. Typical side effects are for instance unwanted additional DNA strands, loss of wanted DNA strands, artifacts, mutations, malformed DNA structures or sequences, impurities, incomplete or unspecific reactions, and unbalanced DNA concentrations. Unfortunately, side effects can sum up in sequences of DNA operations leading to unprecise, unreproducible or even unusable final results [6]. Coping with side effects is to be seen as the main challenge in the research field of experimental DNA computing. We have analyzed processes used in DNA computing at the molecular level in laboratory studies with the aim to specify these processes as detailed as possible. The analysis led to a classification and to a statistical parametric logging of side effects. Based on this knowledge, we have developed a simulation tool of real occurring molecular biological processes considering side effects. The comparison of simulation results with real observations in the laboratory shows a high degree of accordance. Our main objective is to construct error reduced and side effect compensating algorithms. Furthermore, the gap between formal models of DNA

N. Jonoska and N.C. Seeman (Eds.): DNA7, LNCS 2340, pp. 1–13, 2002.
© Springer-Verlag Berlin Heidelberg 2002

computing and implementations in the laboratory should be bridged. A clue to handle side effects in DNA computing can consist in the idea to include them into the definition of DNA operations as far as possible. DNA computing as hardware architecture particularly convinces by its practicability of laboratory implementations based on a formal model of computation.

The simulation tool and continued laboratory studies extend our results presented at DNA6 [4]. Our work focuses a reliable implementation of an optimized distributed splicing system TT6 in the laboratory [8]. Using the simulation tool, prognoses about resulting DNA strands and influences of side effects to subsequent DNA operations can be obtained. The number of strand duplicates reflecting DNA concentrations is considered as an important factor for a detailed description of the DNA computing operations on the molecular level in the simulation. This property allows to evaluate the quantitative balance of DNA concentrations in a test tube. Here, we show the abilities of the simulation using the operations synthesis, annealing, melting, union, ligation, digestion, labeling, polymerisation, PCR, affinity purification, and gel electrophoresis by means of selected examples with comparison to laboratory results.

2   Modelling Molecular Biological Processes

The knowledge about underlying molecular biological processes grows up more and more rapidly. In the meantime, the principles of biochemical reactions are understood very well. Precise descriptions can be found in recent handbooks of genetic techniques like [7]. This pool of knowledge mostly aims at applications in medicine, agriculture, and genetic engineering. Our intention is to use this knowledge and to apply it for approaches in DNA computing.

Biochemical reactions on DNA are generally caused by collisions of the reactants with enough energy to transform covalent or hydrogen bonds. This energy is usually supplied by heating or by addition of instable molecules with a large energy potential. Thus the vis viva of the molecules inside the test tubes increases and they become more moveable. One test tube can contain up to 10^{20} molecules including water dipoles. Which reactive molecules of them will interact indeed? The answer to this question requires to abstract from a macroscopic view. A microscopic approach has to estimate the probability of an inter- or intra-molecular reaction for all combinations of molecules inside the test tube. This can be done by generating a probability matrix whose elements identify all possible combinations how molecules can hit to react together. The probabilities for a reaction between the molecules forming a combination depend on many parameters e.g. chemical properties of the molecules, their closeness and orientation to each other and the neighbourship of other reactive molecules. After creating the matrix of molecular reaction probabilities, a certain combination with acceptable probability > 0 is selected randomly according to the given probability distribution. The molecular reaction is performed and produces a modified contents of the test tube. Using this contents, the subsequent matrix of molecular reaction probabilities is generated and so on. The whole reaction can be understood as a consecutive iterated process of matrix generation, selection of a molecular reaction and its performance. The process stops if all further probabilities for molecular reactions are very low or an equilibrium of the test tube contents occurs. This strategy to model molecular biological processes implies side effects and a nondeterministic behaviour in a natural way. The simulation tool adapts this basic idea to model processes of DNA computing on the molecular level controlled by suitable parameters. A simple annealing example should illustrate the idea how to simulate biochemical reactions closed to the laboratory. Annealing (hybridization) is a process that pairs antiparallel and complementary DNA single strands to DNA double strands by forming hydrogen bonds between opposite orientated complementary bases. Let assume for simplicity that a (very small) test tube contains three different DNA sequences in solution: 10 copies of the DNA single strand 5'-AAGCTCCGATGGAGCT-3', 6 copies of 5'-TGAAGCTCCATCGGA-3', and 7 copies of 5'-GAGCTTATA-3'. Further let assume that these strands are spatially distributed in equipartition and that one molecular reaction affects max. $k = 2$ DNA molecules at once. Figure 1 shows the first iteration of process simulation.

The matrix derived from the test tube contents lists the probabilities for inter- resp. intramolecular collisions that can result in molecular reactions for all combinations of molecules. Subsequently, one combination is selected randomly with respect to the probability distribution. The example uses the collision marked by a grey background. For this selected combination, all possible molecular hybridization products have to be determined.

Two DNA strands can stable anneal to each other if at least approximately 50% of the bases of one participating strand form hydrogen bonds with their complementary counterparts of the other one. A lower bonding rate mostly produces not survivable DNA double strands that melt again. The minimum bonding rate describes the process parameter of annealing. Based on the bonding rate parameter, all possible stable molecular hybridization products from the selected combination are generated. One of these products is selected randomly as performed molecular reaction. The test tube contents is modified accordingly completing one iteration of the process cycle. The modified test tube contents serves as input for the next iteration and so on until no new products can appear.

The annealing example should point out the principle how to model molecular biological processes. Other reactions resp. processes can be considered in a similar way. Our studies include the DNA operations synthesis, annealing, melting, union, ligation, digestion, labeling, polymerisation, PCR, affinity purification, and gel electrophoresis. They affect as follows:

operation	effect
synthesis	generation of DNA single strands (oligonucleotides) up to maximum approximately 100 nucleotides; there are no limitations to the sequence. Most methods use the principle of a growing chain: Fixed on a surface, the DNA single strands are constructed by adding one nucleotide after the other using a special coupling chemistry. Finally, the DNA single strands are removed from the surface and purified.

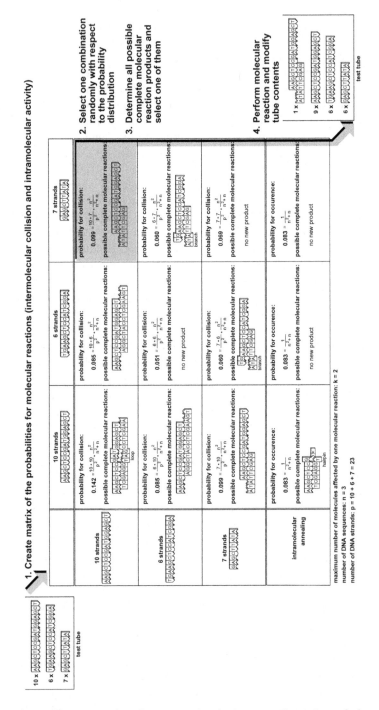

Fig. 1. Annealing example of process simulation, one iteration of the process cycle.

operation	effect
annealing	pairing of minimum two antiparallel and complementary DNA single strands or single stranded overhangs to DNA double strands by forming thermic instable hydrogen bonds; the process is performed by heating above the melting temperature and subsequently slowly cooling down to room temperature. Annealing product molecules can survive if at least 50% of the bases of one participated strand bind to their complementary counterpart.
melting	breaking hydrogen bonds by heating above the melting temperature or by using alkaline environments
union	merging the contents of several test tubes into one common test tube without changes of chemical bonds
ligation	concatenation of compatible antiparallel complementary sticky or blunt DNA double strand ends with 5' phosphorylation; enzym DNA ligase catalyzes the formation of covalent phosphodiester bonds between juxtaposed 5' phosphate and 3' hydroxyl termini of double stranded DNA.
digestion	cleavage of DNA double strands on occurences of specific recognition sites defined by the enzym; all arising strand ends are 5' phosphorylated. Enzym type II restriction endonuclease catalyzes the break of covalent phosphodiester bonds at the cutting position.
labeling	set or removal of molecules or chemical groups called labels at DNA strand ends; enzym alkaline phosphatase catalyzes the removal of 5' phosphates (5' dephosphorylation). Enzym Polynucleotide Kinase catalyzes the transfer and exchange of phosphate to 5' hydroxyl termini (5' phosphorylation). Beyond phosphate, other labels like 5' biotin, fluorescent or radioactive labels can be used in a similar way.
polymerisation	conversion of DNA double strand sticky ends into blunt ends; enzym like vent DNA polymerase (New England Biolabs) catalyzes the completion of recessed 3' ends and the removal of protruding 3' ends.
gel electrophoresis	physic technique for separation of DNA strands by length using the negative electric charge of DNA; DNA is able to move through the pores of a gel, if a DC voltage (usually $\approx 80V$) is applied and causes an electrolysis. The motion speed of the DNA strands depends on their molecular weight that means on their length. After switching off the DC voltage, the DNA is separated by length inside the gel. Denaturing gels (like polyacrylamide) with small pores process DNA single strands and allow to distinguish length differences of 1 base. Non-denaturing gels (like agarose) with bigger pores process DNA double strands with precision of measurement $\approx \pm 10\%$ of the strand length.

operation	effect
polymerase chain reaction (PCR)	cyclic process composed by iterated application of melting, annealing, and polymerisation used for exponential amplification of double stranded DNA segments defined by short ($\approx$ 20 bases long), both-way limiting DNA sequences; these sequences denoted as DNA single strands are called primers. Each cycle starts with melting of the double stranded DNA template into single strands. Subsequently the primers are annealed to the single strands and completed to double strands by polymerisation. Each cycle doubles the number of strand copies. PCR can produce approximately up to 2^{40} strand copies using 40 cycles. Higher numbers of cycles stop the exponential amplification leading to a saturation.
affinity purification	separation technique that allows to isolate 5' biotinylated DNA strands from others; biotin binds very easily to a streptavidin surface fixing according labelled DNA strands. Unfixed DNA strands are washed out and transferred to another tube.

Molecular biological processes annealing and ligation induce interactions between different DNA strands. They are able to produce a variety of strand combinations. The potential and power of DNA computing to accelerate computations rapidly is based on annealing and ligation. Other DNA operations listed above affect the DNA strands inside the test tube independently and autonomously. In this case, interactions are limited to DNA with other reactants or influences from the environment. Union, electrophoresis, and sequencing require modelling as physic processes without reactive collisions between molecules.

3   A Probabilistic Approach to Model DNA Operations with Side Effects

The effect of DNA operations on the molecular level depends on random (non-deterministic) interactions (events) with certain probability. The variety of possible events is specified by biochemical rules and experimental experiences. Only a part of them – but not all – forms the description of formal models of DNA computing. Remaining unconsidered events are subsumed by the term "side effect". Formal models of DNA computing include many significant properties but others are ignored (abstraction). The most commonly used assumptions for abstraction are:

- Linear DNA single or double strands are used as data carrier.
- Information is encoded by DNA sequence (words of formal languages).
- unrestricted approach; arbitrary (also infinite) number of strand copies allowed
- Unique result DNA strands can be detected absolutely reliable.
- All DNA operations are performed completely.
- All DNA operations are absolute reproducible.

Differences from these abstractions are considered as side effects. They can be classified into certain groups with specific common properties. The properties are chosen in a way that the side effect can either be defined by statistical parameters with respect to defaults from the reactants (e.g. mutation error rate of DNA polymerase) or the side effect directly results from the process description. Figure 2 shows a proposal for a classification extending the idea from [1] to the set of frequently used DNA basic operations.

	operations performed with state of the art laboratory techniques	synthesis	annealing	melting	union	ligation	digestion	labeling	polymerisation	PCR	affinity purification	gel electrophoresis
mutations (differences in DNA sequence)	point mutation (% mutation rate)	■							■	■		
	deletion (% deletion rate, max. length of deletion)	■										
	insertion				■							
artifacts (diff. from lin. DNA structure)	loss of linear DNA strands by forming hairpins, bulges, loops, junctions, and compositions of them (% loss rate of tube contents)		■			■			■	■		
failures in reaction procedure (differences from perfect specification of reaction)	incomplete reaction (% unprocessed strands)		■	■		■	■	■	■	■		■
	unspecificity (% error rate, maximum difference)						■				■	■
	supercoils											■
	strand instabilities caused by temperature or pH		■	■		■	■	■	■			■
	impurities by rests of reagences	■				■	■	■	■	■		■
	undetectable low DNA concentration (min. # copies)	■	■	■		■	■	■	■	■		■
	loss of DNA strands (% loss rate of tube contents)				■						■	■

: considered in simulation tool   in brackets: statistical parameters   ■ : significant side effect caused by the operation

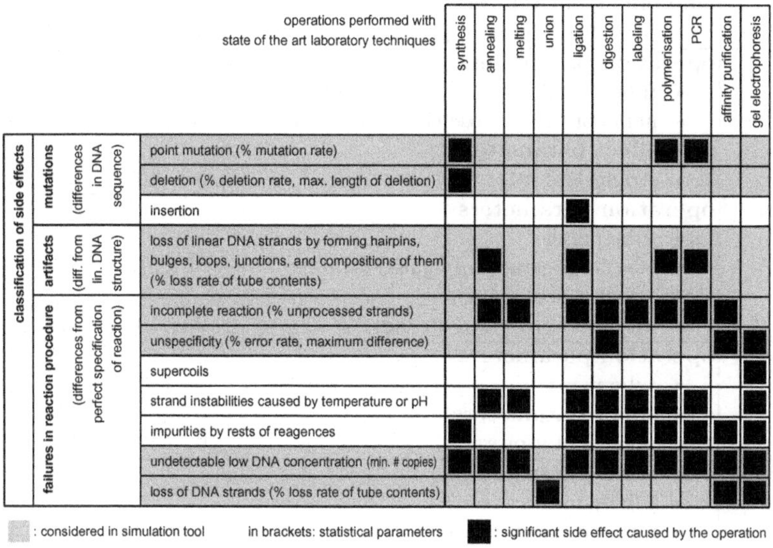

Fig. 2. Significant side effects of frequently used DNA operations

The following table lists the operation parameters and side effect parameters of the considered DNA basic operations. The default values are adapted from laboratory studies. The abbreviation L stands for strand length.

operation	parameter	range	default
synthesis	**operation parameters**		
	• tube name		
	• nucleotide sequence (5'-3')		
	• number of strand copies	$1 \ldots 10^6$	
	side effect parameters		
	• point mutation rate	$0 \ldots 100\%$	5%
	• deletion rate	$0 \ldots 100\%$	1%
	• maximum deletion length	$0 \ldots 100\%$ of L	5%

operation	parameter	range	default
annealing	**operation parameters**		
	• tube name		
	• minimum bonding rate for stable duplexes	$0 \ldots 100\%$	50%
	• maximum length of annealed strands	$1 \ldots 10^6$	
	side effect parameters		
	• base pairing mismatch rate	$0 \ldots 100\%$	$600/L$
	• rate of unprocessed strands	$0 \ldots 100\%$	5%
melting	**operation parameters**		
	• tube name		
	side effect parameters		
	• rate of surviving duplexes	$0 \ldots 100\%$	0.1%
union	**operation parameters**		
	• tube name		
	• name of tube whose contents is added		
	side effect parameters		
	• strand loss rate	$0 \ldots 100\%$	0.5%
ligation	**operation parameters**		
	• tube name		
	• maximum length of ligated strands	$1 \ldots 10^6$	
	side effect parameters		
	• rate of unprocessed strands	$0 \ldots 100\%$	5%
polymeri-sation	**operation parameters**		
	• tube name		
	side effect parameters		
	• point mutation rate	$0 \ldots 100\%$	0.1%
digestion	**operation parameters**		
	• tube name		
	• recognition sequence		
	• restriction site		
	side effect parameters		
	• rate of not executed molecular cuts	$0 \ldots 100\%$	5%
	• rate of star activity (unspecificity)	$0 \ldots 100\%$	5%
	• recognition sequence with wildcard base pairs specifying star activity		
labeling	**operation parameters**		
	• tube name		
	• kind of label (biotin or phosphate)		
	• kind of strand end (3' or 5')		
	• action (set or removal of label)		
	side effect parameters		
	• rate of unprocessed strands	$0 \ldots 100\%$	5%
affinity purifi-cation	**operation parameters**		
	• tube name		
	• kind of extracted strands (with or without biotin label)		
	side effect parameters		
	• rate of false positives (unspecificity)	$0 \ldots 100\%$	8%
	• rate of false negatives (unspecificity)	$0 \ldots 100\%$	8%

operation	parameter	range	default
gel electro- phoresis	**operation parameters** • tube name • minimum number of strand copies with same length, necessary for detection • selection of available length (bands) **side effect parameters** • strand loss rate • rate of strands with forged length • maximum length derivation (forgery)	$1 \ldots 10^6$ $0 \ldots 100\%$ $0 \ldots 100\%$ $0 \ldots 100\%$ of L	 1% 1% 10%

4   Basic Ideas of the Simulation Tool

A simulation tool based on the molecular biological processes from section 2 including optional side effects from section 3 contributes to the experimental setup in the laboratory and is able to explain unexpected results. Our approach extends the idea from [2]. The main features of the simulation tool focus on:

– Specification of DNA operations is set on the level of single nucleotides and strand end labelings using the principle of random probability-controlled consecutive interactions between DNA strands and reactants.
– Number of strand copies is considered to distinguish concentrations of different DNA strands and their influence to the behaviour in the operational process.
– Each DNA operation is processed inside a test tube that collects a set of DNA strands. The simulation tool is able to manage several test tubes.
– Each DNA operation is characterized by a set of specific parameters and side effect parameters that can be stored and load together with all test tube contents as a project.
– Arbitrary sequences of DNA operations including the propagation of side effects can be visualized and logged.

Since a test tube can be considered as a system containing groups of DNA strands and reactants as (autonomous) subsystems, an object-oriented approach for simulation is preferred: Object-oriented simulation means that a system is split into subsystems which are simulated autonomously [5]. A subsystem in this context is named "object" and may contain other objects forming an object hierarchy. An object embeds its own simulation algorithm that can represent both, a small code fragment and an extensive simulator, see figure 3. All implementation details are encapsulated by the object, only an interface allows data exchange and simulation control. The advantage of this approach lies in its flexibility with respect to object combination and exchange. Furthermore, the simulation algorithm can be optimally adapted to the models [9]. The object-oriented simulation approach is suitable for a wide range of applications, e.g. [3].

The implementation uses Java to ensure a wide interoperability to different platforms because of its object-oriented paradigm. The simulation tool requires at least Java Development Kit 2.0.

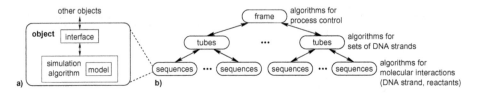

Fig. 3. basic object structure **a)** and hierarchical composition of objects **b)**

5   Comparison of Simulation and Reality

Two examples (PCR and Cut) were selected to confirm simulation results by laboratory experiments. Both examples compare simulation and laboratory experiment and support an explanation of side effects to be seen in the agarose gel photos.

PCR example: A PCR example should illustrate the consequence of deletions and point mutations in synthesized DNA single strands to subsequent iterated PCR cycles. For laboratory implementation, the PCR template was constructed by oligonucleotide synthesis of two complementary DNA single strands named template1 (5'-AGGCACTGAGGTGATTGGCAGAAGGCCTAAAGCTCACTTAAGGGCTACGA-3') and template2 (5'-TCGTAGCCCTTAAGTGAGCTTTAGGCCTTCTGCCAATCACCTCAGTGCCT-3'), both 50 bases (Perkin-Elmer) as well as the primers named primer1 (5'-AGGCACTGAGGTGATTGGC-3') and primer2 (5'-TCGTAGCCCTTAAGTGAGC-3'), both 19 bases (Amersham Pharmacia Biotech). The PCR according to standard protocols was done in four samples using 30 cycles including one sample without Taq-Polymerase as negative control. The PCR product was visualized by agarose gel electrophoresis, see figure 4, box below. Lanes 2 until 4 show the amplified band and below a weaker smear of shorter DNA fragments that has to be comprehended by simulation with side effects.

The simulation uses 1000 copies of template1 considering a point mutation rate of 0.06% and a deletion rate of 0.06%, maximum deletion length 12 bases. These side effect parameters were adapted from properties of oligonucleotide synthesis. Template2 was generated in a same way. 8000 copies from each primer (point mutation rate 0.06%, no deletions) were used. All subsequent DNA operations were assumed to be perfect inside the example. Figure 4 shows a screenshot of synthesized strands after union into one common test tube (box above) and of the simulation result after three PCR cycles affirming unwanted shorter bands (box below). The test tube output lists the DNA strands sorted descendingly by number of copies. Screenshots are truncated to the top of the lists.

The example demonstrates the consequences of point mutations and deletions in synthesized DNA strands to a subsequent PCR decreasing the amount of error free template after three cycles from 8000 expected copies to 5965.

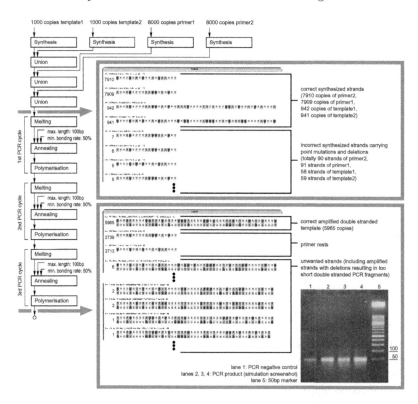

Fig. 4. PCR example: side effect considering simulation result vs. laboratory experiment

Cut example: A cut example should illustrate incomplete and unspecific reactions by digestion of annealed synthesized oligonucleotides. For laboratory implementation, two complementary DNA single strands named oligo1(5'-AGGCACTGAGGTGATTGGCAAGTCCAATCGCGAAAGTCCAAGCTCACTTAAGGGCT-ACGA-3') and oligo2 (5'-TCGTAGCCCTTAAGTGAGCTTGGACTTTCGCGATTGGACTTGCC-AATCACCTCAGTGCCT-3'), both 60 bases (Perkin-Elmer), were synthesized. Aliquots of each were merged and annealed using standard protocols. The subsequent digestion using NruI, a blunt cutter, should cleave all double stranded fragments in the middle producing only 30*bp* strands. The agarose gel photo shows the result of an incomplete reaction, and base pair mismatching supporting unspecific cleavages, see figure 5.

6   Conclusions

The simulation tool represents a restricted and multiset-based model for DNA computing whose operations were specified and adapted directly from the anal-

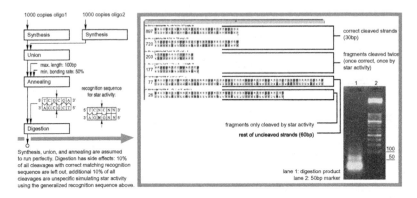

Fig. 5. Cut example: side effect considering simulation result vs. laboratory experiment

ysis of molecular biological processes in the laboratory. In contrast to the most models for DNA computing, the simulation tool also considers the influence of significant side effects. The intensity of side effects can be controlled by suitable statistical parameters in a range from no influence to absolute dominance. The consistent parameterization of DNA operations as well as side effects assigns to the simulation tool a high degree of flexibility and ergonomics. The object-oriented simulation approach supports the modelling of interactions between DNA strands and reactants as autonomous subsystems that are combined to test tubes with frame controlled behaviour. The implementation in Java guarantees interoperability to different platforms. Recently, the simulation tool features by the DNA operations synthesis, annealing, melting, union, ligation, digestion labeling, polymerisation, PCR, affinity purification, and gel electrophoresis. Further studies focus on the extension to additional effects concerning nonlinear DNA structures.

Acknowledgements. This work is a result of an interdisciplinary collaboration between Institute of Theoretical Computer Science, Institute of Computer Engineering, and Department of Surgical Research, Dresden University of Technology, Dresden, Germany.

References

1. K. Chen, E. Winfree. Error correction in DNA computing: Misclassification and strand loss. In E. Winfree, D.K. Gifford, editors. Proceedings 5th DIMACS Workshop on DNA Based Computers, Cambridge, MA, USA, DIMACS Vol. 54, pp. 49–64, 2000
2. M. Garzon, R.J. Deaton, J.A. Rose, D.R. Franceschetti. Soft molecular computing. In E. Winfree, D.K. Gifford, editors. Proceedings 5th DIMACS Workshop on DNA Based Computers, Cambridge, MA, USA, DIMACS Vol. 54, pp. 91–100, 2000

3. U. Hatnik, J. Haufe, P. Schwarz. Object Oriented System Simulation of Large Heterogeneous Communication Systems. Workshop on System Design Automation SDA2000, Rathen, pp. 178–184, March 13–14, 2000

4. T. Hinze, M. Sturm. A universal functional approach to DNA computing and its experimental practicability. PreProceedings 6th International Meeting on DNA Based Computers, University of Leiden, Leiden, The Netherlands, p. 257, 2000

5. J.A. Joines, S.D. Roberts. Fundamentals of object-oriented simulation. In D.J. Medeiros, E.F. Watson, J.S. Carson, M.S. Manivannan, ed., Proceedings of 1998 conference on Winter Simulation, Washington, USA, pp. 141–150, December 13–16, 1998

6. P.D. Kaplan, G. Cecchi, A. Libchaber. Molecular computation: Adleman's experiment repeated. Technical report, NEC Research Institute, 1995

7. F. Lottspeich, H. Zorbas. Bioanalytik. Spektrum Akad. Verlag Heidelbg., Berlin, 1998

8. M. Sturm, T. Hinze. Distributed Splicing of $\mathcal{RE}$ with 6 Test Tubes. Romanian Journal of Information Science and Technology, Publishing House of the Romanian Academy **4(1–2)**:211–234, 2001

9. G. Zobrist, J.V. Leonard. Object-Oriented Simulation – Reusability, Adaptability, Maintainability. IEEE Press, 1997

Towards Optimization of PCR Protocol in DNA Computing

Satoshi Kashiwamura[1], Masashi Nakatsugawa[1], Masahito Yamamoto[1],
Toshikazu Shiba[2], and Azuma Ohuchi[1]

[1] Division of Systems and Information Eng.,
Graduate School of Eng., Hokkaido University
[2] Division of Molecular Chemistry Eng.,
Graduate School of Eng., Hokkaido University
Nishi 8, Kita 13, Kita-ku, Sapporo, Hokkaido, 060-8628, JAPAN
Phone: +81-11-716-2111 (ext. 6498), Fax: +81-11-706-7834
{kashiwa,masashi,masahito,shiba,ohuchi}@dna-comp.org
http://ses3.complex.eng.hokudai.ac.jp/

Abstract. Recently, in the research field of DNA Computing, the improvement of reliability of computation has been expected. Since DNA Computing consists of some chemical reactions, it seems to be clearly important to optimize each reaction protocol in order to improve the reliability of computing. Moreover, the purpose of the reaction in DNA Computing crosses variably. We consider that the optimization of reaction protocol according to each purpose will be necessary. We try to derive positive impact factors on the reaction result from actual experiments by using DOE. From the results of experiments, we show the importance and necessity of optimizing reaction protocol according to each desired reaction.

1 Introduction

Recently, in the research field of DNA Computing, the improvement of reliability of computation (chemical reaction) has been expected. Towards high reliability of DNA Computing, much interests have been mainly directed to sequence designs so far [1,2,3]. However, since DNA Computing consists of some chemical reactions, it seems to be clearly important to optimize each reaction protocol in order to improve the reliability of computing. In this paper, therefore, we focus on optimizing each reaction protocol, in particular, PCR (Polymerase Chain Reaction) protocol.

As a method for optimizing a reaction protocol (particularly PCR), the approaches using Design Of Experiments (DOE) have been proposed [4,5]. In these studies, by finding out the factors which affected a reaction result greatly with the use of DOE, they optimized the reaction protocol and employed PCR was the most general PCR (template is one kind, and the desired reaction is making template amplify as much as possible). Besides, they made light of the interactions between each factor.

N. Jonoska and N.C. Seeman (Eds.): DNA7, LNCS 2340, pp. 14–22, 2002.
© Springer-Verlag Berlin Heidelberg 2002

However, the environment of PCR used in DNA Computing seems to be generally different from one used in the study of molecular biology. For example, in PCR used in DNA Computing, the environment of PCR is almost competitive (two or more kinds of DNA template can be amplified by the same primer). Additionally, in the context of DNA Computing, PCR can be used for different purposes, i.e., desired reactions for PCR will be different according to DNA algorithm. For instance, one of these purposes is to amplify the specific template as much as possible, the other is to amplify holding the ratio of concentrations of template DNAs, and so on. From these considerations, in order to improve the reliability of chemical reactions, optimization of reaction protocol according to each desired reaction will be necessary.

In this paper, we try to derive positive impact factors on the reaction result from actual experiments by using DOE. A lot of experiments are performed in the reaction environment that will be occurred in DNA Computing. From the results of experiments, we show that the importance and necessity of optimizing reaction protocol according to each desired reaction. Additionally, it is said that a chemical reaction involves in a very strong interaction. In this work, since the interactions can be evaluated with DOE, we examine the necessity of taking the interactions into consideration for optimization.

2   PCR

PCR is a standard method used to amplify a template DNA, and the process of amplification is strongly influenced by PCR parameter as shown in Table 1.

Table 1. PCR parameters

Sequence of template	Concentration of template
Sequence of primer	Concentration of primer
Concentration of dNTP	Kind of polymerase
Concentration of polymerase	Concentration of magnesium ion
Reaction buffer	Number of Cycles
Annealing Time	Annealing Temperature
Denaturating Time	Denaturating Temperature
Extension Time	Extension Temperature

2.1   PCR in DNA Computing

PCR is one of the most important operations in DNA Computing because PCR is almost used surely in the algorithm of DNA Computing.

The specific feature of PCR in DNA Computing is to be almost used competitive PCR. In competitive PCR, two or more templates are amplified by the same primer exist together in the same pool[7]. Those templates amplify simultaneously under a reaction.

Additionally, the specific feature of PCR in DNA Computing is that desired reactions of PCR cross variably. Usually, desired reaction of PCR is to make the target DNA amplify as much as possible. But the other desired reactions are also considered. For example, in the DNA Computing model using the concentration control, the amplification holding the initial concentration ratio between templates is important[9].

Optimizations of PCR reaction protocol for making DNA amplify as much as possible have already been performed. In this paper, we show that an optimization of reaction protocol according to each desired reaction is necessary.

3   Setting Experiment

In order to find out the factor which has affected the experimental result greatly, we propose the experiment based on DOE as follows[6].

3.1   The Reaction for Analysis

The reaction for analysis is taken as competitive PCR. In this experiment, two kind of templates made to compete is set, and we call one of each template DNA1(120mer) and another DNA2(80mer). Additionally, we use real time PCR to acquire in-depth experimental data.

3.2   Design of DNA Sequence

Sequences of primer and template are designed at random, fundamentally. However, the sequences in which a mis-annealing tends to occur clearly avoid. Moreover, primer sequence was limited to the sequence which fulfills the minimum conditions which should be fulfilled as a primer [8]. It is shown as Table 2 for details.

3.3   The Orthogonal Array

We plan the experiment using orthogonal array called L_{27} [6] established in DOE. According to DOE, we allocated each parameter to orthogonal array.

3.4   Factors

From a lot of parameters mentioned to Table 1, we focused on three parameters, Concentration of Magnesium ion (Mg), Annealing Time (Time), and Annealing Temperature (Temp). We define these three parameters as factors.

Generally in molecular biology, optimizing PCR is performed with the information of sequences and length of templates and primers. However in DNA

Table 2. Sequences of primer and template

	lenght	Base arrangement	GC contents	Tm
forward primer	18mer	GCCATCATCAGTGGAATC	50	50°C
reverse primer	18mer	CAACTTTGAAGCTGGCTC	50	50°C
DNA1	120mer	GCCATCATCAGTGGAATCCC AGATGCATCTATCGCAATAC CACTTTCGAGCTCAAGTGTT AAGACGCGCCACAGCCGACA GTGACACTATTAGCAGTAAA TGGAGCCAGCTTCAAAGTTG	48.33	≃80°C
DNA2	80mer	GCCATCATCAGTGGAATCTT GCTGTTTGTGCCGAAGCATC GTATAGGCTGACCACCGATT ATGAGCCAGCTTCAAAGTTG	48.75	≃80°C

Computing, since sequences and length of templates are usually unknown, optimizing PCR with the information of them will be impossible. So we expect that we must improve from the most general protocol to optimum protocol by making various PCR parameters change if we are in a situation of DNA Computing. As the first step of optimizing PCR in the situation of DNA Computing, we focused on Mg, Time, and Temp which affected the reaction result of PCR greatly.

Since it was desirable to take the wide range of level values as for the experiment at the first time in DOE, we set up each level as shown in Table 3.

Table 3. Level of each factor

factor	level 1	level 2	level 3
concentration of Mg	1.5mM	4.0mM	10.5mM
Annealing time	1sec	30sec	60sec
Annealing temperature	47°C	52°C	57°C

3.5 Values of Other PCR Parameters

The values of PCR parameters except for above three parameters are set as the most general value [8] as shown in Table 4.

4 Experiments for Detecting Important Factors

From the above setup, we performed the experiment according to L_{27}, and the experimental results were obtained as 27 pieces as shown in Table 5. The values of

Table 4. value of other parameters

parameter	value
size of a reaction system	$25\mu l$
concentration of primers	0.5μM
concentration of NTP	0.2mM
concentration of DNA1	50attomol$/25\mu l$
concentration of DNA2	50attomol$/25\mu l$
polymerase	AmpliTaq Gold
amount of polymerase	2.5U$/100\mu l$
Mg ion	**mM
the concentration of probe1	0.2μM
the concentration of probe2	0.2μM
Denaturing temperature	$94°$C
Denaturing time	30sec
Annealing temperature	$**°$C
Annealing time	**sec
Extension temperature	$72°$C
Extension time	20sec
number of cycles	45

DNA1 and DNA2 in Table 5 are the intensity of light reflecting the concentration of each DNA after amplification.

4.1   Statistical Analysis

Statistical analysis in DOE was applied to the results according to orthogonal array. We focused on the two reaction characteristics for experimental results. The first reaction characteristic is the concentration of DNA1 after amplification. This analysis (*analysis1*) will be performed in order to attain the purpose of obtaining DNA1 as much as possible. The second reaction characteristic is the concentration difference of DNA1 and DNA2 after amplification. In this experiment, the concentration of DNA1 is equaled to DNA2 in an initial condition. If DNA1 and DNA2 are amplified with keeping their concentration ratio, the concentrations of DNA1 will be equaled to that of DNA2 after amplification in this work. So, this analysis (*analysis2*) will be performed in order to attain the amplification in which the initial concentration ratio of each template was made to reflect. Statistical analysis is performed to each reaction characteristic.

4.2   Analysis 1

Here, we regarded the concentration of DNA1 after amplification as attributes, and statistical analysis was performed to the experimental results. Table 6 is a result of statistical analysis of *analysis1*.

Table 5. Results of experiments

Experiment No.	DNA1	DNA2	Mg	Time	Temp
No.1	32.71	0	1.5mM	1sec	47°C
No.2	0	0	1.5mM	1sec	52°C
No.3	24.39	0	1.5mM	1sec	57°C
No.4	0	0	1.5mM	30sec	47°C
No.5	21.89	0	1.5mM	30sec	52°C
No.6	0	0	1.5mM	30sec	57°C
No.7	0	6.39	1.5mM	60sec	47°C
No.8	23.75	0	1.5mM	60sec	52°C
No.9	54.52	0	1.5mM	60sec	57°C
No.10	222.22	82.39	4.0mM	1sec	47°C
No.11	80.77	31.37	4.0mM	1sec	52°C
No.12	38.75	51.94	4.0mM	1sec	57°C
No.13	223.44	206.46	4.0mM	30sec	47°C
No.14	354.78	275.89	4.0mM	30sec	52°C
No.15	239.54	279.43	4.0mM	30sec	57°C
No.16	424.36	292.09	4.0mM	60sec	47°C
No.17	434.06	414.77	4.0mM	60sec	52°C
No.18	308.03	423.36	4.0mM	60sec	57°C
No.19	0	111.23	10.5mM	1sec	47°C
No.20	0	62.51	10.5mM	1sec	52°C
No.21	7.54	42.01	10.5mM	1sec	57°C
No.22	0	91.76	10.5mM	30sec	47°C
No.23	2.95	164.91	10.5mM	30sec	52°C
No.24	0	144.83	10.5mM	30sec	57°C
No.25	0	145.58	10.5mM	60sec	47°C
No.26	0	196.16	10.5mM	60sec	52°C
No.27	31.84	245.01	10.5mM	60sec	57°C

In the statistical analysis of DOE, the factors showing small mean square are pooled to an experimental error. F_0 is calculated for each principal factor to compare with the experimental error. Here, since we obtained the result that the effect of Mg × Time × Temp could be pooled to the experimental error, and since analytic accuracy is improved by enlarging the degree of freedom of error, we performed pooling for
Mg × Time × Temp to the error clause. When the test of the significance level 5% was applied to F_0, we were able to obtain the result that Mg, Time and Mg × Time had the positive impact on a reaction result.

4.3   Analysis 2

Here, we regarded the concentration difference of DNA1 and DNA2 after amplification as attribute. In this case, this attribute is better as small, for it became an improvement in accuracy of initial concentration ratio maintenance. We analyzed the same experiment data as *analysis1*, and treated Mg × Time × Temp as error. Table 7 is a result of statistical analysis in *analysis2*. When the test of the significance level 5% were applied to F_0, we were able to obtain the result that Mg only had the positive impact on a reaction result.

Table 6. Table of analysis of variance 1 **Table 7.** table of analysis of variance 2

	Mean square	F_0
Mg	183520.67	108.44
$Time$	21034.11	12.43
$Temp$	1575.97	0.93
$Mg \times Time$	18223.58	10.77
$Mg \times Temp$	3851.84	2.28
$Time \times Temp$	2397.87	1.41
$Error$	1692.41	-

	Mean square	F_0
Mg	66636.32	29.19
$Time$	6418.49	2.81
$Temp$	4881.76	2.14
$Mg \times Time$	2873.13	1.26
$Mg \times Temp$	7066.35	3.09
$Time \times Temp$	1036.99	0.45
$Error$	2282.88	-

4.4   Discussion

We could get the factors which had a positive impact on an experimental result in both *analysis1* and *analysis2* by using DOE. These results suggest that DOE is effective in optimizing of the PCR protocol in DNA Computing. The following chapter describes about optimization.

   We were able to obtain the result that the factors which had a positive impact on the experimental result were different between two results. In short, the important factors differ by the desired reaction of PCR. In consequence, this suggests that optimizing a reaction protocol had better be performed for each desired reaction. Additionally, we could get the interaction that had a positive impact on the experimental result. Therefore we must take interactions into consideration in favor of optimizing a reaction protocol.

5   Optimizations

In above analyses, we were able to derive the factors which had a positive impact on each reaction characteristic. Here we actually tried to optimize each reaction with using the result obtained above analyses.

5.1   Optimization for Amplifying DNA1

When we tended to obtain DNA1 as much as possible, we should observe only Mg, Time as the result of *analysis1*. In *analysis1*, we found out that the optimum value of Mg was 4mM and the optimum value of Time was 60sec. So we tried to detect more suitable values by searching the neighborhood of these values. We investigated the optimum value of Mg from 3mM to 5mM by 1mM unit and the optimum value of Time from 40sec to 80sec by 20sec unit. Temp was taken as 52°C which was the best value at first experiments.

Furthermore, since the influence of Mg×Time was strong, we evaluated to a total combination of Mg and Time. From the above setup, we performed optimization experiment (*optimization1*) to optimizing the amplification of DNA1, and the experimental results were obtained as 9 pieces. The detailed results are shown in Fig.1.

5.2   Optimization for Decreasing the Concentration Difference

For optimization about *analysis2*, we should observe only Mg as the result of *analysis2*. In *analysis2*, when we focused on the concentration difference of DNA1 and DNA2 purely, the value of Mg which made this attribute the optimum was 1.5mM. However in that case, both DNA1 and DNA2 were not fully amplified. So we regarded 4.0mM as optimum value of Mg in *analysis2*. For more sophisticated result, we tried to detect the neighborhood of 4.0mM. We investigated the optimum value of Mg from 3mM to 6mM by 1mM unit. Time was taken as 60sec, Temp was taken as 52°C.

From the above setup, we performed optimization experiment (*optimization2*) to optimize the decrease of the concentration difference, and the experimental results were obtained as 4 pieces. The detailed results are shown in Fig. 2. These values are absolute values of the concentration difference between DNA1 and DNA2.

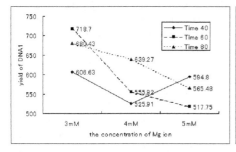   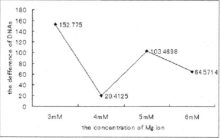

Fig. 1. Results of *optimization1*   **Fig. 2.** Results of *otimization2*

5.3   Discussion

Each reaction characteristics are improved clearly comparing to pre-optimizations. Besides, we found out that these two experimental results fluctuated greatly according to changing important factors, respectively. This indicates the validity of the results of analyses. When we have an interest in the results of *optimization1*, these are scattered intricately. It will be based on effect of interaction and will prove the existence of the effect of the interaction in chemical reactions.

From the results of *optimization1*, we will come to a conclusion that an optimum reaction state is Mg 3mM, Time 60sec, and Temp 52°C when we want to gain DNA1 as much as possible. Additionally, from the result of *optimization2*, we will come to a conclusion that an optimum reaction state is Mg 4mM, Time 60sec, and Temp 52°C when we want to decrease the concentration difference of DNAs. So we obtained the result that optimum reaction protocols were different if desired reactions were different.

6   Concluding Remark

In this paper, we could show the necessity of optimization for each desired reaction that was the purpose of this paper. However PCR is very complicated and there is much characteristic which should be taken into consideration still more. (e.g. specificity, total reaction time etc.) So a lot of evaluations to a lot of other characteristics will be required. Furthermore, we expect that there is a certain regularity between the kind of desired reactions and the kind of important factors. As future work, we are going to clarify this regularity by taking a lot of experimental data. If it succeeds, it will be a guideline of optimizing reaction protocols which are indispensable to improvement in a reliability of calculation in DNA Computing.

References

1. R.J. Deaton, R.C. Murphy, J.A. Rose, M. Garzon, D.R. Franceschetti, S.E. Stevens, Jr.: "A DNA Based Implementation of an Evolutionary Search for Good Encodings for DNA Computation," Proceedings of IEEE International Conference on Evolutionary Computation (ECEC '97), pp.267-271,1997.
2. J.A. Rose, R.J. Deaton, D.R. Franceschetti, M. Garzon, S.E. Stevens, Jr.: "A Statistical Mechanical Treatment of Error in the Annealing Biostep of DNA Computation", Proceedings of GECCO'99 (Genetic and Evolutionary Computation Conference), pp. 1829–1834, 1999.
3. M. Arita, A. Nishikawa, M. Hagiya, K. Komiya, H. Gouzu, K. Sakamoto: "Improving Sequence Design for DNA Computing," Proceedings of GECCO'00 (Genetic and Evolutionary Computation Conference), pp. 875–882, 2000.
4. M.D. Boleda, P. Briones, J. Farres, L. Tyfield, R. Pi: "Experimental Design: A Useful Tool for PCR Optimization," BioTechniques, vol. 21, pp. 134–140, 1996.
5. B.D. Cobb, J.M. Clarkson: "A simple procedure for optimizing the polymerase chain reaction (PCR) using modified Taguchi methods," Nucleic Acids Research, vol. 22, pp. 3801–3805, 1994.
6. G.Taguchi: "Design of Experiments," MARUZEN, 1962.
7. P.D. Siebert, J.W. Larrick: "Competitive PCR," NATURE, vol. 359, pp. 557–558, 1992.
8. M.J. McPHERSON, B.D. HAMES, G.R. TAYLOR: "PCR2 A Practical Approach," IRL PRESS, 1995.
9. Nobuo, M., Masashito, M., Toshikazu, S., Yumi, K., Azuma, O.: "Solutions of Shortest Path Problems by Concentration Control," proceedings of Seventh International Meeting on DNA Based Computers, 2001.

DNASequenceGenerator: A Program for the Construction of DNA Sequences

Udo Feldkamp[1], Sam Saghafi[2], Wolfgang Banzhaf[1], and Hilmar Rauhe[1]

[1] University of Dortmund, Chair of System Analysis, Germany,
{feldkamp, banzhaf, rauhe}@LS11.cs.uni-dortmund.de,
http://ls11-www.informatik.uni-dortmund.de/molcomp/
[2] University of Cologne, Institute of Genetics, Germany

Abstract. In DNA Computing and DNA nanotechnology the design of proper DNA sequences turned out to be an elementary problem [1, 2, 3, 4, 5, 6, 7, 8, 9]. We here present a software program for the construction of sets ("pools") of DNA sequences. The program can create DNA sequences to meet logical and physical parameters such as uniqueness, melting temperature and GC ratio as required by the user. It can create sequences *de novo*, complete sequences with gaps and allows import and recycling of sequences that are still in use. The program always creates sequences that are — in terms of uniqueness, GC ratio and melting temperature — "compatible" to those already in the pool, no matter whether those were added manually or created or completed by the program itself. The software comes with a GUI and a Sequence Wizard. *In vitro* tests of the program's output were done by generating a set of oligomers designed for self-assembly. The software is available for download under http://LS11-www.cs.uni-dortmund.de/molcomp/Downloads/downloads.html.

1 Introduction

The most important requirement for DNA sequences useful for computation is the avoidance of non-specific hybridizations. These can occur between the sequences used in a self-assembly step, in a polymerase chain reaction, in an extraction operation etc. Thus the main purpose for designing DNA sequences is finding a set of sequences as dissimilar as possible, where the dissimilarity usually includes comparison to complementary sequences. Another important aspect is the control of the thermodynamic properties of the sequences, allowing the design of a protocol (the actual application) minimizing the probability of hybridization errors and possibly regarding other constraints given by the application.

There are several approaches to DNA sequence design. Seeman et al. designed sequences using overlapping subsequences to enforce uniqueness [1, 2]. The approach is based on the "repairing" of sequences. Deaton et al. used genetic algorithms to generate a set of unique DNA sequences using the Hamming distance for measuring the uniqueness [3, 4]. Marathe, Condon and Corn chose a dynamic programming approach for DNA sequence design, also using the Hamming distance [5]. They also described a dynamic programming based algorithm

N. Jonoska and N.C. Seeman (Eds.): DNA7, LNCS 2340, pp. 23–32, 2002.
© Springer-Verlag Berlin Heidelberg 2002

for the selection of sequences with a given free energy. Frutos et al. developed a so-called template-map strategy to get a grand number of dissimilar sequences while having to design only a significantly smaller number of templates and maps [6]. They also use a Hamming-like dissimilarity with no shifts of the regarded sequences. Hartemink, Gifford and Khodor designed sequences for the programmed mutagenesis, which demands similar sequences with only a few mismatches [7]. The selection of appropriate sequences is done by exhaustive search, which is feasible for short oligomers. Faulhammer et al. described a designing algorithm for RNA sequences to be used in solving a chess problem [8]. They also use the Hamming distance as measurement for uniqueness and do not construct the sequences but repair them as long as necessary, potentially non-terminating. Baum suggested a method to design unique sequences by avoiding multiple usage of subsequences by restricting the choice of nucleotides at the ends of the sequences [9].

2   Theoretical Background

The program described here uses a concept of uniqueness that, within a pool of sequences, allows any subsequence of a certain (definable) length to occur at most once in that pool. This concept of uniqueness is related to the one described by Seeman et al. [1, 2] but a different approach was chosen. In particular the software described here uses a fully automatic, graph-based approach [10].

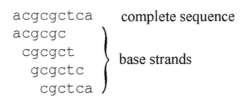

Fig. 1. A sequence of length $n_s = 9$ consisting of $(n_s - n_b + 1) = 4$ overlapping base strands of length $n_b = 6$.

According to this concept a pool of sequences is said to be $n_b - unique$ if any subsequence in the pool of length n_b is unique, i.e. all sequences of the pool have common substrings of maximum length $n_b - 1$. Uniqueness on the other hand is defined as $1 - (n_b - 1)/n_s$, a ratio to measure how much of a set of sequences of length n_s is unique. For example, 20-mers that are 10-unique have common subsequences of at most 9 subsequent nucleotides and a uniqueness of 55%.

The generation algorithm uses a directed graph, where the nodes are base strands (the unique strands of minimal length n_b) and the successors of a node are those four strands that can appear as an (overlapping) successor in a longer sequence (see Fig. 1, 2). Thus, a set of n_b-unique sequences of length n_s corresponds to a set of paths of length $(n_s - n_b + 1)$ through this graph having no node in common.

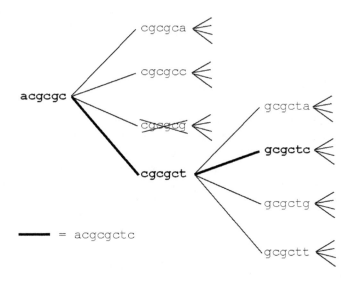

Fig. 2. Graph of base strands. A path of m nodes represents a sequence of length $n_b + m - 1$. The node `cgcgcg` is self-complementary and therefore not used.

The details of this algorithm have been described in more detail earlier [10]. Note that this concept of uniqueness restricts the number of usable sequences strictly. The number of base strands of length n_b is

$$N_{bs}(n_b) = 4^{n_b} \ .$$

(1)

Since complements of already used base strands are not used themselves, self-complementary base strands are not used at all. The number of base strands that can be used in the generation process is

$$N_{useful}(n_b) = \frac{N_{bs}(n_b) - 4^{n_b/2}}{2}$$

(2)

if n_b is even and

$$N_{useful}(n_b) = \frac{N_{bs}(n_b)}{2}$$

(3)

if n_b is odd, because there are no self-complementary base strands of odd length. A sequence of length n_s consists of $n_s - n_b + 1$ base strands. Thus, the maximum number of sequences built with the described algorithm is

$$N_{seqs}(n_\mathrm{s}, n_\mathrm{b}) = \left\lfloor \frac{N_\mathrm{useful}(n_\mathrm{b})}{n_\mathrm{s} - n_\mathrm{b} + 1} \right\rfloor . \tag{4}$$

This estimation is probably a bit too high because it does not include the constraint of the base strands having to overlap. Further requirements such as GC-ratio, melting temperature, the exclusion of long guanine subsequences or of start codons decrease the yield.

3 DNA Sequence Generator

The program DNASequenceGenerator is based on the metaphor of a pool of sequences that can be iteratively filled with sequences which meet the logical and physical requirements. The main window represents a pool of sequences (Fig. 3). The user can add, import, export and print sequences to and from it. E.g., one might import sequences already available in the lab and add new sequences that are compatible to those in terms of uniqueness, melting temperature and GC-ratio.

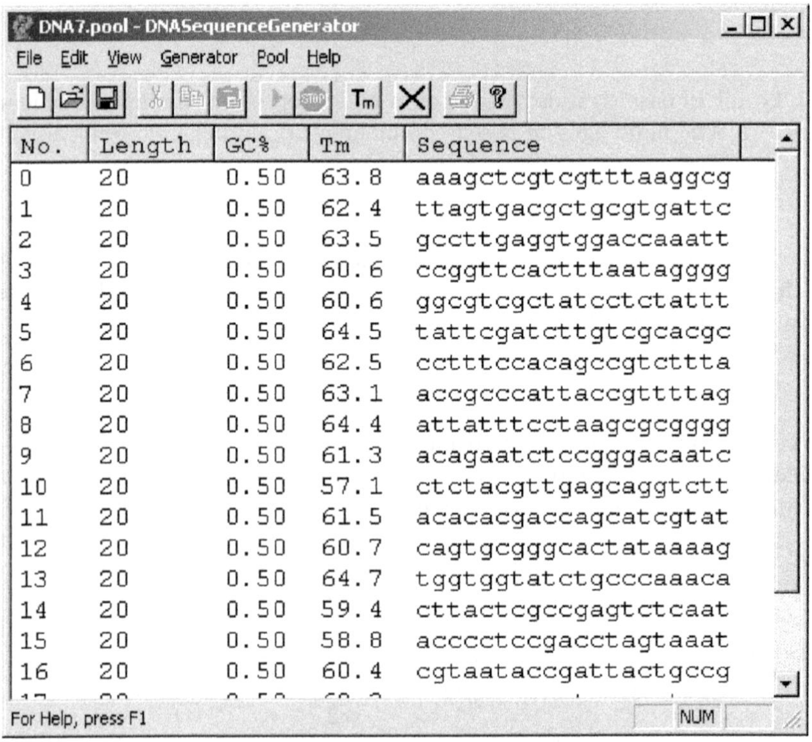
DNA7.pool - DNASequenceGenerator
File Edit View Generator Pool Help
No. Length GC% Tm Sequence
0 20 0.50 63.8 aaagctcgtcgtttaaggcg
1 20 0.50 62.4 ttagtgacgctgcgtgattc
2 20 0.50 63.5 gccttgaggtggaccaaatt
3 20 0.50 60.6 ccggttcactttaatagggg
4 20 0.50 60.6 ggcgtcgctatcctctattt
5 20 0.50 64.5 tattcgatcttgtcgcacgc
6 20 0.50 62.5 cctttccacagccgtcttta
7 20 0.50 63.1 accgcccattaccgttttag
8 20 0.50 64.4 attatttcctaagcgcgggg
9 20 0.50 61.3 acagaatctccgggacaatc
10 20 0.50 57.1 ctctacgttgagcaggtctt
11 20 0.50 61.5 acacacgaccagcatcgtat
12 20 0.50 60.7 cagtgcgggcactataaaag
13 20 0.50 64.7 tggtggtatctgcccaaaca
14 20 0.50 59.4 cttactcgccgagtctcaat
15 20 0.50 58.8 acccctccgacctagtaaat
16 20 0.50 60.4 cgtaataccgattactgccg
For Help, press F1
NUM

Fig. 3. Screenshot of the main window, containing a pool of sequences.

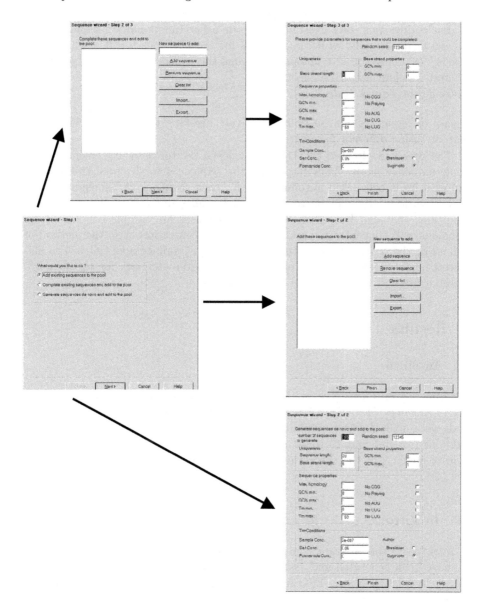

Fig. 4. The different steps of the sequence wizard for the three possible ways to add sequences to the pool.

The process of constructing sequences is controlled by using a "Sequence Wizard" (Fig. 4). The sequence wizard enables the user to:

1. import or manually add existing or strictly required sequences
2. import or manually add sequence templates that are completed to full sequences if possible. Sequence templates use a simple notation: The preset

nucleotides (e.g. of a restriction site or other functional subsequence) are specified normally, while an **n** stands for the positions to fill. E.g., if given the sequence template **nnnnaacgttnnnn**, the generator replaces the leading and rear four **n**s with nucleotides.

3. generate sequences *de novo*.

A pool of sequences can be built iteratively invoking the sequence wizard repeatedly using different parameter sets.

The user can also use the DNASequenceGenerator as a T_m calculator for whole sequence pools. After importing the sequences to the pool the melting temperature is calculated automatically for each sequence. After changing the pool conditions, T_m will be re-calculated for all sequences in the pool. Pool conditions that can be parameterized by the user are sample concentration, monoionic salt concentration and formamide concentration. Also the method of estimating the melting temperature can be chosen (Wallace rule, GC-% formula, nearest-neighbor method) as well as different parameter sets for the nearest-neighbor method.

4   Results

4.1   In Silico

Experiments *in silico* were made to examine the possible yield of n_b-unique DNA sequences (Table 1). Ten runs with different random number generator seeds were made for each combination of n_b- and n_s-values, which ranged from 4 to 7 nt (for n_b) and 10 to 40 nt (for n_s), respectively. Most runs achieved a yield of 80–90 % of the theoretically estimated maximum number of sequences (see (4)).

4.2   In Vitro

In order to test the program's output, oligonucleotides for parallel overlap assembly have been generated and assembled *in vitro*. The single-stranded molecules overlap by 20 nucleotides (E/O sections, see Fig. 5), where the overlap assembly takes place, and a core sequence that stays single stranded in the assembly step and is filled later by the use of polymerase. Additionally, the sequences had to have restriction sites at specified locations.

The DNASequenceGenerator provided the needed sequences, which hybridised which high specificity to form the desired molecules.

Additionally, the melting temperature of a batch of 51 of the generated sequences with a length of 20 bp and a GC ratio of 50 % was analyzed with a Roche LightCycler[TM] (Figs. 6, 7). Since the oligos were selected for their GC ratios rather than their melting temperatures, their melting temperatures ranged between 53 and 63 °C.

Table 1. Yield of n_b-unique sequences averaged over 10 runs, theoretic maximum yield and ratio. This is only an excerpt of all experimental results.

n_s	$n_b = 4$	$n_b = 5$	$n_b = 6$	$n_b = 7$
15	8.1 of 10 (81.0 %)	38.6 of 46 (83.9 %)	165.4 of 201 (82.3 %)	757.8 of 910 (83.3 %)
16	7.3 of 9 (81.1 %)	35.6 of 42 (84.8 %)	150.5 of 183 (82.2 %)	682.1 of 819 (83.3 %)
17	6.8 of 8 (85.0 %)	33.0 of 39 (84.6 %)	138.5 of 168 (82.4 %)	623.4 of 744 (83.8 %)
18	6.1 of 8 (76.3 %)	30.7 of 36 (85.3 %)	127.3 of 155 (82.1 %)	573.3 of 682 (84.1 %)
19	5.8 of 7 (82.9 %)	28.4 of 34 (83.5 %)	118.4 of 144 (82.2 %)	531.1 of 630 (84.3 %)
20	5.6 of 7 (80.0 %)	26.6 of 32 (83.1 %)	111.4 of 134 (83.1 %)	494.8 of 585 (84.6 %)
30	3.3 of 4 (82.5 %)	16.9 of 19 (88.9 %)	67.8 of 80 (84.8 %)	294.0 of 341 (86.2 %)
40	2.0 of 3 (66.7 %)	12.2 of 14 (87.1 %)	48.2 of 57 (84.6 %)	211.1 of 240 (88.0 %)

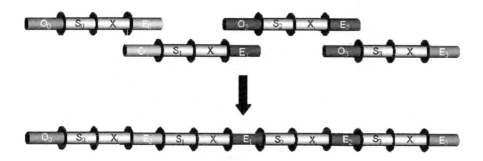

Fig. 5. Overlap assembly of DNA sequences. The O_i- and E_i- subsequences were designed such that E_i is complementary to O_{i+1}.

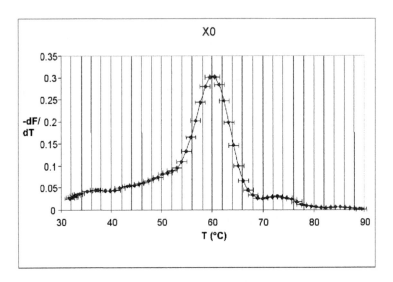

Fig. 6. Sample melting plot of one of the oligos. The y-axis shows the fluorescence absorption, the x-axis shows the current temperature in °C. In this case the oligo's T_m is 60.1 °C.

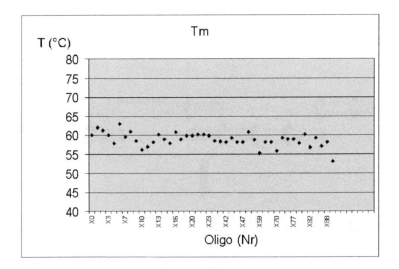

Fig. 7. Distribution of T_m's of 51 oligonucleotides. All oligonucleotides were 20 bp long and had 50 % GC ratio. The distribution of T_m's is in a range from 53 to 63 °C.

5   One Step Further: The DNASequenceCompiler

While the DNASequenceGenerator could, in principle, be used for sticky-end design, its sibling tool, the DNASequenceCompiler, is more specifically suited for this task [10]. It is designed to translate formal grammars directly into DNA molecules representing the rules of the grammar. As these rules determine how terminals are assembled to expressions, they also determine how the DNA molecules self-assemble to larger molecules. Thus, the DNASequenceCompiler provides an interface for the programmable self-assembly of molecules.

It mainly consists of three parts: a parser module for reading the "source code" which contains the symbol sets and the rules of the grammar as well as physical of chemical requirements for the sequences; the generator as a core module for the generation of the sequences; and a coordinating instance controlling the use of the generator while regarding the additional requirements in respect to uniqueness that arise for concatenated sequences.

Currently, a preliminary version of this tool is in use for the design of sequences for algorithmic self-assembly. A user-friendly version is under development. Further enhancements will tackle branched molecules, the consideration of secondary structures occurring on purpose, and a "programming language" on a higher level of abstraction than a formal grammar.

6   Conclusion

Here, a software tool for the design of DNA oligomers useful for DNA computation and DNA Nanotechnology was presented. The software uses a graph-based approach and generates pools of unique DNA sequences automatically according to the user's logical and physical requirements. First experimental results with sequences yielded with the software are encouraging and suggest further investigation. Additional experiments to measure melting temperatures, uniqueness of the sequences and specific hybridization behaviour are currently under investigation. A challenging task will be the development of a benchmark protocol.

The DNASequenceGenerator can be seen as a basic design tool for multiple purposes. The design of sophisticated DNA structures, such as cubes [11] or double- [12] or triple-crossover molecules [13] will require a more specialized program. The DNASequenceCompiler which is currently developed is intended for such a purpose and may help design molecules suitable for algorithmic self-assembly.

7   Acknowledgements

We would like to thank Jonathan C. Howard and the members of his group from the Institute for Genetics, University of Cologne, for their friendly support. The work was supported in parts by the Stifterverband der Deutschen Wissenschaft/Stiftung Winterling Markleuthen.

References

[1] Seeman, N. C., Kallenbach, N. R.: Design of immobile Nucleic Acid Junctions. Biophysical Journal **44** (1983) 201–209

[2] Seeman, N. C.: De Novo Design of Sequences for Nucleic Acid Structural Engineering. Journal of Biomolecular Structure & Dynamics **8**(3) (1990) 573–581

[3] Deaton, R., Murphy, R. C., Garzon, M., Franceschetti, D. T., Stevens Jr., S. E.: Good Encodings for DNA-based Solutions to Combinatorial Problems. Proceedings of the Second Annual Meeting on DNA Based Computers, held at Princeton University (1996) 159–171

[4] Deaton, R., Murphy, R. C., Rose, J. A., Garzon, M., Franceschetti, D. T., Stevens Jr., S. E.: Genetic Search for Reliable Encodings for DNA-based Computation. First Conference on Genetic Programming (1996)

[5] Marathe, A., Condon, A. E., Corn, R. M.: On Combinatorial DNA Word Design. Proceedings of the 5th International Meeting on DNA Based Computers (1999)

[6] Frutos, A. G., Liu, Q., Thiel, A. J., Sanner, A. M. W., Condon, A. E., Smith, L. M., Corn, R. M.: Demonstration of a Word Design Strategie for DNA Computing on Surfaces. Nucleic Acids Research **25**(23) (1997) 4748–4757

[7] Hartemink, A. J., Gifford, D. K., Khodor, J.: Automated Constraint-Based Nucleotide Sequence Selection for DNA Computation. Proceedings of the 4th DIMACS Workshop on DNA Based Computers, held at the University of Pennsylvania, Philadelphia (1998) 227–235

[8] Faulhammer, D., Cukras, A. R., Lipton, T. J., Landweber, L. F.: Molecular Computation: RNA Solutions to Chess Problems Proceedings of the National Acadamy of Sciences USA **97**(4) (2000) 1385–1389

[9] Baum, E. B.: DNA Sequences Useful for Computation. Unpublished, available under http://www.neci.nj.nec.com/homapages/eric/seq.ps (1996)

[10] Feldkamp, U., Banzhaf, W., Rauhe, H.: A DNA Sequence Compiler. Proceedings of the 6th DIMACS Workshop on DNA Based Computers, held at the University of Leiden, The Netherlands (2000) 253. Manuscript available at http://LS11-www.cs.uni-dortmund.de/molcomp/Publications/publications.html

[11] Chen J., Seeman, N. C.: Synthesis from DNA of a Molecule with the Connectivity of a Cube. Nature **350** (1991) 631–633

[12] Winfree, E., Liu, F., Wenzler, L. A., Seeman, N. C.: Design and Self-assembly of Two-dimensional DNA Crystals. Nature **394** (1998) 539–544

[13] Mao, C., LaBean, T. H., Reif, J. H., Seeman, N. C.: Logical Computation Using Algorithmic Self-assembly of DNA Triple-crossover Molecules. Nature **407** (2000) 493–496

DNA Computing in Microreactors

Danny van Noort, Frank-Ulrich Gast, and John S. McCaskill

BioMolecular Information Processing, GMD
53754 Sankt Augustin, Germany
tel.: +49 2241 14 1521/15xx/1526, fax: +49 2241 141511
{danny.van-noort, frank-ulrich.gast, mccaskill}@gmd.de
http://www.gmd.de/BIOMIP

Abstract. The goal of this research is to improve the programmability of DNA-based computers. Novel clockable microreactors can be connected in various ways to solve combinatorial optimisation problems, such as Maximum Clique or 3-SAT. This work demonstrates by construction how one micro-reactor design can be programmed optically to solve any instance of Maximum Clique up to its given maximum size (N). It reports on an implementation of the concept proposed previously [1]. The advantage of this design is that it is generically programmable. This contrasts with conventional DNA computing where the individual sequence of biochemical operations depends on the specific problem. Presently, in ongoing research, we are solving a graph for the Maximum Clique problem with $N = 6$ nodes and have completed the design of a microreactor for $N = 20$. Furthermore, the design of the DNA solution space will be presented, with solutions encoded in customised word-structured sequences.

1 Introduction

DNA computing involves a multidisciplinary interplay between molecular biology, information science, microsystem technology, physical detection methods and evolution. Since the first practical example by Adleman [2] there has been intensive research into the use of DNA molecules as a tool for calculations, simulating the digital information processing procedures in conventional computers. In the short term, however, the main application of DNA computing technology will be rather to perform complex molecular constructions, diagnostics and evolutionary tasks. However, in order to assess the limits of this technology, we are investigating a benchmark computational problem: Maximum Clique, with an NP-complete associated decision problem, chosen because of its limited input information [3]. The step from batch processing in test tubes to pipelined processing in integrated micro-flow reactor networks [1], gives us complete control over the process of information flow and allows operations much faster than in conventional systems. More importantly, it allows the extension to optical programming. Moreover, the proposal differs radically from the surface based DNA computing approach [4] in requiring no problem dependent manual or robotic operations, programming of specific problem instances being completely under

N. Jonoska and N.C. Seeman (Eds.): DNA7, LNCS 2340, pp. 33–45, 2002.
© Springer-Verlag Berlin Heidelberg 2002

light controlled immobilisation techniques. While other publications report on micro-flow reactor networks [5,6,7], this paper describes the first steps towards practical DNA computing in micro-reactors.

2   Benchmark Problem

2.1   Maximum Clique

The decision problem associated with the maximum clique problem becomes rapidly harder to solve as the problem size increases (it is NP-hard). Maximum clique requires finding the largest subset of fully interconnected nodes in the given graph (Fig. 1). To obtain the set of cliques and then determine its largest member using a micro-flow system, an algorithm was devised consisting of a series of selection steps containing three parallel selection decisions.

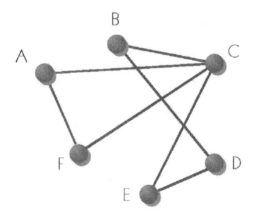

Fig. 1. An $N = 6$ instance of the clique problem. The maximum clique is given by ACF, represented by 101001.

The problem can be divided into two parts: (i) find all the subsets of nodes which correspond to cliques in the graph and (ii) find the largest one. The basic algorithm is simple [1]: for each node i ($i \geq 1$) in the graph retain only subsets either not containing node i or having only other nodes j such that the edges (i, j) are in the graph. This can be implemented in two nested loops (over i and j), each step involving two selectors in parallel. A third selector has been introduced to allow the selector sequences to be fixed independently of the graph instance. Thus the graph dependence is programmed not by which but by whether a sub-sequence selection in the third selector is performed (see Fig. 2). The above procedure is described in more detail by McCaskill [1]. It is important to note that only positive selection for sequences with the desired property is performed, not subtractive selection.

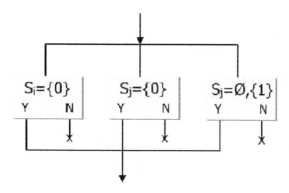

Fig. 2. A flow diagram showing the selection step for node subsets regarding 'cliqueness' at (i, j). The three modules reflect that either node i or node j is absent or the edge (i, j) must be present in the graph.

The edges of the graph, i.e. the connections between the nodes, can be represented by a so called connectivity matrix. The connectivity matrix for the 6-node example shown in Fig. 1 is the 6×6 matrix in Table 1. As Table 1 shows, the matrix is symmetrical over the diagonal, while the diagonal is trivially one, reducing the number of necessary selections from N^2 to $1/2N(N - 1)$.

Table 1. The connectivity matrix for the 6-node graph as shown in Fig. 1. The shaded numbers are trivial selections and don't have to be included in the selection procedure to obtain all the cliques. The boxed positions also do not influence the selection, but are included to allow optical programmability.

2.2   Selection Procedure

Each DNA sequence encodes a binary sequence corresponding to a particular subset of nodes in the graph. Different DNA sub-sequences are used to represent presence (1) or absence (0) at each node. As shown in Fig. 2, each selection step consists of 3 selection modules connected in parallel. After each selection step, the sub-population is passed on to the next selection step. Each selection module is coded with short selection-DNA strand (ssDNA) from a finite set of $2N$ predefined sequences. Following the selection algorithm as described above, the final population of DNA-sequences will consist of all the possible cliques represented in the given graph. For the example given graph in Fig. 1, Table 2 shows all the possible cliques.

Table 2. All the possible cliques from the graph shown in Fig. 1. Note that the nodes themselves are cliques as well.

To determine the maximum clique, a sorting procedure has to follow, to select the DNA-sequence with the largest number of bits with the value 1. As shown in the example, Table 2, this would be 101001 (ACF), which with 3 bits of value 1 represents the maximum clique.

Any graph instance up to the maximum size N can be programmed using the same one microreactor design.

3   DNA Library

Fig. 3. Design of the DNA used for computation.

3.1   Word Design

The DNA consists of 12 bits (V0–V11) and two primer binding sites (PBS), which contain restriction sites for *Bam*HI ($G \downarrow GATCC$) and *Eco*RI ($G \downarrow AATTC$), respectively. (Top) The complete DNA was assembled by hybridization of two oligodeoxynucleotides in bit V6 and polymerization. Only the upper strand (spanning V0–V6) was randomized as described by Faulhammer et al. (2000); for the overlap assembly, two different lower strands (spanning V11–V6) with fixed sequence were used. Cloning and sequencing gave no indication of a nonrandom composition of the library. (Bottom) The sequence of the DNA words used. The bit value (0 or 1) is given in the suffix after the bit position. The sequences do not show significant similarities

The DNA molecules used consist of a series of 12 words which assume a value of either 0 or 1 (Fig. 3). This gives us a certain flexibility in the choice of a optimal set for the case $N = 6$. The word design is a compromise between maximal specificity of the pairing of the DNA words with complementary probes immobilised on magnetic beads and minimal secondary structure of the single-stranded DNA analysed. Since every word not only represents a bit value, but also the bit position, all DNA words must be unique.

In the original work on the Maximal Clique Problem, the length of the 0 and 1 words were different, which helped to identify the maximal clique from a mixture of all cliques by chromatography [3]. A fixed word length with an identical $G+C$ content (50%) was chosen in order to obtain comparable melting points (in contrast, Ouyang et al. [3] used restriction cleavage for selection). A word length of 16 nt is long enough to ensure specific hybridisation and short enough to minimise secondary structure. Unlike the design used by Faulhammer et al. [8] we did not create constant boundary regions between the bits but checked all the overlap regions for non-specific binding. Furthermore, the purification of the library is easier if one uses a fixed length. Since the sorting of the maximal clique occurs in the DNA reactor, no additional molecular properties are needed.

To perform a word design from random sequences, the following criteria had to be fulfilled:

- the difference in base sequence between different words (Hamming distance) including the primer binding sites (PBS), should be maximal in all different registers. Gaps in the sequence alignment were not considered at this time;
- the distance between the sequences should be valid in all possible frames, i.e., the neighbouring sequences (all combinations) should be included in the analysis;
- in the words or at the boundary between words the restriction sites used in the primer binding regions should be present.

3.2   Synthesis of the DNA Library

The DNA synthesis, performed at NAPS (Göttingen, Germany), followed the "mix and split" strategy of Faulhammer et al. The randomised DNA, consisting of PBS1 and bits V0–V6, was then combined by overlap assembly [3,8] with a non-random molecule, consisting of bits V6–V11 (i.e., the overlap was at V6) and a second PBS2. Following PCR amplification and purification, single-stranded DNA molecules could be prepared by linear amplification and purification on non-denaturing gels.

Unambiguous amplification patterns (for two $N = 6$ individual sequences) were detected when a hot start polymerase was used and the reaction was limited to 20 cycles, indicating that the hybridisation of complementary bits is specific and that the DNA words were properly chosen (see Fig. 4). A unique pattern was even obtained in the case of a 10% contamination with a wrong solution. The PCR readout was monitored in real-time using intercalating fluorescent dyes (iCycler, Bio-Rad, CA; SYBR Green I, Molecular Probes, OR). More detailed results of the biochemical analysis will be published elsewhere.

4   DNA-computer Set-up

4.1   Microreactor Structures

The above procedure can be implemented in a network of micro-reactors. To this end we have developed a module which is able to make positive selections from a population of specific DNA sequences. To actively transfer the selected DNA sequences to the appropriate output, they are transferred from one flow to another by moving paramagnetic beads on which single stranded selector-DNA complementary to a nodal sub-sequence is immobilised. The DNA strands in solution hybridise to the selector-strands and are thus transferred to another channel in the micro-flow reactor where they are de-hybridised and passed on to the next selection procedure. To optimise the transfer of only the appropriate sequences, a washing step has to be performed so as to rinse off the non-specific bound DNA sequences. See Penchovsky and McCaskill [9] for an experimental

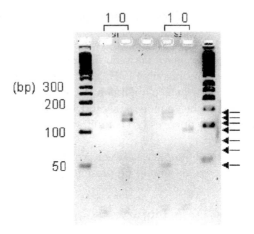

Fig. 4. Typical example of a multiplex PCR analysis. The readout follows Adleman's (1994) strategy, but with a mixture of primers in a single tube, as introduced by Faulhammer et al. (2000). DNAs containing the sequences (bits V0–V6) 0010100 (two left lanes) and 1010011 (two right lanes) were amplified using the constant primer binding to PBS1 (1 μM) and a mixture of reverse primers (hybridizing to V0–V6) with a value of either 0 or 1 (AmpliTaq Gold; 95° C, 10 min, followed by 10 cycles of 95°/47°/72°, 30 s each) and analyzed on a 4% agarose gel (inverted for better readability). If only bands of correct molecular weight are counted (arrows), the correct band pattern (cf. Fig. 3) can be read.

investigation of this process in a separate microreactor design. A typical selection module is shown in Fig. 5.

To prevent the beads from flowing to other STMs, a bead-barrier has been added to the design, which stops the beads from disappearing down the channels. This also provides a straight ledge to move the beads along. Furthermore, by extending the barrier into the channels, a back-flow from the next selection stage is prevented. Experiments and simulations have shown that this design is necessary to ensure a correct flow. More results will be published elsewhere.

Presently we have constructed microflow reactors for $N = 6$ and $N = 20$ nodes and are testing performance firstly with the $N = 6$ version requiring 15 ($= N(N-1)/2$) STMs and 14 ($= 2N - 1 + 3$) inlets. The number of inlets scales linearly and the number of modules quadratically in the problem size.

There are $2N$ selection-strands needed, the 0 and 1 bit representation of the N nodes. This number determines the architecture of the DNA-computer microstructures, while these selection-strands have to be transported to the designated selection and sorting modules. In the $N = 6$ case, there will be 11 programming channels, 2 supply channels and 1 template channel. The programming channels will be used for the de-hybridisation solution during operation.

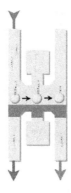

Fig. 5. The Selection Transfer Module (STM). The DNA template enters through channel A. Some of the DNA-strands will hybridise to the selector-strand, which is immobilised on the paramagnetic beads. The beads are moved with a magnet through a wash channel B, to rinse off the unbound strands, into the de-hybridisation channel C from where the selected strands will be transported after a continuos flow neutralisation step, to the next STM. No fluidic switching of flows is required.

Table 3 shows all the ssDNAs necessary to perform the clique selection for a graph of $N = 6$. The selection-strand A_1 is not present due to the fact that it would only be needed to solve the trivial case when comparing node A to itself. The shaded area indicates the positions which must be programmed according to the connectivity matrix. If there is no connection, this field stays empty ($\emptyset$). For the example presented in Fig. 1, the first column, checking the connectivity with node A, would contain beads labelled with ($\emptyset\, C_1\, \emptyset\, \emptyset\, F_1$) successively. It is clear from this table, that the design of the DNA-computer is determined by the problem type and algorithm choice, but not by the problem instance. The design for a 3-SAT problem, for example, would be different. Figure 6 shows the total lay-out for the clique and subset size selection microflow reactor for the case of problems up to $N = 6$.

Table 3. All the necessary selection steps, for a graph of up to $N = 6$ nodes, needed to determine the existence of edges between node i and j. The letter indicates the nodes while the indices denote the bit value (0 or 1). $\emptyset$ is an empty STM. The shaded area is programmable and is determined by the edge between node i and j.

4.2 Sorting Module

The sorting section consist of a parallel selection method to determine the number of bits of value 1 in the DNA-sequence. The algorithm employed is similar in structure to the proposal in Bach et al. [10]. In this design, positive selection for the 1-bits is employed. In the final version of the design (not shown) the 0-bits

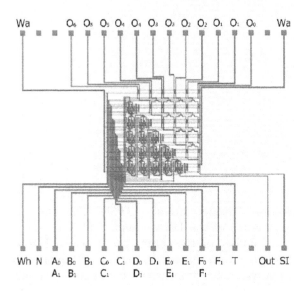

Fig. 6. The complete design of the DNA-computer for graphs with up to 6 nodes. The top right triangle does the length selection, while the other half does the clique selection. The square pads are the input and outputs of the DNA-computer, while the rest of the channels are supply channels. These are connected on the back of the wafer with the reactors (the light grey, horizontal channels). The abbreviations are as follows: Wh is the wash inlet; N is the neutralisation inlet; A_0–F_1 are the ssDNAs programming inlets; T is the input template; Out is the output from the clique sorting modules; SI is the sorting module input channel; Wa is the waste output; Oi are the sorted outputs, with i the clique length. The second row of ssDNAs is for an alternative sorting programming scheme with increasing number of nodes.

will also be actively selected. There are 12 outputs from the selection in which the population is classified (only seven are needed). Table 4 shows the ssDNA needed to perform the sorting procedure as shown in Fig. 6. After step 1, all the strands with A_0 in the sequence will flow down to the step 2, while all the other strands with A_1 move one column to the right and to step 2. This process continues until the output layer is reached.

Table 4. All the selection-strands needed for the length sorting procedure. The output gives the number of bits with value 1.

It can be seen that the path through the sorting module gives the sequence's code. This means that if the DNA strands are fluorescence labelled, their path can be followed optically (e.g. Mathis et al. [11]) and no gel electrophoresis is needed to analyse the sequence.

4.3   Clique Selection Module

The STMs have three supply channels, for the DNA-solution template, the washing solution and de-hybridisation buffer respectively. De-hybridisation is performed by using an alkali solution (NaOH), adjusted in concentration for the common "melting" temperature of the hybridised DNA strands. Because of the change in pH at the de-hybridisation step, a subsequent neutralisation step is necessary after each selection stage before flowing the selected DNA into the next module. This procedure has been successfully applied as shown by Penchovsky et al. [6]. In order to allow the magnetic beads to move uniformly from left to right in all the STMs over the entire microflow reactor, alternate stages are mirrored (see Fig. 7). When the DNA-sequences de-hybridises from the beads, the beads in the next stage will be in the correct place for hybridisation to take place.

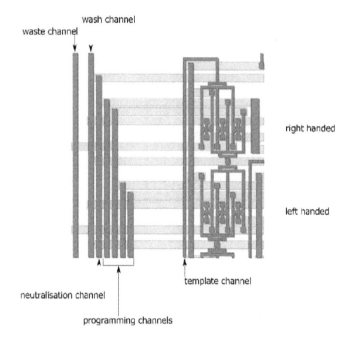

Fig. 7. Two selection steps in sequence as found in the microflow reactor design. Each selection step has 3 selection modules for a node connectivity decision (dark grey) with the template, wash, programming/de-hybridisation, neutralisation and waste channels. The supply channels are connected to the modules by horizontal channels etched on the back of the silicon wafer (light grey).

4.4   Programmability

These Strand Transfer Modules (STM) can be optically programmed as outlined in [1], by means of photo-immobilisation [12], thus creating a programmable micro-fluidic computer. Unlabeled beads are delivered in parallel at the appropriate locations to each STM. UV laser light then determines whether or not a sequence is immobilised on the beads at each STM. If there is no connectivity between a pair of nodes, the corresponding third STM is not immobilised. The other STMs can be preloaded with immobilised beads since they will never have to be changed when re-programming the problem. The immobilisation pattern is directly related to the connectivity matrix, from which a programming mask can be derived. The information flow can be tracked using a sensitive CCD detection system to detect laser-induced fluorescence with intercalating dyes or labelled DNA Because of the fluorescent information from each STM in which a correct DNA strand transfer occurs, it is possible to monitor the solution of the algorithm to its conclusion over time.

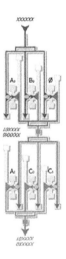

Fig. 8. A close up is shown of the first two clique selection steps. The flows and transfer are shown with the coloured lines. The binary sequences (where $X = 0$ or 1) indicate what selection has taken place.

4.5   Etching Procedure

The whole micro-reactor configuration is photolitho-graphically etched on 4" (100) silicon substrates. The etched wafer is sealed with an anodically bonded pyrex glass wafer. Capillary tubing (0.8 mm diameter) is attached through ultrasound drilled holes in the pyrex wafer. The distance between the holes is 3.5 mm. The supply channels width on the front side are 200 _m, while those on the back are 300 _m. The relatively large width is to reduce the flow resistance and pressure drops in the structure. The channel width in the STMs is 100 _m. To etch through the 220 _m thick silicon wafer, etching pads on the front (200 × 200 _m) and on the back (300 × 300 _m) are made as to obtain holes of 100 × 100 _m. These then have the same width as the channels in the STM.

4.6   Set-up Overview

To set up a DNA-computer in a micro-flow system puts high demands on the control system, an overview of which is presented in Fig. 9.

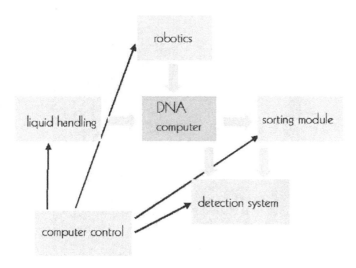

Fig. 9. An overview of the operation of a DNA computer.

To distribute the DNA template and buffer solutions to the wafer, a liquid-handling system is connected. It consists of a pipetting robot and a series of multi-position valves which control the solution distribution. Flow rates will be smaller then 1 _l/min. One of the multi-position valves, together with the micro-structure architecture, makes it possible to address the individual immobilisation sites in the STMs for the selection strand. This can readily be made parallel for full scalability. To programme the computer optically laser light from an EXCIMER laser (308 nm) is used to project either serially trough a microscope or through a static (later dynamic) mask. The wafer is mounted on an xy-translation stage so as to address all the STMs individually. A detection system consisting of a CCD camera for a general overview and a microscope for detailed pictures is in place.

The beads are moved by a magnet which sweeps over the DNA-computer's surface. This sweep clocks the serial steps of the computation, from one stage to the next, which is pipelined to increase throughput.

5   The Next Step: A 20-node DNA-computer

Presently an initial design for a DNA-computer has been constructed to solve a Maximum Clique problem for $N = 20$. The principle layout is the same as for the $N = 6$ case, although it will require a tighter control over the flows. This design has 190 STMs, 210 sorting modules, 43 inlets and 43 outlets. Figure 10 shows the top channel layer for the 20-node computer. The first version has been etched on a 4" wafer in the same fashion as the 6-node computer described above.

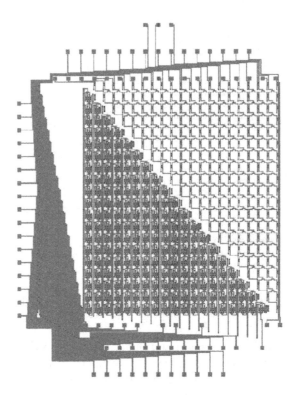

Fig. 10. The complete top layer of the DNA-computer for a 20-node graph. The top right triangle does the length selection, while the other half does the clique selection. The square pads are the input and outputs of the DNA-computer, while the rest of the channels are supply channels.

6   Conclusion

Microflow reactors will prove to be a powerful tool to construct a programmable parallel DNA computer. The main advantage is in the light programmable, integrated system (operating under steady flow) with no problem dependent pipetting steps. Future reactor configurations can be made re-configurable with evolving DNA-populations to obtain a universal programmable DNA-computer [13].

7 Acknowledgement

The authors wish to acknowledge the support of the GMD for DNA computing and the German Ministry of Science (BMBF) for BioMIP's start-up grant (#01SF9952). We would like to thank Marlies Gohlke, Harald Mathis, Robert Penchovsky, Thomas Rücker and Patrick Wagler for their assistance with the word design, biochemical tests, microreactor implementation and set-up.

References

1. McCaskill, J. S. (2001) Optically programming DNA computing in microflow reactors. Biosystems **59** (2), 125–138.
2. Adleman, L. M. (1994) Molecular computation of solutions to combinatorial problems. Science **266**, 1021–1024.
3. Ouyang, Q., Kaplan, P. D., Liu, S. and Libacher, A. (1997) DNA solution to the maximal clique problem. Science **278**, 446–449.
4. Liu, Q., Wang, L., Frutos, A. G., Condon, A. E., Corn, R. M. and Smith, L. M. (2000) DNA computing on surfaces. Nature **403**, 175–179.
5. Gehani, A. and Reif, J. (1999) Micro flow bio-molecular computation. BioSystems **52**, 197–216.
6. Ikuta, K., Maruo, S., Fujisawa, T. and Hasegawa, T. (1999) Biochemical IC family for general micro chemical systems. Transducers '99, June 7–10, 1999, Sendai, Japan, 1046–1049.
7. Suyama, A. (1997) DNA Chips — Integrated Chemical Circuits for DNA Diagnosis and DNA Computers. Proc. of Third International Micromachine Symposium, 7–12.
8. Faulhammer, D., Cukras, A. R., Lipton, R. J. and Landweber, L. F. (2000) RNA solutions to chess problems. Proc. Natl. Acad. Sci. USA **15**, 1385–1389.
9. Penchovsky, R. and McCaskill, J. S. (2001) Cascadable hybridization transfer of specific DNA between microreactor selection modules. Submitted.
10. Bach, E., Condon, A., Glaser, E. and Tanguay, C. (1998) DNA models and algorithms for NP-complete problems. J. of Comp. and Sys. Sci. **57**, 172–186.
11. Mathis, H., Kalusche, G., Wagner, B. and McCaskill, J. S. (1997). Steps towards spatially resolved single molecule detection in solution. Bioimaging **5**, 116–128.
12. Penchovsky, R., Birch-Hirschfeld, E. and McCaskill, J. S. (2000) End-specific covalent photo-dependent immobilisation of synthetic DNA to paramagnetic beads. Nucleic Acids Research **28**, e98, 1–6.
13. McCaskill, J. S. and Wagler, P. (2000) From reconfigurability to evolution in construction systems: spanning the electronic, microfuidic and biomolecular domains. In R. W. Hartenstein and H. Grünbacher (Eds.) FPL 2000, LNCS **1896**, pp. 286–299, Springer-Verlag, Berlin Heidelberg.

Cascadable Hybridisation Transfer of Specific DNA between Microreactor Selection Modules

Robert Penchovsky and John S. McCaskill

GMD-German National Research Center for Information Technology,
Schloss Birlinghoven, D-53754 Sankt Augustin (Bonn), Germany
{McCaskill, Penchovsky}@gmd.de
http://www.gmd.de/BIOMIP.html

Abstract. The paper demonstrates experimentally the basic principle of DNA transfer between magnetic bead based selection stages, which can be used in steady flow microreactors for DNA Computing [McCaskill, J.S.: Biosystems, 59 (2001) 125–138] and molecular diagnostics. Short DNA oligomers, which can be attached covalently to magnetic beads by a light programmable photochemical procedure [Penchovsky et.al.: Nucleic Acids Res., 22 (2000) e98], are used to bind matching ssDNA from a flowing solution. The beads are restrained in two reaction chambers (modules) by etched ledges in a bonded microreactor made of silicon and glass, with the solutions flowing in closed micro-channels. The action of a steady flow network of selection modules is studied in this two chamber microreactor using a succession of different buffer solutions at alternate pH to simulate the transfer between parallel flows in the former system. The pH changes cause successive hybridisation and dissociation of ssDNA to matching sequences on the beads. Detection of DNA is by fluorescence from rhodamine-labelled target DNA. The results demonstrate the successful selection of specific DNA in one module and its subsequent transfer to and pickup on the magnetic beads of a second module. This verifies the biochemical operation of the basic processing step for optically programmable DNA Computing in micro-flow reactors.

1   Introduction

One of the main attractions of DNA-based computation is that it allows a high level of customization as parallel processing hardware. In order for such fully customised approaches to parallel computation to be effective, however, attention must be devoted to ensuring that the DNA computer remains programmable for complex problems. A recent paper by one of the authors has demonstrated that optical programmability is an efficient choice for DNA Computing in conjunction with microreaction technology [1]. Microflow systems can implement a novel kind of dataflow architecture in which a stream of DNA molecules with different sequences are processed in parallel by a network of sequence-discriminating modules. A key example of such modules, as detailed in [1], is called the Strand Transfer Module (STM) and allows the selective transfer of DNA containing

N. Jonoska and N.C. Seeman (Eds.): DNA7, LNCS 2340, pp. 46–56, 2002.
© Springer-Verlag Berlin Heidelberg 2002

a specific subsequence between two flows. In this paper, we demonstrate the ability to select DNA from a continuous stream in one module and pass it on isothermally to be picked up by a second module. On the basis of these experiments, we deduce appropriate chemical conditions for implementing strand transfer modules under constant flow conditions for DNA computations. The technology should also be of interest for molecular diagnostics such as expression and single nucleodite polymorphism analyses.

The transfer is implemented using DNA attached to paramagnetic beads and clocked by an external magnet, in parallel for all modules. A physical realization of such a module was presented previously [2]. Different buffer solutions in neighbouring channel flows are brought into contact by removing a small stretch of intervening wall (to allow the passage between them of beads under a magnetic field). Three different buffer solutions are used to achieve DNA hybridisation to, selective washing on and release from complementary matching oligomers immobilized on the super-paramagnetic beads. In this paper, we will demonstrate that these individual steps can indeed be performed in microflow reactors and will also show the range of buffer conditions necessary for this to be achieved.

A technique for optically programming such modules via the photochemically driven attachment of DNA oligomers to magnetic beads has been developed and tested by the authors [3]. In a parallel paper by our research group, we show how scalable, optically programmable microreaction technology, in steady flow, is being developed using STMs to allow a solution to combinatorial optimisation problems [4].

To quantify the efficacy of DNA selection according to this technique, a special two-chamber microreactor design was employed, allowing independently switchable flows for complete control. This design, shown in Fig. 1, shares the basic feature of STMs in that paramagnetic beads selectively capture DNA and are restrained by ledges positioned across the flow. However, it allows larger bead quantities and manual manipulation to be employed to assist in the quantitative characterization.

This work has direct bearing on our ability to integrate DNA Computing. In contrast with surface-based approaches to DNA Computing [5], the essential computational steps in the flow systems we are proposing are not dictated by off-chip pipetting robots. It can be regarded as a step towards a fully μTAS (micro Total Analysis System) conception of DNA Computing. Such a contribution is beneficial, not simply in terms of automation, but also for the scalable programmability of DNA Computing, since one hardware design is able to deduce the solution to any problem of a given class. We do not address the issue of constructing new solutions in the course of a computation in this paper, although this is the key to overcoming the fundamental limitation to DNA Computing as proposed by Adleman [6]. However, as will be addressed elsewhere, this is possible within the STM framework using evolution.

Microfluidic platforms are being developed in a number of laboratories for various applications including enzyme assays [7], chemical synthesis [8], on chip PCR [9], and DNA analysis [10,11,12]. The research towards labs-on-a-chip is

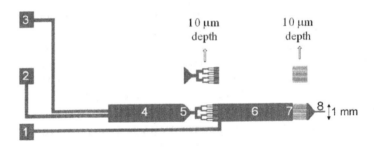

Fig. 1. Overall scheme of the micro-flow reactor employed to quantify DNA selection and transfer. The three inlet channels (1, 2, 3), the outlet channel (8) and the two chambers (4, 6) each with its own bead barrier (5, 7) are shown. Beads are introduced into the first and second chamber and restrained there by the bead barrier. The depth of the channels and the two chambers is 140 μm. The depth of the intervening bead barriers is 10 μm. Different structures for the various bead barriers were used for a functionality not relevant to the current experiments. The microreactor is sealed using anodic bonding of a pyrex wafer, with holes to allow access from polyethylene connecting tubing drilled in the pyrex above the inlets (1, 2, 3).

based on the idea of the integration on a single wafer of different chemical processes, all requiring different reaction conditions. DNA Computing places extreme demands on microfluidic systems, some of which are shared by molecular diagnostics. Recently microfluidic devices have been applied for implementing a parallel algorithm using fluid flow [13]. In contrast to a DNA computing based approach (as presented here) only limited parallelism achievable using only beads to store computational states. In our case we are developing microflow architectures for selection and sorting of DNA from complex populations which could open up new possibilities for integrated diagnostic applications with pipeline and conditional testing. The current work also lays the foundation for such applications.

2   Materials and Methods

2.1   Paramagnetic Beads and Oligodeoxynucleotides

Polystyrene super-paramagnetic beads were purchased from Micromod GmbH (Rostock, Germany). The beads are 15 μm in diameter, carboxyl-coated and essentially monodisperse. Oligomers were obtained from IBA-NAPS (Goettingen, Germany). The DNA was 5' amino-labelled using a C6 linker. All DNA was purified to HPLC grade. The oligomer sequences used are shown in Table 1.

2.2   DNA Immobilisation to Paramagnetic Beads

2.5 nmol of the 35 nt 5' amino-modified oligomer (see Table 1, line 1) was used for the covalent attachment to carboxyl groups on 5 mg beads dissolved in 500 μl 100 mM MES buffer pH 6.1 in the presence of 50 mM EDAC. The reaction was incubated for 3 h at 28°C under continuous shacking. Amino-coated beads could be employed for a light-directed DNA immobilization[3].

Table 1. Oligonucleotides used

N	Type of 5' modification	Sequence 5' - 3'	Length [nt]
1	Amino C6	TTTTTTTTTTTTTTTTACAGACAAGCTGTGACCGTC	35
2	Rhodamine 6G	GACGGTCACAGCTTGTCTGTA	21
3	————	TACTGTCGCAGCTTGTCTGTATTTTT	26
4	Rhodamine 6G	TACTGTCGCAGCTTGTCTGTA	21

2.3   Microflow Reactor for Quantifying DNA Selection and Transfer

The microflow reactor used in these experiments was a two serially coupled chamber design with larger chambers than in the standard STMs to assist in quantitation, see Fig. 1. The ledges to restrain beads under fluid flow have a more complicated form than necessary, to allow the microreactor to be used in other contexts not discussed here. The construction of the microreactors is in the same materials (100 oriented silicon wafers bonded to pyrex glass) and micro-machining procedure employed in the full microflow DNA Computer [2,4].

2.4   DNA Selection and Transfer Protocol on Beads in Microflow Reactor and Its Quantification

Beads with immobilised DNA (as described above) were incorporated in a microreactor using a precision syringe pump (Model 260, World Precision Instruments Inc., Sarasota, FL). The hybridisation was carried out in 500 mM tris-acetate buffer pH 8.3 and 50 mM NaOH in the presence of 1 μM 5' rhodamine 6G labelled DNA oligomers (see Table 1, line 2) at 45°C under flow conditions. After each hybridisation step, the beads were washed with 150 mM tris-acetate buffer pH 8.3 at the same temperature. Hybridised DNA on the beads was denatured using 100 mM NaOH. The denaturation solution is reversed to a hybridising solution by adding an equal volume of 1 M tris-acetate buffer pH 8.3. The capacity of the buffer is not exhausted by the NaOH solution, so that no change in the pH of the tris-acetate buffer upon mixing was observed.

2.5   Fluorescence Quantification of DNA In Situ

An inverted microscope (model Axiovert 100 TV, Carl Zeiss, Jena, Germany) was applied for detection of the hybridisation of 5' rhodamine 6G labelled DNA oligomers to immobilised DNA to polystyrene carboxyl-coated beads incorporated into a microflow reactor. The microscope was connected by an optical fiber to an argon-ion laser (model 2080-15S, Spectra-Physics Lasers, Inc., CA). The fluorescent images were detected by a CCD camera (CH250-KAF1400, Photometrics, Tucson, AZ) connected to the microscope by a C-Mount adapter. An emission fluorescent filter opening at 540 nm and closing at 580 nm (Oriel Instruments, Stratford, CT) was used. The images were obtained and analysed by PMIS image processing software (GKR Computer Consulting, Boulder, CO). The average fluorescence signal of regions of images containing beads was measured, subtracting background (dark plus bias) images.

3   Results and Discussion

3.1   Efficient Buffers for DNA Selection, Washing, and Release

Rather than moving the beads, as will be done in integrated applications using the STMs, we keep the beads stationary to facilitate better comparative quantification and switch the composition of the flow instead. A theoretical background for such selective hybridisation kinetics has been presented previously [1]. We postpone the important analysis of the efficiency of DNA selection to the following section. There are several other issues which must first be resolved experimentally.

- the ability to capture DNA on beads from a flowing solution
- the ability to discriminate mismatched DNA by washing
- the ability to dissociate hybridised DNA efficiently with a reversible buffer
- the ability to bind matching DNA selectively in the presence of mismatched DNA

The experiment presented in this section was designed to address these issues in practice. One chamber of the microreactor shown in Fig. 1 was filled with beads immobilized with DNA (see Table 1, line 1). The reactor was held at a temperature of 45°C while a series of different buffer solutions were pumped through the chambers as described below (see Fig. 2). The selective hybridisation in flow involved firstly exactly matching DNA (1 μM ssDNA, see Table 1, line 2) fluorescently labelled with rhodamine 6G in the hybridising buffer solution (500 mM tris-acetate pH 8.3 and 50 mM NaOH). This buffer was chosen since it is the solution in which DNA released from a previous module will be immersed. It arises in a previous module through neutralization of the dehybrising solution (100 mM NaOH) by mixing with an equal volume flow of buffer (1 M tris-acetate pH 8.3). In a separate measurement, it was confirmed that the pH change of the buffer on mixing is negligible. The series of fluorescence measurements, monitoring the concentration of hybridised DNA on the beads, is shown in Fig.2.

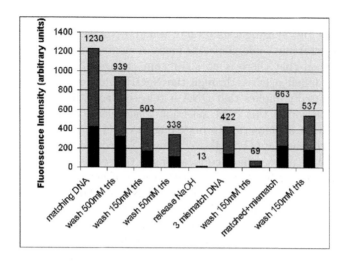

Fig. 2. DNA hybridisation fidelity analysis on beads placed in a microflow reactor. The average fluorescence intensity (minus background signal) of the region of the microreactor containing beads (cf. Fig. 4) is shown under various flow conditions. A single chamber of the microreactor shown in Fig. 1 was employed in this experiment. The measurements are averages over several images, after allowing time for reaction. For details, see text. The main features are successful hybridisation (column 1), release (column 5), the ability to separate matching and weakly mismatching DNA (cf columns 3 vs 7) with the appropriate buffer (tris 150 mM), and the ability to obtain high yield selective hybridisation of matched DNA in the presence of mismatched DNA (columns 8 and 9).

After switching to the hybridising solution, a rapid increase in signal (cf[1]), compatible with time scale induced by partial mixing lengthwise along the inlet tubes (upon buffer change from water), was followed by a slower increase towards saturation as more DNA is flowed past the beads. The DNA hybridisation on beads is taking place in less than a half of minute.

The beads were then washed with an identical buffer solution, not containing DNA, to remove unbound DNA and the fluorescence signal from the free solution. Washing continued for 40 min. The reduction in fluorescence signal (Fig. 2) was consistent with just removing the signal from DNA in the surrounding free solution. Next, the buffer solution was changed to 150 mM tris-acetate, which at lower ionic strength provides a more stringent test of DNA binding than the 500 mM solution. This buffer concentration was suggested as a wash solution by batch experiments in Eppendorf tubes (data not shown) and, as we shall see below, allows the discrimination of imperfectly matching DNA. The data reported in Fig. 2 confirms that significant signal is retained, with the exactly matching DNA, after a period of 30 min washing with this solution (Decreasing the buffer concentration further to 50 mM tris caused a steady depletion of

DNA, 50% over the next 30 min). Next the remaining DNA was dehybridised completely (see Fig. 2) using a solution of 100 mM NaOH. The fluorescent signal returned to its starting value. Separate measurements in a fluorimeter (data not shown) confirmed that there is no significant change of fluorescence signal of rhodamine-6G labelled DNA through the NaOH solution, so that the removal of fluorescence really implies that DNA is released from the beads.

After completion of this basic selection-wash-release cycle, a second round was initiated with an imperfectly matching DNA (1 μM, see Table 1, line 4) labelled with rhodamine 6G as above. This DNA contained three mismatches over a length of 21 nt, at separated locations in the sequence. A smaller amount of DNA bound to the beads was observed (even after 10 min hybridisation at 45°C) than in the case of perfectly matched DNA, as shown in Fig. 2. Washing with the discriminating buffer (150 mM tris-acetate, see above) was sufficient to remove the imperfectly matched DNA from the beads (see Fig. 2), so this buffer provides a good candidate for the intermediate wash channel in the STM modules in [1,2].

Finally, the ability of the wash buffer to discriminate in cases of simultaneous competitive binding between matched and imperfectly matched DNA was tested using a solution of hybridising buffer as at the outset but with 0.5 μM each of the two types of DNA (Table 1, lines 2 and 3), this time only the perfectly matching DNA being fluorescently labelled. The fluorescent signal on the beads rose to about 50% of its value in the first phase of the experiment (with 1 μM of labelled matching DNA), consistent with successful competitive binding of the matched form. After washing with the discriminating buffer (150 mM tris-acetate), the signal reduction was consistent with a removal of free labelled DNA as in the first phase of the experiment (see Fig. 2b). These results demonstrate that selective DNA hybridisation and release can be performed with appropriate buffer solutions and temperature, compatible with a steady flow implementation of DNA Computing as proposed in [1].

3.2   Demonstration of Transfer of DNA between Two Selection Modules with Reversible Chemistry

In a second experiment, the ability to select, restrain, transfer and pick up DNA between two modules was examined. The experimental procedure is similar to that in the previous section (but at 40°C) except that both chambers are filled with beads and separate buffer solutions are used at the inlets to the two chambers at various stages. Whereas in the integrated STMs, operating under steady flow, switching buffers is induced by magnetic bead transfer, here the flows are switched in larger chambers to facilitate quantitation of the transfer yields with constant bead locations.

The series of events for a complete transfer round from one module to the next is:

1) Hybridisation of perfectly matching DNA (see Table 1, line 2) to beads in both chambers. The hybridising buffer (500 mM tris pH 8.3 and 50 mM

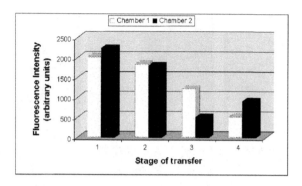

Fig. 3. DNA hybridisation transfer from beads in one module to those in a second module. Average fluorescence intensity (including the background signal of 490 units) images of two regions of the microreactor (see Fig. 1) containing beads (see Fig. 4) are shown under various flow conditions. White bars: fluorescence intensity from the first chamber of the micro-reactor. Black bars, fluorescence intensity from the second chamber. 1. DNA hybridisation on the beads in the first and second chamber. 2. Washing step of DNA hybridised on the beads in the both micro-reactor chambers. 3. Continuous washing step on the beads in the first micro-reactor chamber and denaturation step on the beads in the second chamber. 4. Denaturation of DNA from the beads in the first chamber and pick up of the same DNA by hybridisation on the beads placed in the second micro-flow reactor chamber. For details, see text.

NaOH) containing the DNA is pumped in parallel (4 μl/min) into all three inlets of the microreactor shown in Fig. 1.

2) Washing of the beads in both chambers (500 mM tris pH 8.3 and 50 mM NaOH without DNA), using inlets 1, 2 and 3.

3) Releasing DNA from the beads in the second downstream chamber and washing of the beads in the first chamber. NaOH solution at 100 mM is pumped through inlet 1 and 150 mM tris pH 8.3 through inlets 2 and 3.

4) Simultaneous release of DNA from the beads in the first chamber and pickup of the same DNA in the second chamber. 1 M tris pH 8.3 is pumped into the second chamber (inlet 1) and 100 mM NaOH into the first chamber (inlets 2 and 3). The outflow from the first chamber is neutralised by the tris buffer entering the second chamber from the side and then the DNA hybridises to the beads in the second chamber.

The results are summarized in Fig. 3, showing the quantitation of DNA on the beads in the two chambers at the four stages of the procedure. Figure 4 shows images of the fluorescence from DNA hybridisation to the beads in the two chambers at various stages (2, 3 and 4) of the above transfer procedure. Moderately high flow rates are employed, so that diffusive mixing is not perfect as indicated by the slightly inhomogeneous pickup (Fig. 4, image F). As seen in

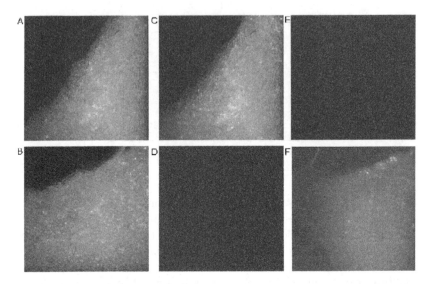

Fig. 4. Fluorescence images of DNA hybridisation on beads in micro-flow reactor showing the course of transfer from one module to the next. DNA hybridisation on the beads placed in the first chamber (A) and second chamber (B); Washing of the beads in the first chamber (C) and DNA denaturation on the beads in the second chamber (D); DNA denaturation from the beads in the first chamber (E) and pick up of the same DNA on the beads placed in the second chamber.

Fig. 3, about 50% of the DNA bound in chamber 1 is successfully picked up in chamber 2 in the transfer step (4, above).

4   Conclusions

This paper demonstrates that the basic processing and linkup steps of steady flow DNA Computing in microreactors can be experimentally verified. The key step of reversible chemical hybridisation and dehybridisation has been demonstrated under flow in microreactor modules similar to the integrated devices to be used in the complete DNA Computer [4]. Once a single complete round of selection, release and pickup on new beads has been achieved, as shown in the previous section, the procedure can be cascaded to many processors in series and in parallel. It should be noted that the problem of increasing salt concentration, on successive rounds of hybridisation and dehybridising by chemical means, does not occur for STMs [1] because the buffers are only used in one complete cycle before the beads are transferred to a fresh flow [1].

Experiments in Eppendorf tubes (data not shown) suggest that the discrimination between matched and non-matched DNA is significantly better with members of the DNA library [4]. This is because they have a modular structure, with unwanted hybridisation involving more than 3 mismatches over a length of

16 nt, so the experiments reported here provide an underestimate of fidelity. This issue will be addressed more completely with the full library in other work. The present paper establishes that a suitable wash buffer of 150 mM tris-acetate pH 8.3, compatible with the reversible hybridisation-dehybridisation solutions, can be employed in the steady flow integrated DNA Computer to enhance discrimination and hence fidelity of matching. The concentration of the denaturating solution (100 mM NaOH) could be reduced down to 50 mM for denaturation of small oligomers so that the concentration of the buffering solution of tris could be reduced to 500 mM instead of 1 M. It could be used a hepes buffer as a buffering solution as well (data not shown). Additional results (data not shown) imply that an alkali DNA denaturation to beads is more efficient than using temperature. This requires reducing to a minimum the non-specific attachment of DNA to the beads. Carboxyl-coated beads show less non-specific DNA attachment than amino-coated ones[3].

The efficiency of DNA transfer from one module to the next of 50% (see Fig. 3, stage of transfer 2 and 4) is not yet optimised, with values greater than 90% being sought. The flow rates employed in the current series of experiments were too high to allow sufficient time for either complete mixing or completed binding to beads. Optimisation of these parameters will be carried out using the network of STM microreactors actually to be employed in the computations[4]. We would like to stress that the current paper represents the first results in which selective bead to bead transfer of DNA with release and pickup has been quantitatively investigated, and that there is significant room for improvement.

Acknowledgements: The transfer and pickup experiments were quantified in a larger microreactor designed previously (to test switchable magnetic mixers) by Kristina Schmidt and reconstructed at the GMD by Marlies Gohlke. The authors wish to thank Jonathan Howard for helpful discussions. Assistance with the fluorescence imaging setup by Harald Mathis, Thomas Kirner and Danny van Noort was appreciated. The support of the GMD for DNA computing and German Ministry of Sience (BMBF) for BioMIP's start-up grant (#01SF9952) is gratefully acknowledged.

References

1. McCaskill, J.S.: Biosystems, 59 (2001) 125–138.
2. McCaskill, J., Penchovsky, R., Gohlke, M., Ackermann, J. and Ruecker, T.: DNA6 Sixth International Meeting on DNA Based Computers- Preliminary Proceedings. Leiden Center for Natural Computing, The Netherlands (2000) 239–246.
3. Penchovsky, R., Birch-Hirschfeld, E. and McCaskill, J.S.: Nucleic Acids Res., 22 (2000) e98.
4. Van Noort, D., Gast, F.U. and McCaskill, J.S.: DNA Computing in Microreactors, DNA7 Conference Proceedings (2001).
5. Frutos, A. G, et.al.: Nucleic Acids Res., 25 (1997) 4748–4757.
6. Adleman, L.M.: Science, 266 (1994) 1021–1024.
7. Hadd, A.G., Raymound, D.E., Hailiwell, J.W., Jacobson, S.C., Ramsey, J.M.: Anal. Chem., 69 (1997) 3407–3412.

8. Hossein, S., Tang, T., Harrison, D.J.: J. Am. Chem. Soc., 119, 8716–8717 (1997).

9. Kopp, M.U., de Mello, A. J. and Manz, A.: Science, 280, 1046-1048 (1998).

10. Burns, M.A., Johnson, B.N., Brahmasandra, S.N., Handique, K., Webster, J.R., Krishnan, M., Sammarco, T.S., Man, P.M., Jones, D., Heldsinger, D., Mastragelo, C.H., Burke, D.T.: Science, 282 (1998) 484–487.

11. Han, J. and Craighead, H. G.: Science, 288 (2000) 1026–1029.

12. Hugh Fan, Z., Mangru, S., Granzow, R., Heaney, P., Ho, W., Dong, Q. and Kumar, R.: Anal. Chem., 71 (1999) 4851–4859.

13. Chui, D.T., Pezzoli, E., Wu, H., Stroock, A.D. and Whitesides, G.M.: PNAS, 98 (2001) 2961–2966.

Coding Properties of DNA Languages*

Salah Hussini[1], Lila Kari[2], and Stavros Konstantinidis[1]

[1] Department of Mathematics and Computing Science, Saint Mary's University,
Halifax, Nova Scotia, B3H 3C3, Canada
s.konstantinidis@stmarys.ca
[2] Department of Computer Science, University of Western Ontario
London, Ontario, N6A 5B7 Canada
lila@csd.uwo.ca
http://www.csd.uwo.ca/~lila

Abstract. The computation language of a DNA-based system consists of all the words (DNA strands) that can appear in any computation step of the system. In this work we define properties of languages which ensure that the words of such languages will not form undesirable bonds when used in DNA computations. We give several characterizations of the desired properties and provide methods for obtaining languages with such properties. The decidability of these properties is addressed as well. As an application we consider splicing systems whose computation language is free of certain undesirable bonds and is generated by nearly optimal comma-free codes.

1 Introduction

DNA (deoxyribonucleic acid) is found in every cellular organism as the storage medium for genetic information. It is composed of units called nucleotides, distinguished by the chemical group, or base, attached to them. The four bases, are *adenine, guanine, cytosine* and *thymine*, abbreviated as A, G, C, and T. (The names of the bases are also commonly used to refer to the nucleotides that contain them.) Single nucleotides are linked together end–to–end to form DNA strands. A short single-stranded polynucleotide chain, usually less than 30 nucleotides long, is called an *oligonucleotide*. The DNA sequence has a *polarity*: a sequence of DNA is distinct from its reverse. The two distinct ends of a DNA sequence are known under the name of the $5'$ end and the $3'$ end, respectively. Taken as pairs, the nucleotides A and T and the nucleotides C and G are said to be complementary. Two complementary single–stranded DNA sequences with opposite polarity are called *Watson/Crick complements* and will join together to form a double helix in a process called *base-pairing*, or *hybridization* [9].

In most DNA computations, there are three basic stages which have to be performed. The first is encoding the problem using single-stranded or double-stranded DNA. Then the actual computation is performed by employing a suc-

---

* All correspondence to Lila Kari. Research partially supported by Grants R2824A01 and R220259 of the Natural Sciences and Engineering Research Council of Canada.

N. Jonoska and N.C. Seeman (Eds.): DNA7, LNCS 2340, pp. 57–69, 2002.
© Springer-Verlag Berlin Heidelberg 2002

cession of *bio-operations* [9]. Finally, the solutions, i.e. the result of the computation are sequenced and decoded.

In many proposed DNA-based algorithms, the initial DNA solution will
contain some oligonucleotides which represent single "codewords", and some
oligonucleotides which are strings of catenated codewords. This leads to two
main types of possible undesirable hybridizations. First, it is undesirable for any
DNA strand representing a codeword to form a hairpin structure, which can
happen if either end of the strand binds to another section of that same strand.
Second, it is undesirable for any DNA strand representing a codeword to bind to
either another codeword strand or to the catenation of two codeword strands. If
such undesirable hybridizations occur, they will in practice render the involved
DNA strands useless for the subsequent computations.

To formalize these problems, let us introduce some notations. An alphabet
X is a finite non-empty set of symbols (letters). We denote by $\Delta = \{A, C, G, T\}$
the DNA alphabet. A word u over the alphabet X is a sequence of letters and
$|u|$ will denote the length of u, that is, the number of letters in it. By 1 we
denote the empty word consisting of zero letters. By convention, a word u over
the DNA alphabet Δ will denote the corresponding DNA strand in the 5′ to
3′ orientation. We denote by $\overleftarrow{u}$ the Watson-Crick complement of the sequence
u, i.e., its reverse mirror image. For example, if $u = 5′ - AAAAGG - 3′$ then
$\overleftarrow{u} = 5′ - CCTTTT - 3′$ will be the Watson-Crick complement that would
base-pair to u. By X^* we denote the set of all words over X, and by X^+ the
set of all non-empty words over X. A language (over X) is any subset of X^*.
For a language K, we denote by K^* the set of words that are obtained by
concatenating zero or more words of K; then, $1 \in K^*$. We write K^+ for the
subset of all non-empty words of K^*.

Let now $K \subseteq \Delta^*$ be a set of DNA codewords. Several types of undesirable
situations can occur with these DNA strands encoding initial data.

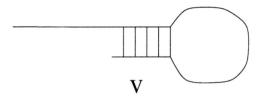

Fig. 1. Intramolecular hybridization I: $uv \in K$, $\overleftarrow{v}$ being a subword of u.

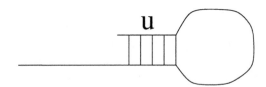

Fig. 2. Intramolecular hybridization II: $uv \in K$, $\overleftarrow{u}$ being a subword of v.

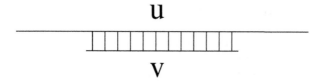

Fig. 3. Intermolecular hybridization I: $u, v \in K$, $\overleftarrow{v}$ being a subword of u.

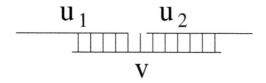

Fig. 4. Intermolecular hybridization II: $u_1, u_2, v \in K$, $\overleftarrow{v}$ being a subword of $u_1 u_2$.

Several attempts have been made to address this issue by trying to find sets of codewords which are unlikely to form undesired bonds in this manner [2], [5], [4]. For example genetic algorithms have been developed which select for sets of DNA sequences that are less likely to form undesirable bonds [3], and combinatorial methods have been used to calculate bounds on the size of a set of uniform codewords (as a function of codeword length) which are less likely to mishybridize [11].

In this paper, we continue the approach in [10] where the notion of θ-compliance has been defined, where θ is an arbitrary morphic or antimorphic involution. In the particular case when θ is the Watson-Crick complement, a language is strictly θ-compliant (respectively strictly prefix θ-compliant, suffix θ-compliant) if situations like the ones depicted in Figure 3 (respectively Figures 1, 2) cannot occur. [10] studied properties of such θ-compliant languages, i.e. languages consisting of words with good coding properties. Here we define the notion of θ-freedom: if θ is the Watson-Crick complement, a language is θ-free iff situations like the ones depicted in Figure 4 cannot occur.

Section 3 studies θ-freedom and its relation to θ-compliance. It turns out that, under some conditions, languages that are θ-free are also θ-compliant.

Section 4 studies the question of whether or not, given a set of codewords, we can constructively decide if the set is θ-compliant or θ-free, i.e., if it has desirable coding properties.

Section 5 studies under what conditions, if we start with an initial set of "good" codewords, θ-compliance and θ-freedom are preserved by all intermediate strands that occur during computations.

With these studies, we attempt to gain a deeper insight into encoding information in DNA, which will hopefully assist solving this otherwise difficult problem.

2 Definitions and Notations

For a set S, we denote by $|S|$ the cardinality of S, that is, the number of elements in S. The set of non-negative integers is denoted by $\mathbf{N}$. Let X^* be the free monoid generated by the finite alphabet X. A mapping $\alpha : X^* \to X^*$ is called a *morphism* (*anti-morphism*) of X^* if $\alpha(uv) = \alpha(u)\alpha(v)$ (respectively $\alpha(uv) = \alpha(v)\alpha(u)$) for all $u, v \in X^*$. A bijective morphism (anti-morphism) is called an *isomorphism* (*anti-isomorphism*) of X^*.

An involution $\theta : S \to S$ of S is a mapping such that θ^2 equals the identity mapping, i.e., $\theta(\theta(x)) = x$ for all $x \in S$. It follows then that an involution θ is bijective and $\theta = \theta^{-1}$. The identity mapping is a trivial example of involution.

If Δ^* is the free monoid generated by the DNA-alphabet Δ then two involutions can be defined on Δ^*: the mirror involution μ which is an anti-morphism, and the complement involution γ which is a morphism ([10]).

Indeed, the mapping $\gamma : \Delta \to \Delta$ defined by $\gamma(A) = T, \gamma(T) = A, \gamma(C) = G, \gamma(G) = C$ can be extended in the usual way to a morphism of Δ^* that is also an involution of Δ^*. Obviously, γ is an involution as the complement of the complement of a sequence equals the sequence itself. This involution γ will be called the *complement involution* or simply the *c-involution* of Δ^*.

Let $\mu : \Delta^* \to \Delta^*$ be the mapping $\mu(u) = v$ defined by

$$u = a_1 a_2 \ldots a_k, \quad v = a_k \ldots a_2 a_1, a_i \in \Delta, 1 \leq i \leq k.$$

The word v is called the mirror image of u. Since μ^2 is the identity mapping, μ is an involution of Δ^* which will be called the *mirror involution* or simply the *m-involution* of Δ^*. The m-involution is an anti-morphism as $\mu(uv) = \mu(v)\mu(u)$ for all $u, v \in \Delta^*$.

It is easy to see that γ and μ commute, i.e. $\gamma\mu = \mu\gamma$, and hence $\gamma\mu$ which will denoted by τ is also an involution of Δ^*. Furthermore τ is an anti-morphism which corresponds to the notion of Watson-Crick complement of a DNA sequence and will therefore sometimes be called the DNA involution.

Instead of γ, μ, τ we sometimes use the alternative notation

$$\gamma(u) = \bar{u}, \ \mu(u) = \tilde{u}, \ \tau(u) = \gamma\mu(u) = {}^{\leftharpoonup}\overline{u}.$$

Following [10], if $\theta : X^* \to X^*$ is a morphic or antimorphic involution, a language $L \subseteq X^*$ is said to be θ-*compliant*, (prefix θ-compliant, suffix θ-compliant) iff, for any words $x, y, u \in X^*$,

$$u, x\theta(u)y \in L \ (\theta(u)y \in L, x\theta(u) \in L) \ \Rightarrow \ xy = 1.$$

If, in addition, $L \cap \theta(L) = \emptyset$ then the language L is called strictly θ-compliant (strictly prefix θ-compliant, strictly suffix θ-compliant, respectively).

Note that if the involution θ is the DNA involution, i.e., it represents the Watson-Crick complement, then a language L being strictly θ-compliant (strictly prefix θ-compliant, strictly suffix θ-compliant) amounts to the fact that situations of the type depicted in Figure 3 (respectively Figure 1, Figure 2) do not occur. Such languages with good coding properties have been studied in [10].

We close the section with some terminology on codes [8]. A code K is a subset of X^+ satisfying the property that, for every word w in K^+, there is a unique sequence $(v_1, v_2, \ldots, v_n)$ of words in K such that $w = v_1 v_2 \cdots v_n$. A bifix code K is a prefix and suffix code; that is, $K \cap KX^+ = K \cap X^+K = \emptyset$.

Every bifix code is indeed a code. An infix code, K, has the property that no word of K is properly contained in another word of K, that is, $K \cap (X^+KX^* \cup X^*KX^+) = \emptyset$. Every infix code is a bifix code. A comma-free code K is a language with the property $K^2 \cap X^+KX^+ = \emptyset$. Every comma-free code is an infix code.

3   Involution-Freedom and Involution-Compliance

In this section we define the notion of an involution-free language which formalizes the situation depicted in Figure 4. Moreover, this notion extends the concept of comma-free code, [13], in the same fashion that involution-compliance extends the concept of infix code.

We define the notion of a θ-free language (Definition 1) and establish relations between θ-freedom and θ-compliance (Lemma 1). We also establish some properties of θ-free languages. For example, Proposition 1 states that if a set of codewords K is strictly θ-free then the set consisting of arbitrary catenations of codewords from K will be strictly θ-free as well. Proposition 2 states that if a language is dense (it contains all possible sequences in X^* as subsequences of some of its strands), then it cannot be strictly θ-compliant or θ-free.

To aid the intuition, Propositions 3, 6, 7, 8 are proved for the particular case of the complement involution even though the results hold for the more general case of an arbitrary morphic involution.

After some examples of complement-free languages, Propositions 3, 6 bring together the notions of complement-compliance and complement-freedom. As it turns out, under some conditions, languages that avoid undesirable bindings of the type in Figures 1–3 avoid also situations of the type in Figure 4, and vice-versa.

Definitions 3 and 4 introduce the notions of complement-reflective respectively anti complement-reflective languages, which are generalizations of the notions of reflective and anti-reflective languages, [13], and relates them to complement-freedom.

This paves the way to methods of constructing complement-free languages as the one described in Proposition 7.

Proposition 8 gives a sufficient condition under which the catenation of two languages is complement-free.

Propositions 9, 10 address similar problems but this time for the anti-morphic involution case which is the type that the DNA Watson/Crick involution belongs to. Because of the additional "flipping" complication that such an involution entails, the results are slightly different.

Definition 1. *Let θ be an involution of X^*. A language K is called θ-free if $K^2 \cap X^+\theta(K)X^+ = \emptyset$. If, in addition, $K \cap \theta(K) = \emptyset$ then K is called strictly*

θ-free. For convenience, we agree to call K strictly $θ$-free when $K \setminus \{1\}$ is strictly
θ-free.

As an example, consider the DNA involution τ of Δ^* and the language $ACC\Delta^2$. Then, $ACC\Delta^2 ACC\Delta^2 \cap \Delta^+(\Delta^2 GGT)\Delta^+ = \emptyset$. Hence, $ACC\Delta^2$ is τ-free and, in fact, strictly τ-free. The same language is strictly free for the complement involution as well. On the other hand, $A\Delta^2 A$ is strictly τ-compliant but not τ-free, as $ACTAATAA \in (A\Delta^2 A)^2 \cap \Delta^+ \tau(A\Delta^2 A)\Delta^+$.

Lemma 1. *Let $θ$ be a morphic or antimorphic involution and let K be a non-empty subset of X^+.*
  (i) If K is $θ$-free then both K and $θ(K)$ are $θ$-compliant.
  (ii) If K is strictly $θ$-free then $K^2 \cap X^* θ(K) X^* = \emptyset$.

Proposition 1. *Let $θ$ be a morphic or antimorphic involution and let K be a non-empty subset of X^+. If K is strictly $θ$-free then also K^+ is strictly $θ$-free.*

It is shown next that the properties of strict $θ$-compliance and strict $θ$-freedom impose restrictions on the words of a language in terms of density and completeness. A language L is called dense, [1], if every word is a subword of some word of L; that is, $L \cap X^* w X^* \neq \emptyset$ for every $w \in X^+$. The language L is complete if L^* is dense, [1].

Proposition 2. *Let $θ$ be a morphic or antimorphic involution and let K be a non-empty subset of X^+.*

(i) *If K is dense then K is not strictly $θ$-compliant.*
(ii) *If K is complete then K is not strictly $θ$-free.*

To make results more intuitive, we will prove the following four propositions for the particular case where $θ$ is the complement involution. Note that the results can be generalized to refer to any arbitrary morphic involution.

There are many examples of complement-free languages. $L = AC^+ A$ is an infinite complement-free language as

$$\{AC^{n_1} AAC^{n_2} A \mid n_1, n_2 > 0\} \cap \Delta^+ TG^+ T\Delta^+ = \emptyset.$$

For any $L_1 \subseteq AC^*$ and $L_2 \subseteq C^+ A$, $L_1 L_2$ is a complement-free language. Indeed, $L_1 L_2 = AC^+ A$ which is complement-free.

Every subset of a complement-free language is itself complement-free.

Let $L \subseteq \Delta^+$ be a finite language and let $m = \max\{|u| \mid u \in L\}$. Then $L' = AC^m LAC^m$ is complement-free. Indeed, assume $L'^2 \cap \Delta^+ \overline{L'} \Delta^+ \neq \emptyset$. Then, $AC^m w_1 AC^m AC^m w_2 AC^m = \alpha_1 TG^m \overline{w_3} TG^m \alpha_2$ for some $w_1, w_2, w_3 \in L, \alpha_1, \alpha_2 \in \Delta^+$. This implies TG^m is a subword of w_1 or w_2 – contradiction as $|w_1|, |w_2| \leq m$.

An example of a language L satisfying $L \cap \overline{L} = \emptyset$ is $\Delta^* \{A, C\}$.

For a language $L \subseteq X^+$ denote by

$$L_{pref} = \{x \in X^+ \mid xy \in L \text{ for some } y \in X^+\}$$

$$L_{suff} = \{y \in X^+ \mid xy \in L \text{ for some } x \in X^+\}$$

the language of non-empty proper prefixes of L and the language of non-empty proper suffixes of L respectively.

Proposition 3. *Let $L \subseteq X^+$, $L \neq \emptyset$, be a language. The following statements are equivalent:*

(1) *L is complement-free.*
(2) *L is c-compliant and $\overline{L} \cap L_{suff}L_{pref} = \emptyset$.*
(3) *L is c-compliant and $L^2 \cap L_{pref}\overline{L}L_{suff} = \emptyset$.*
(4) *$\overline{L}$ is complement-free.*

Definition 2. *Let $\theta : X^* \longrightarrow X^*$ be a morphic or antimorphic involution. For a non-empty language $L \subseteq X^+$ define*
$$L_s = \{z \in X^+ \mid ux \in L, xz \in \theta(L) \text{ for some } u, x \in X^+\},$$
$$L_{is} = \{x \in X^+ \mid wx \in \theta(L), xv \in L \text{ for some } w, v \in X^+\},$$
$$L_p = \{x \in X^+ \mid xv \in \theta(L), vy \in L \text{ for some } y, v \in X^+\},$$
$$L_{ip} = \{x \in X^+ \mid ux \in L, xv \in \theta(L) \text{ for some } u, v \in X^+\}.$$

Note that in the particular case, where we talk about a morphic involution we have that $L_{is} = \theta(L_{ip})$. Indeed $x \in L_{is}$ means $wx \in \theta(L), xv \in L$ for some $w, v \in X^+$ which means that $\theta(w)\theta(x) \in L, \theta(x)\theta(v) \in \theta(L)$ which, in turn, means that $\theta(x) \in L_{ip}$, that is, $x \in \theta(L_{ip})$.

Proposition 4. *Let $L \subseteq X^+$ be a non-empty language and $\theta : X^* \longrightarrow X^*$ be an antimorphic involution. Then $\theta(L)_s = \theta(L_p)$ and $\theta(L)_p = \theta(L_s)$.*

Proposition 5. *Let $L \subseteq X^+$ be a non-empty language and $\theta : X^* \longrightarrow X^*$ be a morphic involution. Then $\theta(L)_s = \theta(L_s)$ and $\theta(L)_p = \theta(L_p)$.*

Take the particular case where θ is the complement involution.

Proposition 6. *Let $L \subseteq X^+$ be a non-empty language. Then L is complement-free if and only if L is complement-compliant and one of the following conditions hold:*

(1) *$L_s \cap L_{is} = \emptyset$,*
(2) *$L_p \cap L_{ip} = \emptyset$,*
(3) *$L \cap \overline{L_p}L_p = \emptyset$,*
(4) *$L \cap L_s\overline{L_s} = \emptyset$.*

Note that Proposition 6 holds also if we replace the complement involution with an arbitrary morphic involution.

Definition 3. *A language $L \subseteq X^+$, $L \neq \emptyset$ is called complement-reflective iff for all $x, y \in X^+$ we have that $xy \in L$ implies $\overline{yx} \in L$.*

Definition 4. *A language $L \subseteq X^+$, $L \neq \emptyset$ is called anti complement-reflective iff for all $x, y \in X^+$ we have that $xy \in L$ implies $\overline{yx} \notin L$.*

Note that every complement-free language is anti complement-reflective. Indeed, if for a complement-free language L we would have $xy \in L$ and $\overline{yx} \in L$ for some $x, y \in X^+$ then

$$(\overline{yx})(\overline{yx}) = \overline{yxyx} = \overline{y}(\overline{xy})\overline{x} \in X^+ \overline{L} X^+ \cap L^2$$

which violates the condition of complement-freedom.

However, one can find anti complement-reflective languages that are not complement-free. Indeed, take $L = \{A^n T^{2n} \mid n \geq 1\}$ over the alphabet $\Delta = \{A, C, G, T\}$ with the usual complement function. Then L is anti complement-reflective but not complement-free. Indeed, $xy \in L$ implies $xy = A^n T^{2n}$ for some $n \geq 1$. For $\overline{yx}$ to belong to L, it must be the case that $x \in A^+$ and $y \in T^+$. Then $\overline{yx} = A^{2n} T^n \notin L$ therefore L is anti complement-reflective.

To show that L is not complement-free, take $u_1 u_2 \in L$, that is, $u_1 u_2 = A^n T^{2n} A^m T^{2m}$ for some $n, m \geq 1$. As it is possible to find some indices n, m, p such that $v \in L$, $v = A^p T^{2p}$ and $\overline{v} = T^p A^{2p}$ a subword of $u_1 u_2$, it results that L is not complement-free.

In general the concatenation of two complement-free languages may not be complement- free. For example, take $L_1, L_2 \subseteq \Delta^+$,

$$L_1 = \{CACA, AC\}, \quad L_2 = \{TGTG, GT\}.$$

Both L_1 and L_2 are complement-free. On the other hand $L_1 L_2$ contains the word $u = CACATGTG$ therefore we have that $uu = CACATGTGCACATGTG \in (L_1 L_2)^2$ and the word $v = ACGT \in L_1 L_2$ has the property that $\overline{v} = TGCA$ is a subword of uu. This means that $L_1 L_2$ is not complement-free.

Nevertheless, for a given finite language $L \subseteq \Delta^+$ we can always find a word $u \in \Delta^+$ such that uL is complement-free.

Indeed, if $L = \{u_1, u_2, \ldots, u_n\}$ is a finite language and $m = \max\{|u| \mid u \in L\}$ then for the word $u = A^{m+1} C$, $m \geq 1$ the language uL is complement-free.

Take $\alpha_1 \alpha_2 \in uLuL$. Then $\alpha_1 \alpha_2 = A^{m+1} C u_i A^{m+1} C u_j$ for some $1 \leq i, j \leq n$. Consider $\alpha \in uL$. Then $\overline{\alpha} = T^{m+1} G \overline{u_k}$ for some $1 \leq k \leq n$. The only possibility for $\overline{\alpha}$ to be a subword of $\alpha_1 \alpha_2$ would be that $T^{m+1} G$ is a subword of u_i or of u_j which is impossible because of the way m has been defined. Consequently, uL is complement-free.

Proposition 7. *For any finite language $L \subseteq \Delta^+$ there exists an infinite language $R \subseteq \Delta^+$ such that RL is a complement-free language.*

The next question to address is: given two languages $A, B \subseteq X^+$, when is AB a complement-free language?

Proposition 8. *Let* $A, B \subseteq X^+$ *be two non-empty languages. Assume that* $A \cap \overline{B} = \emptyset$ *and that* $A \cup B$ *is complement compliant. If* $A_s \cap \overline{B_p} = \emptyset$ *then* AB *is a complement-free language.*

Corollary 1. *Let* $A, B \subseteq X^+$ *be two non-empty languages. If* $A \cup B$ *is strictly complement compliant and* $A_s \cap \overline{B_p} = \emptyset$ *then* AB *is complement free.*

Remark that the preceding proposition and corollary hold also for any arbitrary morphic involution, not only for the complement involution.

The case of antimorphic involutions, i.e., of involutions of the type of the DNA involution, is slightly different.

Proposition 9. *Let* $\theta : X^* \longrightarrow X^*$ *be an antimorphic involution and* $\emptyset \neq L \subseteq X^+$ *be a language. Then* L *is* θ-free *if and only if* L *is* θ-compliant *and one of the following conditions hold:*

(1) $L_s \cap L_{is} = \emptyset$,
(2) $L_p \cap L_{ip} = \emptyset$,
(3) $L \cap \theta(L_s)L_p = \emptyset$,
(4) $L \cap L_s\theta(L_p) = \emptyset$.

Proposition 10. *Let* $A, B \subseteq X^+$ *be two non-empty languages and* $\theta : X^* \longrightarrow X^*$ *be an antimorphic involution. Assume, furthermore, that* A *and* B *are strictly* θ-compliant *and* $A \cup B$ *is* θ-compliant. *If* $B_s \cap \theta(A_s) = A_p \cap \theta(B_p) = \emptyset$ *then* AB *is a* θ-free *language.*

4 Decidability Issues

Given a family of sets of codewords, we ask if we can construct an effective algorithm that takes as input a description of a set of codewords from the given family and outputs the answer yes or no depending on whether or not the set is θ-free or θ-compliant (has good encoding properties). If such an algorithm exists, the question is deemed "decidable," otherwise it is "undecidable." This section considers the problem of deciding τ-compliance and τ-freedom where τ is the DNA involution. As in the case of infix and comma-free codes — see [8] — it turns out that these properties are decidable for regular languages but undecidable for context-free languages. We use regular expressions to represent regular languages and context-free grammars for context-free languages. If E is a regular expression (context-free grammar), $L(E)$ denotes the language represented by (generated by) E.

Proposition 11. *Let* τ *be the DNA involution. The following two problems are decidable.*

1. *Input: A regular expression* E.
   Output: YES or NO depending on whether $L(E)$ *is* τ-free.

2. *Input: A regular expression E.*
   Output: YES or NO depending on whether $L(E)$ is τ-compliant.

Proposition 12. *Let τ be the DNA involution. The following problems are undecidable:*

1. *Input: A context-free grammar F.*
   Output: YES or NO depending on whether or not $L(F)$ is τ-free.
2. *Input: A context-free grammar F.*
   Output: YES or NO depending on whether or not $L(F)$ is τ-compliant.

5 Splicing Systems That Preserve Good Encodings

The preceding sections studied conditions under which sets of DNA codewords have good encoding properties, like DNA compliance and τ-freedom, where τ is the DNA involution. The next question is to characterize initial sets of codewords having the additional feature that the good encoding properties are preserved during *any* computation starting out from the initial set. As a computational model we have chosen the computation by splicing [7,12,9]. Proposition 14 states that if the splicing base (initial set of codewords) is strictly θ-free then all the sequences that may appear along any computation will not violate the property of θ-freedom. Proposition 15 is a stronger result stating that for any computational system based on splicing one can construct an equivalent one with "good," i.e. θ-compliance and θ-freedom properties, being preserved during any computation. Finally, Proposition 16 provides a method of construction of a strictly θ-free code that can serve as splicing base, i.e. initial soup, for a computation. Moreover, it is proved that from the point-of-view of the efficiency of representing information, the constructed code is close to optimal.

A multiset M over an alphabet Σ is a collection of words in Σ^* such that a word can occur in M more than once. More formally, a multiset M is a mapping of Σ^* into $\mathbf{N}$ such that $M(w)$ is the number of copies of the word w in the multiset. For a multiset M, we write $\mathrm{supp}\,(M)$ to denote the set of (distinct) words that occur in M; that is, $\mathrm{supp}\,(M) = \{w \in \Sigma^* \mid M(w) > 0\}$. A splicing system — see [9] — is a quadruple $\gamma = (\Sigma, T, A, R)$ such that Σ is an alphabet, T is a subset of Σ, the set of terminal symbols, A is a multiset over Σ, the initial collection of words, and R is a set of splicing rules of the form $\alpha_1 \# \beta_1 \$ \alpha_2 \# \beta_2$ with $\alpha_1, \alpha_2, \beta_1, \beta_2$ being words in Σ^*. For two multisets M and M' we write $M \Longrightarrow_\gamma M'$, if there is a splicing rule $\alpha_1 \# \beta_1 \$ \alpha_2 \# \beta_2$ in R and two words in M of the form $x_1 \alpha_1 \beta_1 y_1$ and $x_2 \alpha_2 \beta_2 y_2$ such that M' is obtained from M by replacing those words with $x_1 \alpha_1 \beta_2 y_2$ and $x_2 \alpha_2 \beta_1 y_1$ – see [9]. The language, $L(\gamma)$, *generated* by a splicing system $\gamma = (\Sigma, T, A, R)$ is the set $\{w \in T^* \mid \exists M, A \Longrightarrow_\gamma M, w \in \mathrm{supp}\,(M)\}$. The *computation language of* γ is the set of words over Σ that can appear in any computation step of γ; that is, the language

$$\{w \mid w \in \mathrm{supp}\,(M) \text{ and } A \Longrightarrow_\gamma^* M, \text{ for some multiset } M\}$$

Note that the language generated by γ is a subset of the computation language of γ.

When the computation language of a splicing system involves only symbols of the alphabet Δ, with the usual functionality of these symbols, one wants to prevent situations like the ones described in the introduction and this requires choosing languages with certain combinatorial properties. In this section, we consider a type of splicing systems in which only symbols of the alphabet Δ are used and require that the language of computation of such systems is strictly θ-free, where θ is a morphic or antimorphic involution. To this end, we utilize Proposition 1 by requiring that the computation language is of the form K^* for some strictly θ-free subset K of Δ^+. Moreover, we require that K is a splicing base, as defined below, to ensure that, after performing a splicing operation between two words in K^*, the resulting words are still in K^*

Definition 5. *A splicing base is a set of words K such that, for all $x_1, x_2, y_1, y_2 \in \Delta^*$ and for all $v_1, u_1, v_2, u_2 \in K^*$ with $|v_1 u_1| > 0$ and $|v_2 u_2| > 0$, if $x_1 v_1 u_1 y_1$, $x_2 v_2 u_2 y_2 \in K^*$ then $x_1 v_1 u_2 y_2$, $x_2 v_2 u_1 y_1 \in K^*$.*

In the following lemma it is shown that the splicing base condition is equivalent to the following:

(C_K) For all $x \in \Delta^*$ and for all $u \in K$, if xu is a prefix of K^* then $x \in K^*$ and if ux is a suffix of K^* then $x \in K^*$.

Remark: A splicing base is not necessarily a code. For example, $K = \{A, A^2\}$ is not a code but it is a splicing base.

Lemma 2.

(i) *Condition (C_K) implies the following: For all $x \in \Delta^*$ and for all $v \in K^+$, if xv is a prefix of K^* then $x \in K^*$, and if vx is a suffix of K^* then $x \in K^*$.*
(ii) *A set of words K is a splicing base if and only if (C_K) holds.*
(iii) *Every splicing base which is a code is a bifix code.*
(iv) *Every comma-free code is a splicing base.*

We define now a special type of splicing systems that involve only symbols of the alphabet Δ.

Definition 6. *Let K be a splicing base. A K-based splicing system is a quadruple $\beta = (K, T, A, R)$ such that $T \subseteq K$, A is a multiset with $\mathrm{supp}\,(A) \subseteq K^*$, and R is a subset of $K^* \# K^* \$ K^* \# K^*$ with the property that $v_1 \# u_1 \$ v_2 \# u_2 \in R$ implies $|v_1 u_1| > 0$ and $|v_2 u_2| > 0$.*

For two multisets M and M', the relationship $M \Longrightarrow_\beta M'$ is defined as in the case of ordinary splicing systems – see [9]; that is, M' is obtained from M by applying a splicing rule to two words in $\mathrm{supp}\,(M)$. The language $L(\beta)$ *generated* by a K-based splicing system β is the set $\{w \in T^* \mid \exists M, A \Longrightarrow_\beta^* M$ and $w \in \mathrm{supp}\,(M)\}$. The *computation language* of the system β is the set $\{w \in \Delta^* \mid \exists M, A \Longrightarrow_\beta^* M$ and $w \in \mathrm{supp}\,(M)\}$. As shown next, this language is a subset of K^*.

Proposition 13. *For every splicing base K and for every K-based splicing system β, the computation language of β is a subset of K^*.*

Proposition 14. *Let θ be a morphic or antimorphic involution and let β be a K-based splicing system, where K is a splicing base. If K is strictly θ-free then the computation language of β is strictly θ-free.*

In the rest of the section we consider the question of whether an arbitrary splicing system γ can be translated to a K-based splicing system β over Δ such that the computation language of β is strictly θ-free and the languages generated by β and γ are isomorphic.

Definition 7. *Let θ be a morphic or antimorphic involution of Δ^* and let $\mathbf{K}$ be an infinite family of splicing bases such that every splicing base in $\mathbf{K}$ is a strictly θ-free code. The involution-free splicing class $\mathbf{C}_{\theta,\mathbf{K}}$ is the set of all K-based splicing systems, where $K \subseteq K'$ for some $K' \in \mathbf{K}$.*

Proposition 15. *Let $\mathbf{C}_{\theta,\mathbf{K}}$ be an involution-free splicing class and let Σ be an alphabet. For every splicing system $\gamma = (\Sigma, T, A, R)$ there is a K-based splicing system $\beta = (K, F, B, P)$ in $\mathbf{C}_{\theta,\mathbf{K}}$ and an isomorphism $f : T^* \to F^*$ such that the computation language of β is strictly θ-free and $L(\beta) = f(L(\gamma))$.*

We close the section with a simple construction of an infinite comma-free code, $K_{m,\infty}$, which is strictly free for the DNA involution τ. By Lemma 2 (iv), the code is a splicing base as well. Moreover, we define an infinite family $\{K_{m,n} \mid n \in \mathbf{N}\}$ of finite subsets of $K_{m,\infty}$ whose information rate tends to $(1 - 1/m)$. The involution τ and the family $\{K_{m,n} \mid n \in \mathbf{N}\}$ define a τ-free splicing class which can be used to 'simulate' every splicing system according to Proposition 15. The fact about the information rate of the codes $K_{m,n}$ concerns the efficiency (or redundancy) of these codes when used to represent information. In general, the information rate of a finite code K over some alphabet X is the ratio

$$\frac{\log_{|X|}|K|}{(\sum_{v\in K}|v|)/|K|}$$

When that rate is close to 1, the code is close to optimal.

Proposition 16. *Let m and n be non-negative integers with $m > 1$, let $K_{m,\infty} = A^m T(\Delta^{m-1}T)^*C^m$, and let $K_{m,n} = \cup_{i=0}^n A^m T(\Delta^{m-1}T)^iC^m$. Then,*

(i) *$K_{m,\infty}$ and, therefore, $K_{m,n}$ are comma-free codes and strictly τ-free, where τ is the DNA involution*

(ii) *The information rate of $K_{m,n}$ tends to $(1 - 1/m)$ as $n \to \infty$.*

References

1. J. Berstel and D. Perrin. *Theory of Codes*, Academic Press, Inc., Orlando, Florida, 1985.
2. R. Deaton, R. Murphy, M. Garzon, D.R. Franceschetti, and S.E. Stevens. Good encodings for DNA-based solutions to combinatorial problems. *DNA-based computers II*, in AMS DIMACS Series, vol. 44, L.F. Landweber and E. Baum, Eds., 1998, 247–258.
3. R. Deaton, M. Garzon, R. Murphy, D.R. Franceschetti, and S.E. Stevens. Genetic search of reliable encodings for DNA-based computation, *First Conference on Genetic Programming GP-96*, Stanford U., 1996, 9–15.
4. M. Garzon, R. Deaton, P. Neathery, D.R. Franceschetti, and R.C. Murphy. A new metric for DNA computing. *Proc. 2nd Genetic Programming Conference*, Stanford, CA, 1997, Morgan-Kaufmann, 472–478.
5. M. Garzon, R. Deaton, L.F. Nino, S.E. Stevens, Jr., and M. Wittner. Genome encoding for DNA computing. *Proc. 3rd Genetic Programming Conference*, Madison, WI, 1998, 684–690.
6. T. Harju and J. Karhumäki. Morphisms. In *Handbook of Formal Languages,* vol. 1, G. Rozenberg, A. Salomaa, Eds., Springer-Verlag, Berlin, 1997, 439–510.
7. T. Head. Formal language theory and DNA: an analysis of the generative capacity of recombinant behaviors. *Bulletin of Mathematical Biology* 49 (1987), 737–759.
8. H. Jürgensen and S. Konstantinidis. Codes. In *Handbook of Formal Languages*, vol. 1, G. Rozenberg and A. Salomaa, Eds., Springer-Verlag, Berlin, 1997, 511–607.
9. L. Kari. DNA computing: arrival of biological mathematics. *The Mathematical Intelligencer*, vol. 19, no. 2, Spring 1997, 9–22.
10. L. Kari, R. Kitto, and G. Thierrin. Codes, involutions and DNA encoding. Workshop on Coding Theory, London, Ontario, July 2000. To appear.
11. A. Marathe, A. Condon, and R. Corn. On combinatorial DNA word design. *DNA based Computers V*, DIMACS Series, E. Winfree, D. Gifford, Eds., AMS Press, 2000, 75–89.
12. G. Paun, G. Rozenberg, and A. Salomaa. *DNA Computing: New Computing Paradigms*, Springer-Verlag, Berlin, 1998.
13. H.J. Shyr, *Free Monoids and Languages*, Hon Min Book Company, Taichung, Taiwan, R.O.C., 1991.

Boundary Components of Thickened Graphs

Nataša Jonoska and Masahico Saito

Department of Mathematics
University of South Florida
Tampa, FL 33620
{jonoska,saito}@math.usf.edu

Abstract. Using linear DNA segments and branched junction molecules many different three-dimensional DNA structures (graphs) could be self-assembled. We investigate maximum and minimum numbers of circular DNA that form these structures. For a given graph G, we consider compact orientable surfaces, called thickened graphs of G, that have G as a deformation retract. The number of boundary curves of a thickened graph G corresponds to the number of circular DNA strands that assemble into the graph G. We investigate how this number changes by recombinations or edge additions and relate to some results from topological graph theory.

1 Introduction

In biomolecular computing, often, DNA linear segments are used as computational devices. As DNA strands form 3-dimensional (3D) structures [10], it was proposed in [4,5] (see also [17,18]) that 3D DNA structures be used as computational devices. The idea was to solve computational problems related to graphs by actually assembling the graphs by DNA segments, linear DNA double stranded molecules and k-branched junction molecules. The circular strands of these complex DNA structures form knots and links. Knot structures of DNA have been studied [14] and methods of detection of knotted DNA strands have been developed [13]. Such molecules can be detected using electron microscopy or gel-electrophoresis techniques [3,16]. As DNA structures made of single circular DNA strand can be potentially easier to amplify, clone and in general use in laboratory experiments, it is of interest to know what kind of DNA structures can be made by such molecules. Therefore, use of a single stranded knotted DNA that appear as a double stranded version of the graph constructed are preferable. This problem was addressed from chemical point of view by N. Seeman [11] where it was suggested that by allowing portions of "free" single stranded DNA segments, not considered as parts of the structure, the entire DNA structure could be made by a single strand. In this article we consider the case when such single stranded parts, without their Watson-Crick complements, are not used. As the size and molecular weight of such molecules that represent a graph structure will be in question, it is natural to ask how many strands, i.e. components it can have (not always a single component for a given graph), thus the problem investigated herein.

N. Jonoska and N.C. Seeman (Eds.): DNA7, LNCS 2340, pp. 70–81, 2002.
© Springer-Verlag Berlin Heidelberg 2002

Specifically, the changes in the number of DNA strands that make up these DNA graph structures, when such structures are changed by recombinations or edge additions, are of interest from point of view of applications. In fact, in topological graph theory it has been observed that the number of DNA strands (which corresponds to the number of boundary components of a thickened graph, see Section 2) is related to the (maximal and minimal) genus of the graph, and a number of results are known. The purpose of this paper is to observe how the number of DNA strands changes under possible recombinations of DNA, explain how these numbers are related to other notions in topological graph theory, and lead the reader to some relevant mathematical references.

The paper is organized as follows. Definitions of a thickened graph and double strand numbers are given in Section 2. The effects of local changes of a graph on the double strand numbers, such as edge additions, are addressed in Section 3. Some simple ways of computations of double strand numbers using planar diagrams are given in Section 4. We end the paper by relating the double strand numbers to commonly known concepts in topological graph theory.

2   Thickning Graphs and Double Strand Numbers

The mathematical model that we consider for a DNA graph structure is a thickened graph. We imagine that a strip of surface is laid along the Watson-Crick base pairing of the molecule and the phosphodiester bonds are the boundaries of these strips. The number of boundary components for the thickened graph corresponds to the number of DNA strands that make up the DNA graph structure. More precisely, we have the following definition.

Definition 1. Let G be a finite graph. A *thickened graph* $F(G)$ (of G) is a compact orientable surface (2-dimensional manifold) such that G is topologically embedded (as a 1-complex) in $F(G)$ as a deformation retract (see [7] for example for the definition of a deformation retract).

Note that $F(G)$ is not necessarily uniquely determined by G, even up to homeomorphism. The notation $F(G)$ simply indicates that we choose and fix one of such surfaces.

Definition 2. Let G be a finite graph. The number of boundary components (curves) of a thickened graph $F(G)$ is denoted by $\#\partial F(G)$. The maximum of $\#\partial F(G)$ over all thickened graphs $F(G)$ of the given graph G is called the *maximal double strand number* of G and denoted by $DS(G)$. The *minimal double strand number* is similarly defined and is denoted by $ds(G)$.

When we construct a graph by DNA, the boundary curves $\partial F(G)$ are realized by single stranded DNA molecules, and each edge is a double stranded molecule, which motivates the names for ds and DS. Note that the orientability of the surfaces is required, as DNA strands paired by WC complementarity are oppositely oriented segments. Figure 1 shows how two distinct thickened graphs can be constructed by DNA strands, for the same graph. The first thickened graph (a) consists of five distinct strands, and hence $\#\partial F(G)$ for this thickened graph is 5. In the second case (b), it is shown that a single DNA strand can form a double stranded version of the graph. In that case $\#\partial F(G) = 1$ and in fact, for this graph $DS(G) = 5$ and $ds(G) = 1$.

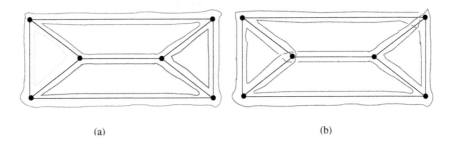

(a)   (b)

Fig. 1. One graph made by DNA strands in two ways.

Example 1. Let O_n denote the set of disjoint circles. Then $F(O_n)$ is homeomorphic to n copies of annuli (and in this case unique up to homeomorphism). Therefore $DS(O_n) = 2n = ds(O_n)$.

In general, we observe from the definition that if $G = G_1 \cup G_2$ is a disjoint union of two graphs G_1 and G_2, then $DS(G) = DS(G_1) + DS(G_2)$ and $ds(G) = ds(G_1) + ds(G_2)$. Therefore, we focus on connected graphs from now on. Since a vertex with degree more than three can be perturbed to a combination of 3-valent vertices, we focus on 3-valent graphs. Such graphs are also known as *three-regular graphs*. It is easy to see that ds and DS numbers do not change by this perturbation.

It turns out that ds and DS are closely related to the *Betti number* of graphs. Intuitively, the first Betti number $\beta_1(G)$ is the number of "independent" loops in a given graph G. More details will be given in the last section. Here we note the following two properties of $\beta_1(G)$ which will be explained in the last section.

(a) For any connected 3-valent graph G, we have

$$ds(G) \leq DS(G) \leq \beta_1(G) + 1.$$

(b) Let G be a 3-valent graph, and let $F(G)$ be a thickened graph. Then $\#\partial F(G) \equiv \beta_1(G) + 1 \pmod 2$. Consequently, $ds(G)$, $DS(G)$, and $\beta_1(G) + 1$ are all even or all odd.

The following problems on double strand numbers are of interest for potential applications.

- When a new graph is constructed from old ones, find a formula relating their double strand numbers. In particular, constructions that occur in self assembling of DNA graph structures are of interest.
- Find estimates of the double strand numbers by means of numerical graph invariants, such as Tutte polynomials.
- Characterize the positive integers k, m, n such that there exists a finite connected 3-valent graph G with $\mathrm{ds}(G) = k$, $\mathrm{DS}(G) = m$, and $\beta_1(G)+1 = n$, where β_1 denotes the first Betti number.
- Determine how are DS and ds of a cover of a graph related to the DS and ds of the graph itself.
- Characterize the graphs of given numbers for ds, DS, and $\beta_1(G) + 1$.

Some of these problems have been considered in topological graph theory within different settings. In the last section we will explain how the double strand numbers can be interpreted in terms of standard terminologies in topological graph theory, and lead the readers to a few relevant references for guidance.

3   Changes in Graphs and Double Strand Numbers

In this section we point out how the changes like edge additions in graphs affect the DS and ds numbers.

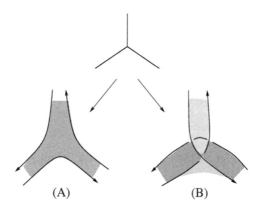

Fig. 2. Neighborhood of a vertex

For 3-valent graphs, the following description is convenient for the thickened graphs. At a 3-valent vertex, consider the portions of an orientable surface depicted in Fig. 2. The arrows of the curves in the figure indicate that the surfaces are orientable. These thick arrowed curves in the figure will become the boundary of $F(G)$ constructed. There are three segments corresponding to the ends of edges of a vertex, where the portions of the surface end. Two of these segments of the surfaces corresponding to two vertices v_1 and v_2 are glued and identified with each other when an edge connects v_1 and v_2 in G. This situation is depicted in Fig. 3. The top of the figure represents two vertices connected by an edge. The middle describes how to glue the patches together, and the bottom figure describes the resulting surface.

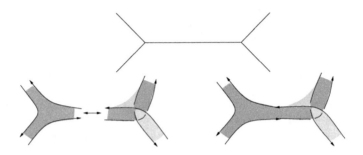

Fig. 3. Gluing patches together

Note that by replacing the surface between (A) and (B) of Fig. 2, the way the boundary curves are connected change. Therefore, it can be regarded that the choices of these surfaces at 3-valent vertices produce variety of homeomorphism types of $F(G)$.

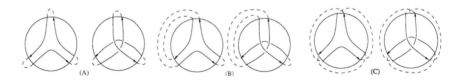

Fig. 4. How boundary curves change

Next Lemma shows how the DS and ds numbers change under such replacements.

Lemma 1. *The replacement at a vertex between Fig. 2 (A) and (B) changes the number of components involved between $1 \leftrightarrow 1$, $2 \leftrightarrow 2$, or $1 \leftrightarrow 3$.*

In other words, if the number of boundary curves belong to a single component, then the number either remains unchanged or changes to three under the replacement. If the number is 2, then the number remains unchanged. If the number is three, then the number changes to one.

Proof. When a replacement is performed at a vertex, the number of boundary components changes depending on how they are connected outside the neighborhood, and all such cases are depicted by dotted lines in Fig. 4. By counting the number of components for each case, we obtain the result. □

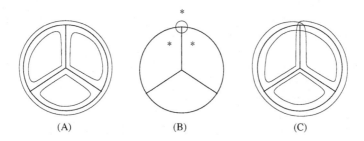

(A)   (B)   (C)

Fig. 5. Re-connections for planar graphs

Definition 3. Let G be a finite 3-valent graph. A finite 3-valent graph G' is said to be obtained from G by an *edge addition* if there are two points v_1 and v_2 on the interior of edges e_1 and e_2 (possibly $e_1 = e_2$ but $v_1 \neq v_2$) of G such that G' is obtained from G by attaching a (new) edge e from v_1 to v_2 (so that $\partial e = v_1 \cup v_2$).

Conversely, it is said that G is obtained from G' by an *edge deletion*.

Lemma 2. *Let G' be a graph obtained from a graph G by an edge addition. Then $\mathrm{DS}(G') = \mathrm{DS}(G) \pm 1$, and $\mathrm{ds}(G') = \mathrm{ds}(G) \pm 1$.*

Proof. Suppose that a thickened graph $F(G')$ is obtained from a $F(G)$ by adding a band along the boundary. If the band is attached to a single component of $\partial F(G)$, then the number of components increases by 1. If the band is attached to distinct components, the components decreases by 1. Hence $\#\partial F(G') = \#\partial F(G) \pm 1$. In the first case, the band is cutting the component, and in the second case, the band is gluing two components. A similar argument applies when a band is deleted.

If we take $F(G)$ which gives $\mathrm{ds}(G) = \#\partial F(G)$, then the above argument shows that $\mathrm{ds}(G') \leq \mathrm{ds}(G) + 1$. On the other hand, $F(G)$ is obtained from $F(G')$ by a band deletion. Hence if we take $F(G')$ to be a thickened graph such that $\mathrm{ds}(G') = \#\partial F(G')$, then it follows that $\mathrm{ds}(G) \leq \mathrm{ds}(G') + 1$. Hence $\mathrm{ds}(G') = \mathrm{ds}(G) \pm 1$. Similarly for DS. □

Lemma 3. *Let G_1 and G_2 be finite connected graphs, and v_i is a point on an edge of G_i ($i = 1, 2$). Let G be the graph obtained from G_1 and G_2 by adding an edge e connecting v_1 and v_2, so that the end points of e are v_1 and v_2. Then $\mathrm{DS}(G) = \mathrm{DS}(G_1) + \mathrm{DS}(G_2) - 1$ and $\mathrm{ds}(G) = \mathrm{ds}(G_1) + \mathrm{ds}(G_2) - 1$.*

Proof. This is proved by considering $F(G_i)$ ($i = 1, 2$) that realizes $\mathrm{DS}(G_i)$ and $\mathrm{ds}(G_i)$, and Observation 2. □

4 Some Computations Using Planar Projections

Consider a graph G embedded in the plane (i.e., G is a planar graph). Let $N(G)$ denote the regular neighborhood of G in the plane $\mathbf{R}^2$. Thus $N(G)$ consists of neighborhoods for edges and vertices. Neighborhoods of edges are bands, and neighborhoods of 3-valent vertices are portions of the surfaces depicted in Fig. 2 (A), and all neighborhoods are patched together smoothly. Hence $N(G)$ defines one of the $F(G)$s. In this case, the boundary $\partial N(G)$ consists of the boundary curves of the complementary disk regions $\mathbf{R}^2 \setminus \text{Interior}(N(G))$ (one of them is actually the region at infinity, R_∞, instead of a disk). See Fig. 5 (A).

Let G be the graph depicted in Fig. 5. When the neighborhood of the vertex on the top is replaced by the portion depicted in Fig. 2, the surface $F(G)$ changes to the one depicted in Fig. 5 (C). We see that three circles corresponding to three regions in Fig. 5 (A) were connected to a single circle by this operation. We represent this operation and the new situation diagrammatically by Fig. 5 (B). Here the vertex at which the replacement was performed is depicted by a small circle around the vertex, and three regions whose boundaries are connected together are marked by $*$. The three curves are connected to form a single curve by Lemma 1. Note that this operation can be continued for every remaining vertex at which three boundary curves meet. When another operation is performed for a large graph in a distant vertex, three other curves are connected, forming a distinct new component. In such a case, the new circles should be marked by a different marker, such as $*_1$ for the first and $*_2$ for the second, to indicate that new curves are distinct components. However, if the next operation is performed nearby, then all components involved can be connected. In this case, the same marker $*$ is used as long as the components are the same.

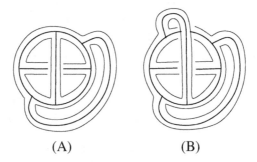

(A)   (B)

Fig. 6. An example of planar presentations of two thickened graphs for the same non-planar graph

More generally, planar presentations of non-planar thickened graph can be obtained by taking bands "parallel" to the plane into which the graph is projected. Examples are depicted in Fig. 6. The graph in this figure is non-planar,

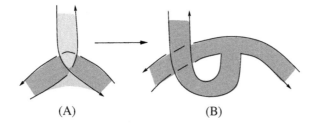

(A)   (B)

Fig. 7. Two planar representations of a band connection at a vertex

but projections of two of its thickened graphs are depicted with "crossing points" between edges. The under-arc at a crossing is broken as in knot diagrams to indicate relative height of the edges in space. In Fig. 7, the same type of connection at a vertex (depicted in Fig. 2) is specified by two planar presentations. Thus, Fig. 6 (A) and (B) represent different thickened graphs of the same graph using planar presentations of two types of connections at one vertex.

Remark 1. (1) Let K be a knot or a link. Then there is a 3-valent graph G and a thickened graph $F(G)$ such that $\partial F(G) = K$. (2) For any thickened graph $F(G)$ of a 3-valent graph G, there is a planar presentation.

Proof. It is well-known in knot theory [9] that any link K bounds an orientable surface S, called a Seifert surface. For any such S, there is a 3-valent graph G in S which is a deformation retract. Thus S can be isotoped to a neighborhood of G and can be regarded as $F(G)$. Then $\partial F(G) = K$.

For (2), we recall a standard argument in knot theory (see [6] for example). Since $F(G)$ is orientable, there are two sides of the surface, one positive, the other negative (say). Try to put the surface on the plane in such a way that only the positive sides are visible. First we arrange the surface at every vertex in such a way, using alternative diagrammatic method of representing connections as depicted in Fig. 7 (B). Then the rest are twisted edges, and arrange them following the over-under information. When the thickened edges are twisted, arrange them in small links as depicted in Fig. 8 (this is a standard trick in knot theory which can be found, for example, in [6]). □

In [12,13], methods of constructing knotted DNA molecules have been reported. A theoretical description for obtaining any single stranded DNA knot experimentally is given in [12]. In [4], construction of graph structures by DNA is proposed, and a plastic model of a building block was physically constructed. Thus the above remark indicates that variety of knots and links may be constructed by DNA molecules by constructing thickened graphs, giving an alternate approach (to [12]) of constructing knotted DNA structures. As DNA molecules form a right-handed double helix every 10.5 base pairs, the twists of thickened edges are not freely given but rather are determined by the lengths of edges (as DNA molecules). Hence the above remark does not imply that any DNA knot

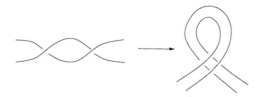

Fig. 8. Arranging a twist on the plane

can be generated by this method, and it is an interesting problem to determine which knots can indeed be constructed by this method.

When the double strand numbers satisfy certain relations, it is possible to characterize graphs with such conditions. The planar presentation is useful for such computations. For example, we have the following.

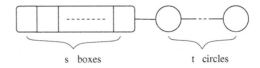

s   boxes   t   circles

Fig. 9. The graph $G(s,t)$ of a chain of circles

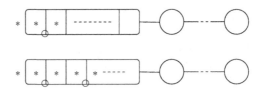

Fig. 10. Connecting boundary curves

Consider the graph $G(s,t)$ depicted in Fig. 9, where non-negative integers s and t indicate the numbers of squares and circles. Note that $\#\partial N(G(s,t))$ is exactly $\mathrm{DS}(G(s,t))$.

In Fig. 10, the replacements at vertices are successively performed. The left bottom vertex is marked in the top of the figure, and the curves corresponding to the region at infinity, the left-most and second left-most squares are connected to a single boundary circle by the replacement. These regions are marked by $*$. The third left bottom vertex (if any) is chosen next, as indicated in Fig. 10 bottom, and the curves corresponding to the newly marked regions are connected.

Inductively, all s squares (possibly but one, depending on whether s is even or odd) from the left are marked. When we try to proceed to the further right, we notice that every vertex now has two boundary components coming together. Hence we are not able to perform the replacements to reduce the number of boundary curves.

By induction on $s + t$, the above computations combined with Observations 2, 3 imply

Lemma 4. *For the graph $G(s,t)$ with non-negative integers s and t with $s+t > 0$, if s is even, $\mathrm{DS}(G(s,t)) = s+t+1$ and $\mathrm{ds}(G(s,t)) = t+1$, and if s is odd, $\mathrm{DS}(G(s,t)) = s+t+1$ and $\mathrm{ds}(G(s,t)) = t+2$.*

Corollary 1. *Let k, n be integers such that (1) $k \leq n$, (2) $k \equiv n \pmod 2$, and (3) if $k = 1$ or 2, $k < n$. Then there is a graph G with $\mathrm{ds}(G) = k$ and $\mathrm{DS}(G) = n$.*

5   Relations to Genera and Euler Characteristics

In this section we give a short overview of how ds and DS are related to commonly known concepts in topological graph theory. Details can be found, for example, in [1,2].

Any orientable surface (compact, without boundary) is constructed from a sphere by attaching some number of handles in an orientable way. For example, the surface of a doughnut, called a torus, is topologically a sphere with a single handle. The number of such handles is called the *genus* of a surface. The genus of a sphere is 0, and the genus of a torus is 1.

The (orientable) *minimal genus* $\gamma(G)$ of a graph G is the minimum among genera of all orientable compact surfaces S in which G can be embedded. The (orientable) *maximal genus* $\gamma_M(G)$ of a graph G is the maximum among genera of all orientable compact surfaces S in which G can be embedded in such a way that the complement $S - G$ consists of open 2-cells (such an embedding is called *cellular*). Here, an open 2-cell means a disk with its boundary points excluded.

Let χ denote the Euler characteristic. For a surface S divided into some polyhedral regions (for example into triangles), the Euler characteristic is defined by $\chi(S) = \#V - \#E + \#F$, where $\#V$, $\#E$, $\#F$ are the numbers of vertices, edges, and faces, respectively. For example, from a tetrahedron, we see that $\chi(S^2) = 2$, where S^2 denotes a sphere.

For a graph G, let $F(G)$ be its thickened graph. To each component (a circle) of $\partial F(G)$, attach a disk along its boundary to cap it off. After attaching disks to all boundary components, we obtain a surface F without boundary. Clearly the embedding of G in F is cellular and $\#\partial F(G)$ is equal to the number of 2-cells in the complement $F - G$. Each time a disk is attached, 1 is added to the number of faces. Thus we compute $\chi(F) - \#\partial F(G) = \chi(F(G)) = \chi(G)$. On the other hand, we know that $\chi(F) = 2 - 2$ genus (F) [8]. Hence we have

$$2 \text{ genus } (F) = 2 - \chi(F) = 2 - \chi(G) - \#\partial F(G).$$

A thickened graph $F(G)$ which realizes $DS(G)$ gives the inequality

$$\text{genus } (F) \leq (1/2)(2 - \chi(G) - DS(G)).$$

In fact we have the following.

Lemma 5 ([15]). *Let G be a finite connected graph. Then*

$$\gamma(G) = (1/2)(2 - \chi(G) - DS(G)).$$

Let $\beta_i(G)$ denote the i-th Betti number of a graph G, where i is a non-negative integer. Geometrically, $\beta_0(G)$ is the number of connected components of G and $\beta_1(G)$ is the "number of loops" in G. More specifically, $\beta_1(G)$ remains unchanged by homotopy (continuous deformations), and any graph is homotopic to a bouquet of circles. The number of such circles is equal to $\beta_1(G)$. Since $\chi(G) = \beta_0(G) - \beta_1(G)$ (see [7]), if G is connected, we have $\beta_1(G) = 1 - \chi(G)$.

The above lemma implies that DS can be computed from $\gamma(G)$ by

$$DS(G) = 2 - \chi(G) - 2\gamma(G) = (\beta_1(G) + 1) - 2\gamma(G),$$

and $\gamma(G)$ has been studied extensively in topological graph theory (see [1], for example). Similarly, we have

$$ds(G) = 2 - \chi(G) - 2\gamma_M(G) = (\beta_1(G) + 1) - 2\gamma_M(G).$$

Lemma 6. *For any connected 3-valent finite graph G, the following inequalities hold:*

$$ds(G) \leq DS(G) \leq \beta_1(G) + 1.$$

Any finite 3-valent graph is obtained from a disjoint circles by a finite sequence of edge additions. The disjoint circles O_n have $\chi(O_n) = 0$ and $DS(O_n) = ds(O_n) = 2n$. By adding an edge, $\chi(G)$ decreases by 1, so Observation 2 implies the following fact.

Lemma 7. *Let G be a finite 3-valent graph, and $F(G)$ be its thickened graph. Then $\#\partial F(G) \equiv \chi(G) \pmod 2$.*

Consequently, $ds(G)$, $DS(G)$, and $\beta_1(G) + 1$ are all even or all odd.

Example 2. Let $G = K_{3,3}$ be the complete bipartite graph of type $(3,3)$. Then $ds(G) = 1$, $DS(G) = 3$, and $\beta_1(G) + 1 = 5$.

In fact, $K_{3,3}$ is depicted in Fig. 6 with examples of its thickened graphs. It is computed that $\beta_1(G) + 1 = 5$. It is well known that this graph is non-planar and $g(G) = 1$, and so $DS(G) = \beta_1(G) + 1 - g(G) = 3$. By performing a replacement at the right-top corner vertex of (A) (or (B)), we realize $ds(G) = 1$. □

Acknowledgment. This research is partially supported by NSF grants EIA-0086015 and EIA-0074808 (Jonoska) and DMS 9988101 (Saito).

References

1. D. Archdeacon, *Topological graph theory, a survey*, Congressus Numerantium **115** (1996), 5–54.
2. G. Chartrand and L. Lesniak, *Graphs and Digraphs*, Third Edition, Chapman and Hall, London, 1996.
3. N.R. Cozzarelli, *The structure and function of DNA supercoiling and catenanes*, The Harvey Lecture Series, **87** (1993) 35–55.
4. N. Jonoska, S. Karl, and M. Saito, *Creating 3-dimensional graph structures with DNA*, in DNA computers III, eds. H. Rubin and D. Wood, AMS DIMACS series, vol. 48, 123–135, (1999).
5. N. Jonoska, S. Karl, M. Saito, *Three dimensional DNA structures in computing*, BioSystems, **52**, 143–153, (1999).
6. L. H. Kauffman, *Knots and Physics*, World Scientific, Series on knots and everything, vol. 1, 1991.
7. J. R. Munkres, *Topology, a first course*, Prentice-Hall, 1975.
8. J. R. Munkres, *Elements of Algebraic Topology*, Addison-Wesley Publ. Co., 1994.
9. Rolfsen, D., *Knots and Links*. Publish or Perish Press, (Berkley 1976).
10. N.C. Seeman, *Physical Models for Exploring DNA Topology*. Journal of Biomolecular Structure and Dynamics **5**, 997–1004 (1988).
11. N.C. Seeman, *Construction of Three-Dimensional Stick Figures From Branched DNA*, DNA and Cell Biology **10** (1991), No. 7, 475–486.
12. N.C. Seeman, *The design of single-stranded nucleic acid knots*, Molecular Engineering **2**, 297–307, (1992).
13. N. C. Seeman, Y. Zhang, S.M. Du and J. Chen, *Construction of DNA polyhedra and knots through symmetry minimization*, Supermolecular Stereochemistry, J. S. Siegel, ed., (1995), 27–32.
14. D.W. Sumners, ed., *New Scientific Applications of Geometry and Topology*, Proc. of Symposia in Applied Mathematics, Vol. 45, A.M.S., (1992).
15. J.W.T. Youngs, *Minimal Imbeddings and the genus of a graph*, J. of Math. Mech., vol. 12, (1963), 303–315.
16. S. A. Wasserman and N. R. Cozzarelli, *Biochemical Topology: Applications to DNA recombination and replication*. Science, vol. 232, (1986), 951–960.
17. E. Winfree, X. Yang, N.C. Seeman, *Universal Computation via Self-assembly of DNA: Some Theory and Experiments*. in DNA computers II, eds. L. Landweber and E. Baum, AMS DIMACS series, vol. 44, 191–214 (1998).
18. Winfree, F. Liu, L. A. Wenzler, and N.C. Seeman, *Design and Self-Assembly of Two-Dimensional DNA Crystals*, Nature **394** (1998), 539–544.

Population Computation and Majority Inference in Test Tube

Yasubumi Sakakibara*

Department of Information Sciences, Tokyo Denki University,
Hiki-gun, Saitama 350-0394, Japan
`yasu@j.dendai.ac.jp`

Abstract. We consider a probabilistic interpretation of the test tube which contains a large amount of DNA strands, and propose a population computation using a number of DNA strands in the test tube and a probabilistic logical inference based on the probabilistic interpretation. Second, in order for the DNA-based learning algorithm [4] to be robust for errors in the data, we implement the weighted majority algorithm [3] on DNA computers, called *DNA-based majority algorithm via amplification* (DNAMA), which take a strategy of "amplifying" the consistent (correct) DNA strands while the usual weighted majority algorithm decreases the weights of inconsistent ones. We show a theoretical analysis for the mistake bound of the DNA-based majority algorithm via amplification, and imply that the amplification to "double the volumes" of the correct DNA strands in the test tube works well.

1 Introduction

Population computation is to combine a number of functions (algorithms) in the pool and obtain a significantly good result of computation. The concept of "population computation" has been introduced in the context of machine learning research [2]. A similar approach is the weighted majority algorithm [3] for on-line predictions and to make the learning algorithms robust for errors in the data. In order to introduce such population computations into DNA computing, we propose the following interpretation of the test tube:

- A single DNA strand represents a computable function,
- The test tube which contains a large amount of DNA strands represents a population (set) of functions,
- The volume (number) of copies of each DNA strand represents a weight assigned to the encoded function,
- The output of (population) computations in the test tube is taken as a weighted majority of the outputs of functions encoded by DNA strands in the test tube. More precisely, in the case of the binary output (that is, $\{0, 1\}$), we compare the total volume of the DNA strands outputting 0 to the total

---

* This work is supported in part by "Research for the Future" Program No. JSPS-RFTF 96I00101 from the Japan Society for the Promotion of Science.

© Springer-Verlag Berlin Heidelberg 2002

volume of the DNA strands outputting 1, and produce the output according to the larger total.
- The test tube is trained (learned) based on the sample data and the learning strategy is that when the output of the test tube makes a mistake, the volume of the DNA strands that output the correct value are amplified.

More concretely, our scenario is as follows. First, by using some encoding method, we assume that each individual DNA strand represents a computable function. Second, the initial test tube contains a large number of randomly generated DNA strands which represent various functions (random pool of various functions), and the volume of copies of each DNA strand represents a weight assigned to the encoded function. This is illustrated in the lefthand side of Figure 1. Third, the output of computations with the test tube is taken as a weighted majority of the outputs of DNA strands (the outputs of encoded functions) in the test tube. This is illustrated in the righthand side of Figure 1. Forth, we have some training data to tune the test tube so that after the training, the test tube contains desirable volumes of each DNA strands and turns to be the best majority function. Here, a training data is a set of correct input-output pairs of the goal (target) function, and the training (learning) in the test tube is to change the volumes of DNA strands inside the test tube. Our strategy to train (learning) the test tube is "amplifying" DNA strands encoding correct (consistent) functions with the training data. This is illustrated in Figure 1.

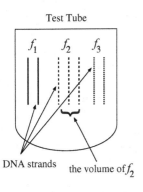 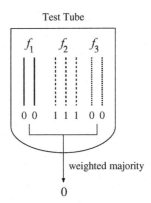

Fig. 1. (left:) A random pool of DNA strands in the test tube, and (right:) the weighted majority of the outputs of functions encoded by DNA strands in the test tube.

Thus, the "entire" test tube containing a number of DNA strands is regarded as a single entity of a "function" and trained (learned) on a sequence of examples.

In this paper, we consider each individual DNA strand represents a Boolean formula and computes a Boolean function for the input of truth-value assignment. In our previous work [4], we have proposed a DNA-based learning algorithm for Boolean formulae from the training data without errors. We have

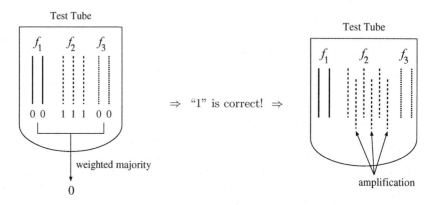

Fig. 2. "Amplifying" DNA strands encoding correct (consistent) functions with the training data.

further applied this DNA-based learning algorithm to DNA chip technologies [5]. The application is called *intelligent DNA chip* and employs the DNA-based learning algorithm to extract useful informations from gene expression profile data. Our new method proposed in this paper enlarges these algorithm and application to be robust for errors in data. We start with this method in our previous work [4] which will be briefly explained in the next section.

2   Evaluation of Boolean Formulae

We assume there are n Boolean variables (or attributes) and we denote the set of such variables as $X_n = \{x_1, x_2, \ldots, x_n\}$. A *truth-value assignment* $a = (b_1, b_2, \ldots, b_n)$ is a mapping from X_n to the set $\{0, 1\}$. A *Boolean function* is defined to be a mapping from $\{0, 1\}^n$ to $\{0, 1\}$. *Boolean formulae* are useful representations for Boolean functions. Each variable x_i $(1 \leq i \leq n)$ is associated with two *literals*: x_i itself and its negation $\neg x_i$. A *term* is a conjunction of literals. A Boolean formula is in *disjunctive normal form* (DNF, for short) if it is a disjunction of terms. For any constant k, a k-term DNF formula is a DNF Boolean formula with at most k terms.

First, we encode a k-term DNF formula β into a DNA single-strand as follows:

Let $\beta = t_1 \vee t_2 \vee \cdots t_k$ be a k-term DNF formula.

(1) For each term $t = l_1 \wedge l_2 \wedge \cdots l_j$ in the DNF formula β where l_i $(1 \leq i \leq j)$ is a literal, we use the DNA single-strand of the form:

$$5' - \textbf{stopper} - \textbf{marker} - seqlit_1 - \cdots - seqlit_j - 3'$$

where $seqlit_i$ $(1 \leq i \leq j)$ is the encoded sequence for a literal l_i. The **stopper** is a *stopper sequence* for the polymerization stop that is a technique developed by Hagiya et al. [1]. The **marker** is a special sequence for a extraction used later at the evaluation step.

(2) We concatenate all of these sequences encoding terms t_j ($1 \leq j \leq k$) in β. Let denote the concatenated sequence encoding β by $e(\beta)$.

For example, the 2-term DNF formula $(x_1 \wedge \neg x_2) \vee (\neg x_3 \wedge x_4)$ on four variables $X_4 = \{x_1, x_2, x_3, x_4\}$ is encoded as follows:

$$5' - \mathbf{marker} - x_1 - \neg x_2 - \mathbf{stopper} - \mathbf{marker} - \neg x_3 - x_4 - 3'$$

Second, we put the DNA strand $e(\beta)$ encoding the DNF formula β into the test tube and do the following biological operations to evaluate β for the truth-value assignment $a = (b_1, b_2, \ldots, b_n)$.

Algorithm $B(T, a)$:
(1) Let the test tube T contain the DNA single-strand $e(\beta)$ for a DNF formula β.
(2) Let $a = (b_1, b_2, \ldots, b_n)$ be the truth-value assignment. For each b_i ($1 \leq i \leq n$), if $b_i = 0$ then put the Watson-Crick complement $\overline{x_i}$ of the DNA substrand encoding x_i into the test tube T, and if $b_i = 1$ then put the complement $\overline{\neg x_i}$ of $\neg x_i$ into T.
(3) Cool down the test tube T for annealing these complements to complementary substrands in $e(\beta)$.
(4) Apply the primer extension with DNA polymerase to the test tube T with these annealed complements as the primers. As a result, if the substrand for a term t_j in β contains a literal lit_i and the bit b_i makes lit_i 0, then the complement $\overline{seqlit_i}$ of the substrand $seqlit_i$ has been put at Step (2) and is annealed to $seqlit_i$. The primer extension with DNA polymerase extends the primer $\overline{seqlit_i}$ and the subsequence for the marker in the term t_j becomes double-stranded, and the extension stops at the stopper sequence. Otherwise, the subsequence for the marker remains single-stranded. This means that the truth-value of the term t_j is 1 for the assignment a.
(5) Extract the DNA (partially double-stranded) sequences that contains single-stranded subsequences for markers. These DNA sequences represent the DNF formulae β whose truth-value is 1 for the assignment a.

We have already verified the biological feasibilities for this evaluation method of Boolean formulae. Yamamoto et al. [6] have done the following biological experiments to confirm the effects of the evaluation algorithm $B(T, a)$:

1. For a simple 2-term DNF Boolean formula on three variables, we have generated DNA sequences encoding the DNF formula by using DNA ligase in the test tube,
2. The DNA sequences are amplified by PCR,
3. For a truth-value assignment, we have put the Watson-Crick complements of DNA substrands encoding the assignment, applied the primer extension with DNA polymerase, and confirmed the primer extension and the polymerization stop at the stopper sequences,
4. We have extracted the DNA sequences encoding the DNF formula with magnetic beads through biotin at the 5'-end of primer and washing.

3   Probabilistic Interpretation of the Test Tube

In our probabilistic interpretation, we simply represent a "probability (weight)" by the volume (number) of copies of a DNA strand which encodes the probabilistic attribute. Approximately, 2^{40} DNA strands of length around several hundreds are stored in $1.5ml$ of a standard test tube, and by considering the test tube of $1.5ml$ as the unit, we can represent the probabilistic values using the quantities of DNA strands with precision up to 2^{40}.

In probabilistic logic, the logical variable "x" takes a real truth-value between 0 and 1. Further, when the value of the variable x is c ($0 \leq c \leq 1$), the negation "$\neg x$" of the variable x takes the value $1 - c$. Hence, for the logical variable x and its value c, we put the DNA strands encoding x with volume c of the unit (entire) volume for x, and simultaneously we put the DNA strands for $\neg x$ (the negation of x) with volume $1 - c$ into the test tube. Combined with the methods for representing and evaluating Boolean formulae, we can execute the following probabilistic inference:

1. We extend the truth-value assignment $a = (b_1, b_2, \ldots, b_n)$ to the probabilistic truth-value assignment $a' = (c_1, c_2, \ldots, c_n)$ where each c_i is a real value between 0 and 1 to represent the probability that the variable x_i becomes 1.
2. We execute a modified algorithm $B'(T, a')$ for the probabilistic truth-value assignment $a' = (c_1, c_2, \ldots, c_n)$ such that for each c_i ($1 \leq i \leq n$) and the unit (entire) volume Z of the test tube, we put $(1 - c_i)Z$ amount of the Watson-Crick complements $\overline{\neg x_i}$ of the DNA strand for the negation $\neg x_i$ into the test tube T and put $c_i Z$ amount of the complement $\overline{x_i}$ of x_i into T.

Example 1. We consider two Boolean variables $\{x, y\}$ and a Boolean formula $x \wedge \neg y$ which is encoded as follows:

$$5' - \mathbf{marker} - x - \neg y - 3'$$

We prepare enough amount Z of copies of this DNA strand and let the probabilistic assignment be $a' = (c_1, c_2)$.

- Case $c_1 = 0.2$ and $c_2 = 0.0$: (illustrated in the lefthand side of Figure 1). The execution of the algorithm $B'(T, a')$ implies that 20% amount (that is, $0.2Z$) of DNA strands for $x \wedge \neg y$ have single-stranded markers and hence the probability that the truth-value of $x \wedge \neg y$ becomes 1 is 0.2.
- Case $c_1 = 0.2$ and $c_2 = 0.7$: (illustrated in the righthand side of Figure 1). The execution of the algorithm $B'(T, a')$ implies that the amount between 0% and 20% of DNA strands for $x \wedge \neg y$ have single-stranded markers (the expected amount is 6%) and hence the probabilistic truth-value of $x \wedge \neg y$ is between 0.0 and 0.2.

Example 2. We consider a Boolean formula $x \vee \neg y$ which is encoded as follows:

$$5' - \mathbf{marker} - x - \mathbf{stopper} - \mathbf{marker} - \neg y - 3'$$

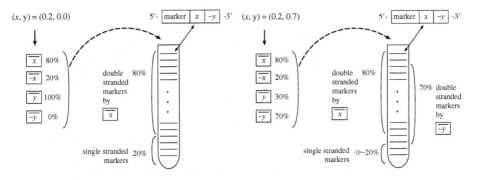

Fig. 3. Probabilistic logical inferences with a Boolean formula $x \wedge \neg y$.

- Let the probabilistic assignment be $a' = (0.2, 0.7)$.

  The execution of the algorithm $B'(T, a')$ implies that 80% left markers of DNA strands representing $x \vee \neg y$ become double-stranded and 70% right markers become double-stranded, and hence the amount between 30% and 50% of DNA strands have at least one single-stranded marker. Thus the probabilistic truth-value of $x \vee \neg y$ is between 0.3 and 0.5.

4   Learning by Amplification and the Majority Inference

In our previous work [4] for DNA-based learning of Boolean formulae from the training data without errors, we have proposed a learning method of "selecting" DNA strands of consistent Boolean formulae and "eliminating" inconsistent ones, which is illustrated in Figure 4. In this paper, for robust learning of Boolean formulae, we take the strategy of "amplifying" DNA strands encoding consistent Boolean formulae.

First, we briefly explain the *weighted majority algorithm* [3] that is a general and powerful method for on-line predictions and to make the learning algorithms robust for errors in the data. Next, we propose a method to implement the weighted majority algorithm on DNA computers and extend the DNA-based consistent learning algorithm to being robust for errors in the data. A fundamental strategy of our method is the "amplification" of the consistent (correct) DNA strands while the weighted majority algorithm decreases the weights of inconsistent ones. This is a obvious difference of our method from the weighted majority algorithm in order to be adequate for implementing on DNA computers.

The weighted majority algorithm [3] is a general and powerful method for on-line prediction problems in the situation that a pool of prediction algorithms are given available to the master algorithm and the master algorithm uses the predictions of the pool to make its own prediction. The on-line prediction problem is generally defined as follows: Learning proceeds in a sequence of trials, and in each trial the prediction algorithms receive an input instance from some fixed domain and produce a binary prediction. At the end of the trial, the prediction

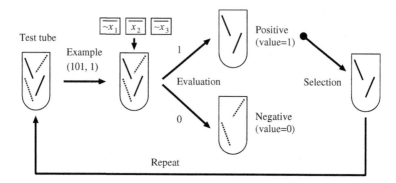

Fig. 4. DNA-based learning algorithm by "selecting" DNA strands of consistent Boolean formulae and "eliminating" inconsistent ones, for a training example $(101, 1)$ where the correct value is 1.

algorithms receive a binary *label*, which can be viewed as the correct prediction for the instance. The goal of the on-line prediction problems is to design a prediction algorithm which makes mistakes as few as possible. The fundamental ideas of the weighted majority algorithm is to assign a weight to each prediction algorithm in the pool and take the *weighted voting* for the prediction of the master algorithm. When it makes a mistake, the weighted majority algorithm updates the weights of the prediction algorithms which made the mistake by multiplying by a fixed constant less than 1 (that is, decreasing the weights). This strategy has been proved to work very well for the on-line prediction problems.

Now, we implement the weighted majority algorithm on DNA computers, called *DNA-based majority algorithm via amplification* (DNAMA), as follows (illustrated in Figure 4).

1. (Representation:) We first encode each (prediction) function into a DNA strand and represent the weight of a function by the volume of DNA strands encoding the function in the test tube. The set of DNA strands encoding the functions in the "test tube" correspond to the set of functions in the "pool".
2. (Prediction:) The master algorithm compares the total volume of the DNA strands predicting 0 to the total volume of the DNA strands predicting 1, and predicts according to the larger total (taking the weighted majority).
3. (Update:) When the master algorithm makes a mistake, the volume of the DNA strands that agreed with the label of the given example are amplified (multiplied) by some fixed γ such that $\gamma > 1$.

In the next section, we show a mistake bound that the DNA-based majority algorithm via amplification makes $O(\log N + m)$ mistakes, where m is the number of mistakes of the best function in the test tube and N is the number of different functions in the test tube.

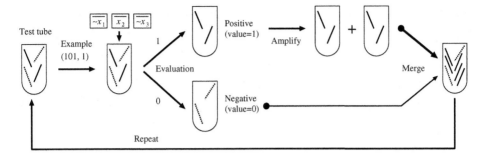

Fig. 5. DNA-based learning (weighted majority) algorithm by "amplifying" DNA strands of consistent Boolean formulae.

5   Mistake Bound for Updating by Amplification

To prove the theorem for a mistake bound of DNAMA, we prepare some notations and definitions. Assume that there exist a set of N prediction functions encoded by DNA strands in the test tube and let w_i denote the volume (weight) of i-th function in the test tube. For a given trial, let q_0 and q_1 denote the total volume of the functions in the test tube that predict 0 and 1, respectively. Then DNAMA updates the volumes by amplifying the volumes of the functions which made the correct prediction on the trial by some fixed γ such that $\gamma > 1$. Let w_{init} denote the total initial volume of all functions in the test tube and w_{fin} denote the total final volume after all examples have been processed.

Theorem 1. *Assume that T is a set of N prediction functions encoded by DNA strands in the test tube and that m_i is the number of mistakes made by i-th function in the test tube on a sequence S of instances (examples) with binary labels. DNAMA makes $O(\log N + m_i \log \gamma)$ mistakes on the sequence S when DNAMA is applied to the test tube T with equal initial volumes.*

Proof. Let w_{pre} and w_{after} denote the total volume of all functions in the test tube before a given trial and after the trial, respectively, and m be the number of mistakes made by DNAMA on the sequence S.

We first assume that $w_{pre} \cdot u \geq w_{after}$ holds for some fixed constant $1 < u < \gamma$ when DNAMA makes a mistake. Then it obviously holds that $w_{init} \cdot u^m \geq w_{fin}$ and $w_{fin} \geq w_i \cdot \gamma^{m-m_i}$. By a simple calculation, we have

$$\left(\frac{\gamma}{u}\right)^m \leq \frac{w_{init} \cdot \gamma^{m_i}}{w_i}.$$

By putting $w_{init} = N$ and $w_i = 1$, we have a bound

$$m \leq \frac{\log N + m_i \log \gamma}{\log\left(\frac{\gamma}{u}\right)}.$$

Now we show $w_{pre} \cdot u \geq w_{after}$ for some fixed $1 < u < \gamma$. Without loss of generality, in this trial we assume that the DNAMA's prediction is 0 and hence

$q_0 \geq q_1$, and the correct label is 1. Then, we need to show that $(q_0 + q_1)u \geq q_0 + q_1\gamma$. By a simple trick [3], it holds that $(q_0 + q_1)(1 + \beta) \geq 2\beta q_0 + q_1$ for $0 < \beta < 1$. When we put $u \geq 1 + \beta$ and $2\beta q_0 + 2q_1 \geq q_0 + q_1\gamma$, it must hold that

$$\beta \geq \frac{1}{2} + \frac{q_1}{q_0}\left(\frac{\gamma}{2} - 1\right)$$

$$u \geq \frac{3}{2} + \frac{q_1}{q_0}\left(\frac{\gamma}{2} - 1\right)$$

$$\gamma \geq u.$$

Therefore, $2 \leq \gamma < 2 + q_0/q_1$, $(\gamma - 1)/2 \leq \beta < 1$, and $(\gamma + 1)/2 \leq u \leq \gamma$ satisfy the required conditions.   □

For examples, $\gamma = 2$, $u = 3/2$ and $\beta = 1/2$ derive a mistake bound

$$m \leq \frac{\log N + m_i \log 2}{\log\left(\frac{4}{3}\right)}.$$

Hence, the amplification to "double the volumes" of the correct DNA strands (that is, DNA strands consistent with the given example) works well and the DNAMA makes $O(\log N)$ additional mistakes to the mistakes m_{best} of the best prediction function in the test tube. If $m_{best} = 0$, that means the errors contained in the data is 0, the DNAMA makes at most $O(\log N)$ mistakes.

6   Taking Account of PCR Replication Errors

In our DNA-based majority algorithm via amplification, we assume that the amplification by PCR must be precisely done and make a correct copy of DNA strands. However, this is not a practical assumption and in a real laboratory experiment, some errors on replications can occur with some probability. Hence, we need to incorporate the PCR replication errors into our method. The DNAMA has some tolerance for replication errors in the current form with some expense of additional mistakes. Here, we provide a rough analysis of mistake bounds of the DNAMA with the replication errors.

When we execute the PCR amplification to make a copy of some DNA strand which encodes some function, say i-the function, the replication errors on the PCR amplification will cause two possible results:

(1) the DNA strand produced by the amplification with errors encodes no meaningful function (that is, it does not satisfy the encoding syntax), and
(2) the produced DNA strand encodes other function, say j-th function.

In any case, the condition $w_{pre} \cdot u \geq w_{after}$ in the proof of Theorem 1 can hold. However, the condition $w_{fin} \geq w_i \cdot \gamma^{m-m_i}$ does not hold in both cases. For a rough analysis, we assume that in ϵ fraction ($0 \leq \epsilon < 1$) of the total number m_i of amplifications (updates of volumes), the replication errors will occur and fail to make a correct copy of i-th DNA strand (i-th function). In this case, we have the following inequality:

$$w_{fin} \geq w_i \cdot \gamma^{(m-m_i)\cdot(1-\epsilon)}.$$

By a simple calculation, we have

$$m \leq \frac{\log N + (1-\epsilon)m_i \log \gamma}{\log \left(\frac{\gamma^{1-\epsilon}}{u}\right)}.$$

Then it must hold that $\gamma^{1-\epsilon} > u \geq (\gamma+1)/2$. In the case of $\gamma = 2$, $\epsilon \leq 1/3$ satisfies the condition. Hence, $1/3$ fraction of the total number of amplifications is allowed to have the replication errors and fail to make a correct copy. The expense of additional mistakes is roughly $O(\epsilon \log N/(1-\epsilon))$.

7   Application to Intelligent DNA Chip

The intelligent DNA chip [5] applies the DNA-based learning algorithm to medical diagnoses from gene expression profile data. An advantage of it is that the DNA-based computation directly accepts gene expression profile as input in the form of molecules. We consider to apply our new learning method, DNAMA, to the intelligent DNA chip.

If we put a gene expression profile data as a training data to the DNAMA, the DNAMA trains the test tube to represent some characteristic rule for the gene expression profile. For example, when the gene expression profile is specific to a disease, the trained test tube containing a number of DNA strands for Boolean formulae could be used to diagnose a test data as the disease state. Thus, the DNAMA will make a "magic water" inside the test tube for a diagnosis of the specific disease.

References

1. M. Hagiya, M. Arita, D. Kiga, K. Sakamoto, and S. Yokoyama. Towards parallel evaluation and learning of Boolean μ-formulas with molecules. In *Proc. of Third Annual Meeting on DNA Based Computers*, 105–114, 1997.
2. M. Kearns, H. Sebastian Seung. Learning from a population of hypotheses. *Machine Learning*, 18, 255–276, 1995.
3. N. Littlestone and M. K. Warmuth. The weighted majority algorithm. *Information and Computation*, 108, 212–261, 1994.
4. Y. Sakakibara. Solving computational learning problems of Boolean formulae on DNA computers, *Proc. 6th International Meeting on DNA Based Computers*, 193–204, 2000.
5. Y. Sakakibara and A. Suyama. Intelligent DNA chips: Logical operation of gene expression profiles on DNA computers, *Genome Informatics 2000 (Proceedings of 11th Workshop on Genome Informatics)*, Universal Academy Press, 33–42, 2000.
6. Y. Yamamoto, S. Komiya, Y. Sakakibara, and Y. Husimi. Application of 3SR reaction to DNA computer, *Seibutu-Buturi*, 40(S198), 2000.

DNA Starts to Learn Poker

David Harlan Wood[1*], Hong Bi[2*], Steven O. Kimbrough[3**], Dong-Jun Wu[4],
and Junghuei Chen[2*]

[1] Computer Science, University of Delaware, Newark DE, 19716
[2] Chemistry and Biochemistry, University of Delaware, Newark DE, 19716
[3] The Wharton School, University of Pennsylvania, Philadelphia, PA, 19104
[4] Bennett S. Lebow College of Business, Drexel University, Philadelphia,PA, 19104

Abstract. DNA is used to implement a simplified version of poker. Strategies are evolved that mix bluffing with telling the truth. The essential features are (1) to wait your turn, (2) to default to the most conservative course, (3) to probabilistically override the default in some cases, and (4) to learn from payoffs. Two players each use an independent population of strategies that adapt and learn from their experiences in competition.

1 Introduction

The long-term goal is to use DNA to construct special purpose computers. Their special purpose is to learn game-playing strategies adapting to the strategies of opponents, even while opponent's strategies are also changing and adapting. It is clear that many real-world problems have this nature—and it is equally clear that no general solution method is known for these problems. The ultimate payoff for our research is a method of searching for adaptive game-theoretic strategies.

The ultimate aim is to use DNA to encode game strategies that improve over time and adapt to the strategies of other players. In the long term, this is to be addressed for the game of poker. In the medium term, a simplified 3-person poker [1] that has no "equilibrium" is to be addressed.

In the near term, this paper demonstrates the necessary DNA laboratory techniques for an example from a textbook on game theory [2]. This game is a simplified version of poker, but it still involves probabilistic strategies of bluffing versus truth-telling and calling versus folding. The essential features are to wait your turn, to default the most conservative course, and to probabilistically override this default in some contexts, and to learn from the payoffs obtained. Each of two players competes using a large population of strategies that adapt and learn from their experiences in competition.

We employ laboratory evolution of DNA [3,4,5,6,7], DNA hairpin extension [8,9,10], and the evolutionary computation paradigm from conventional computing [11,12,13]. All three of these techniques been used before, but have never been combined.

* Partially supported by NSF Grant No. 9980092 and DARPA/NSF Grant No. 9725021.
** Partially supported by NSF Grant No. 9980092

N. Jonoska and N.C. Seeman (Eds.): DNA7, LNCS 2340, pp. 92–103, 2002.
© Springer-Verlag Berlin Heidelberg 2002

1.1   The Advantages of DNA for Computing

Computations of evolving strategies seem particularly well suited to DNA implementation.

1. Estimated answers for a particular problem can be encoded in DNA molecules using binary representation.
2. Selection by fitness, and breeding via mutation and crossover, can be implemented by laboratory procedures, as demonstrated in [4,5].
3. Evolutionary computation, like natural evolution, benefits from tolerance of error [14,15], requiring only that selection be correlated with fitness.
4. Massive parallel processing of up to 10^{18} independent bytes of data is a characteristic of DNA laboratory processes (about one milligram). This is comparable to projected next-generation silicon computers [16].
5. A very large amount of information storage is available using DNA. For example, the entire Internet contains about the same amount of data as a milligram of DNA [17].
6. DNA laboratory procedures can multiplex many simultaneously evolving populations at no extra cost. Multiplexing permits large-scale sampling of the distribution of possible population evolutions.

2   Where Do Game Strategies Come from?

A game is a situation in which two or more players make moves (or plays). The reward received by a player for its moves depends in part on the moves made by the other player(s). The broad applicability of game theory not only ensures its importance, but also explains why game theory is unlikely to produce general methods for finding good strategies. While we admire the accomplishments of game theory, we regret that equilibrium and hyper-rationality are so often unrealistically assumed [18].

Playing competitive poker, for example, seems to be (1) a dynamic process of adapting one's strategies while (2) exploiting the mistakes of opponents. Regrettably, neither of these two features is usual in game-theoretic analysis. We depart from what has been the mainstream of game theory in that we focus on the dynamics of play and strategy creation, rather than on the statics and the various equilibrium concepts. See [19].

A recent commentary in Nature nicely captures our perspective:

> "Of course, the main problem with Nash equilibria is still there: they may exist, but how does one reach them? . . . .We are in a situation akin to the beginning of mechanics: we can do the statics, but we don't have the dynamics."

Some of our prior research provided application of evolutionary computation to competitions [20,21,22,23]. We are led to using evolutionary computation because: (1) it is a general paradigm for exploring large search spaces. (2) its robustness under change or uncertainty is important since the very meaning of "good strategies" dynamically changes as opponents evolve their own strategies.

2.1   Complexity of Seeking Game Strategies

The complexity of the problem of finding good strategies can be indicated in the following way. Roughly speaking, all interesting games have exponentially many possible strategies. As for finding good strategies, no definite procedures are known which can consistently outperform simple enumeration. This is analogous to the complexity of seeking solutions or approximations for NP-complete problems, plus extra difficulties arising from dynamically changing situations.

Even in the case of finding Nash equilibrium, which we regard as regrettably static, no polynomial algorithm is known. In fact, Papadimitriou says [24],

> "...the complexity of finding a Nash equilibrium is in my opinion the most important concrete open question on the boundary of P today."

3   Simplified Poker via DNA

In this paper, we use an very simplified version of poker taken from a game theory textbook [2]. Even so, it incorporates bluffing, calling, and folding—all of which must be done with varying probabilities, if good payoffs are to be achieved.

1. There is a Dealer and a Player. Each contributes $1 into the pot to start one hand of play. The Dealer deals a single card, an Ace or a 2, so that only the Player can see it.
2. If the card it is an Ace, the Player must add $1 and say "Ace." If the card is a 2, the Player may say "2," losing the hand, or may add $1 and bluff by saying "Ace."
3. If the Player has said "Ace," it becomes the game continues an it is the Dealer's turn. The dealer may choose to fold, losing the hand and ending the game, or the Dealer may add $1 and call.
4. If called, the Player must show the card. Player wins the hand if the card is an Ace, and loses if it is a 2.

3.1   Strategies Are Learned by Playing Trillions of Simultaneous Hands Using DNA

Our DNA based implementation of playing poker is organized as shown in Fig. 1. This figure gives a broad overview of three independent but linked processes. The overall approach of selection by fitness, adding variation by crossover, is an extension of *in-vitro* evolution [25].

At the top, differing strategies compete, and the resulting histories of play are separated by outcomes.

In the middle, the many dealer-strategies are evaluated and selected by using a procedure based on payoffs achieved. This must be done carefully. Many selection criteria are possible. Considerable care is needed. For example, it is foolish to insist on consistently high payoffs because such strategies become predictable, and therefore exploitable by one's opponent. Population size is restored by making many copies of the selected strategies. Then, crossover can be used

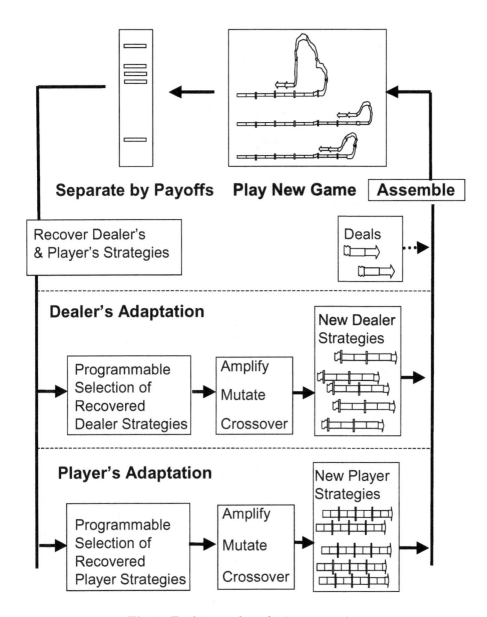

Fig. 1. Evolving poker playing strategies.

to induce variation within the population of strategies. Finally, these new dealer strategies are entered into another round of competition.

At the bottom of Fig. 1 the other player uses a similar process, but with an independent method of selection.

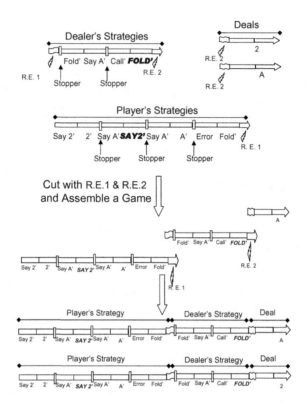

Fig. 2. A Player's strategy and a Dealer's strategy join for one hand of poker.

3.2   Encoding Strategies in DNA Strands

The Player's strategies are encoded in single-stranded DNA as shown at the top of Fig. 2. This DNA strand consists of four pairs of labeled regions, with pairs separated by "stoppers." The roles of the various regions will be explained shortly.

It is important to note that all of the Player's strategy strands are identical except in one variable region, labeled *SAY 2'*. This region will vary throughout the Player's population of strategies. Its purpose, as we will see, is to implement diverse probabilities of bluffing. The Dealer's strategies are similarly encoded, having one variable region labeled *FOLD'* to implement various probabilities of calling. Of course, the variable regions are not predictable in advance. These variable regions will probabilistically determine the course of the play.

The cards to be dealt are Ace and 2, shown with unlabeled spacers at their left (labeled Deals in Fig. 2.)

Restriction enzymes, R.E. 1 and R.E. 2, cut DNA strands at specific locations, where indicated. This facilitates first joining each Dealer strategy strand to one Player strategy strand and to a Deal strand. The restriction enzymes are later used to sever these strands so they can be individually recovered (separated by length using gel electrophoresis).

Two situations can result, as shown at the bottom of Fig. 2. They differ in that their right hand ends have an Ace or a 2.

3.3   The DNA Sequences Used

The specific DNA sequences used are shown in Table 1. These sequences are based on [9] where they were used in DNA hairpin extension. The hairpin extension of [9] is similar to the play of one hand of simplified poker, except using two fixed non-probabilistic strategies.

Table 1. DNA sequences encoding simplified poker.

Names	Size	Sequences
A (Ace)	15-mer:	5' CCGTCTTCTTCTGCT 3'
A'	15-mer:	5' CCGTCTTCTTCTGCT 3'
2	15-mer:	5' TTCCCTCCCTCTCTT 3'
2'	15-mer:	5' AAGAGAGGGAGGGAA 3'
Say A'	15-mer:	5' CGTCCTCCTCTTGTT 3'
Say 2	15-mer:	5' CCCCTTCTTGTCCTT 3'
SAY 2'	15-mer:	Random with T,G, & C
Fold	15-mer:	5' TGCCCCTCTTGTTCT 3'
FOLD'	15-mer:	Random with T,G, & C
Call'	20-mer:	5' CTCCTCTTCCTTGCTCTTCTCCCTT 3'

3.4   DNA Hairpin Extension Plays a Hand of Poker

At the top of Fig. 3 below, two strategies are combined with a dealt Ace. The rest of the figure shows how the play of this one hand can result in two possible outcomes, depending on whether the Dealer decides to fold or call. The play of a hand when a 2 is dealt is similar.

Having been dealt an Ace, the Player must say "Ace." This is accomplished using DNA in the following way. The sequence encoding A at the end of the DNA strand strongly pairs with its Watson-Crick complementary sequence A'. This enables DNA polymerase to extend the strand by appending the sequence Say A'. Extension halts at a "stopper." To continue extension into a stopper region (encoded with 4 A-bases) would require dTTP, which is withheld from the reaction. Raising the temperature disrupts interstrand pairing. Recooling begins the Dealer's turn.

The Dealer must decide to call or fold. This is an IF-THEN-ELSE type decision, but we implement it in the form, "By default, fold, but if the probability

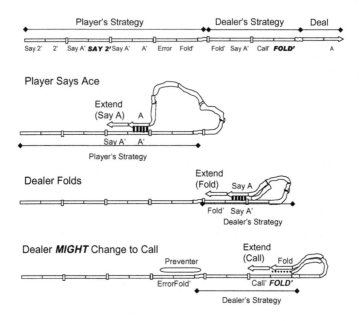

Fig. 3. Play of one hand of simplified poker in the case when an Ace is dealt.

is large enough, change your mind and call." That is, in the part of Fig. 3 labeled Dealer Folds, the Dealer's strategy encodes this situation and extends the DNA strand with the Fold sequence.

After heating followed by cooling, the FOLD sequence may or may not pair with the *FOLD'* sequence, as at the bottom of Fig. 3. If and only if pairing occurs, the DNA strand is extended by the Call sequence, essentially changing the Dealer's decision from fold to call.

Success of pairing depends on the *FOLD'* sequence, which is generally different for different Dealer strategy strands. The *FOLD'* sequences in the initial population of strategies are initially randomized during the synthesis of Dealer's strategy strands. Therefore, the population of Dealer's strategies will generally produce some fold outcomes and some call outcomes. These outcomes are later used to select strategies by payoffs. Thus, it is the *FOLD'* sequences within Dealer's strategy strands that adapt by learning from outcomes.

3.5 Experimental Results

In this experiment, about a million distinct Dealer strategies are used. (Each one is likely to occur with many duplicates of itself.) This large variety of strategies is present because the *FOLD'* labeled region is randomized during the DNA synthesis. Similarly, about a million independent Player strategies are present.

The experiment consists of about a million million distinct hands of poker all being played at one time, with and Ace being dealt in all cases. Each hand to be played pairs up one Dealer strategy with one Player strategy.

Each hand played is some one strand of DNA, which is extended as shown in Fig. 3. Because of the randomization in the *FOLD'* region, some Dealer strategies will fold and some will call, as seen in Fig. 3.

Figure 4 shows the result of laboratory reactions that enable extensions of the poker hands encoded in DNA. The gel band at 282 shows that in some cases a dealer strategy has called. The band at 262 are the cases where the dealer has folded. The band at 247 contains games yet to be completed, and the band at 232 consists of hands that have failed to extend.

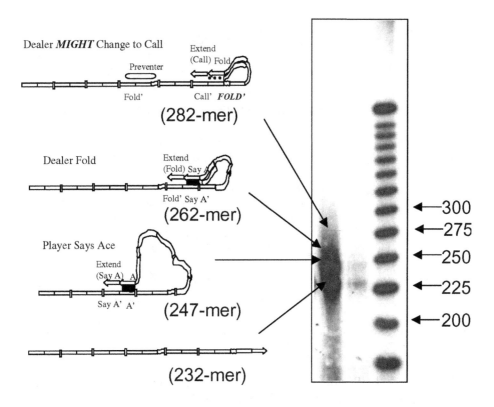

Fig. 4. The extensions illustrated in Fig. 3 are carried out in this experiment.

3.6 The Dealer and Player Independently Evolve Their Strategies

So far, we have explained how DNA laboratory techniques are able to pair off Dealer strategies and Player strategies along with a deal of an Ace or a 2. The result is: each such DNA strand is extended so that it records the entire history of the play of one hand of simplified poker. So far, we have elaborated on the

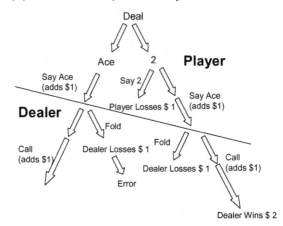

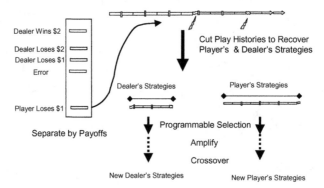

Fig. 5. Each branch of the game tree produces different length DNA strands.

top part of Fig. 1. We now go on to explain how the Dealer and the Player can independently evolve their populations of strategies.

Fig. 5A (the extensive form of the game of simplified poker, plus an error output) contains all five possible game histories, along with their payoffs, positive or negative, for the Dealer. The left side of the game tree corresponds to Fig. 3. Figure 5A also indicates how different final lengths of the DNA strands encode each of the possible histories.

Differing lengths make it convenient to physically separate histories using denaturing gel electrophoresis. Readout is provided by quantifying the amounts of DNA in each band of the gel. Other techniques could also be used, for example the 2d-DGGE techniques that we have used in evolutionary computations [4,5].

Physical separation by length via gel electrophoresis is indicated on the left of Fig. 5B. Each band of the gel corresponds to a different payoff.

What follows is that for each possible payoff the Dealer receives a quantified sample of the strategies that led to the given payoff. These samples are obtained by literally cutting the bands from the gel and extracting the DNA from them. The Dealer is then able to recombine strategies in various dilutions of her own choosing. Using this freedom of choice, and a chosen amount of crossover to explore further variations, the Dealer produces a new generation for strategies that will hopefully improve her net payoff. Improvement cannot be guaranteed, of course, because the Player is independently striving for the opposite outcome.

3.7   Regular Poker Would Use Similar Techniques

In regular poker decisions are similar but somewhat more complex. For example, the Dealer's decision in Fig. 3 would become the following. "By default, I fold, but append a copy of the hand I have been dealt. If my hand is good enough to match an evolved criterion, I can change my mind and call. But I append another copy of my hand and if it is good enough, I make a small raise." An additional comparison can result in a larger raise, etc.

A "wait your turn" feature is also shown in the last step in Fig. 3. Strictly speaking, this feature is not needed in simplified poker, but we wish to test it because it is needed in other games where players may take several turns. In essence, the Player's strategy is prepared to react to folding, but must not react before there is a chance for the Dealer to change from fold to call. Thus, as we cool the DNA we include a Preventer stand that preferentially (at higher temperature) pairs as shown in the Player's strategy. Should prevention fail, we would detect the presence of the Error sequence in some outcomes.

4   Anticipated Directions

We wish to address some game theoretic questions on the evolution of strategies for simplified poker. For simplified poker the questions can also addressed by analytic means, and by computer simulation. However, computer simulation would be difficult for populations as large as when using DNA.

The main outcome sought is to gain confidence in the DNA encodings and techniques that could be applied to more challenging games, especially poker. We will address questions such as the following. Is equilibrium maintained once it is induced? If one party uses an equilibrium strategy, will the other party evolve to equilibrium? If one party does a poor job of learning strategies, does the other party exploit this? What are good choices for programmable selection in evolving strategies? Will they result in obtaining equilibrium? If so, how fast? How much does crossover help? What crossover rates are best?

Thus, we have cited many more questions than answers. However, we hope to provide a technique for answering such questions—namely, taking advantage of the massive parallelism of DNA computing to test huge numbers of strategies in competition and to improve them based on their outcomes.

References

1. Nash, J., Shapley, L.S.: A simple three-person poker game. In Kuhn, H.W., Tucker, A.W., eds.: Contributions to the Theory of Games. Annals of Mathematics Studies. (1950) 105–116
2. Thomas, L.C.: Games, Theory and Applications. Ellis Horwood, Ltd., West Sussex, England (1984)
3. Chen, J., Wood, D.H.: Computation with biomolecules. Proceedings of the National Academy of Sciences, USA **97** (2000) 1328–1330 Commentary.
4. Wood, D.H., Chen, J., Antipov, E., Lemieux, B., Cedeño, W.: A design for DNA computation of the OneMax problem. Soft Computing **5** (2001) 19–24
5. Wood, D.H., Chen, J., Antipov, E., Lemieux, B., Cedeño, W.: *In vitro* selection for a OneMax DNA evolutionary computation. In Gifford, D., Winfree, E., eds.: DNA Based Computers V: DIMACS Workshop, June 14-15, 1999. Volume 54 of DIMACS series in discrete mathematics and theoretical computer science., Providence, American Mathematical Society (2000) 23–37
6. Deaton, R., Murphy, R.C., Rose, J.A., Garzon, M.H., Franceschetti, D.R., Jr., S.E.S.: A DNA based implementation of an evolutionary search for good encodings for DNA computation. In: Proceedings of the 1997 IEEE International Conference on Evolutionary Computation, Indianapolis, Indiana, IEEE Press (1997) 267–271
7. Bäck, T., Kok, J.N., Rozenberg, G.: Evolutionary computation as a paradigm for DNA-based computing. In Landweber, L., Winfree, E., Lipton, R., Freeland, S., eds.: Preliminary Proceedings DIMACS Workshop on Evolution as Computation, DIMACS, Piscataway NJ (1999) 67–88 Available on request from DIMACS. Paper found at URL: http://www.wi.LeidenUniv.nl/~joost.
8. Sakamoto, K., Gouzu, H., Komiya, K., Kiga, D., Yokoyama, S., Yokomori, T., Hagiya, M.: Molecular computation by DNA hairpin formation. Science **288** (2000) 1223–1226
9. Komiya, K., Sakamoto, K., Gouzu, H., Yokoyama, S., Arita, M., Nishikawa, A., Hagiya, M.: Successive state transitions with I/O interface by molecules. In Condon, A., Rozenberg, G., eds.: DNA Computing: Revised papers / 6th International Workshop on DNA-Based Computers, DNA 2000, Leiden, The Netherlands, June 13–17, 2000. Volume 2054 of Lecture Notes in Computer Science., Berlin, Springer-Verlag (2000) 17–26
10. Winfree, E.: Whiplash PCR for $O(1)$ computing. Unpublished manuscript availble from http://dope.caltech.edu/winfree/Papers/pcr.ps (1998)
11. Bäck, T., Fogel, D.B., Michalewicz, Z., eds.: Handbook of Evolutionary Computation. Institute of Physics Publishing, Philadelphia (1997)
12. Goldberg, D.E.: Genetic Algorithms in Search, Optimization and Machine Learning. Addison Wesley (1989)
13. Vose, M.D.: The Simple Genetic Algorithm. MIT Press, Cambridge (1999)
14. Goldberg, D.E., Deb, K., Clark, J.H.: Genetic algorithms, noise, and the sizing of populations. Complex Systems **6** (1992) 333–362
15. Goldberg, D.E., Miller, B.L.: Genetic algorithms, selection schemes, and the varying effects of noise. Evolutionary Computation **4** (1996) 113–131
16. High end architectures. (2000) National Coordination Office for Computing, Information, and Communications.
    Web page at http://www.ccic.gov/pubs/blue00/hecc.html#architectures
17. Lawrence, S., Giles, C.: Accessibility of information on the web. Nature **400** (1999) 107—109

18. Von Neumann, J., Morgenstern, O.: Theory of Games & Economic Behavior. Princeton University Press, Princeton, NJ (1953) ISBN:0691003629.
19. Maynard Smith, J.: Evolution and the Theory of Games. Cambridge University Press, Cambridge, Great Britain (1982) ISBN: 0-521-28884-3.
20. Kimbrough, S.O., Dworman, G.O., Laing, J.D.: On automated discovery of models using genetic programming: Bargaining in a three-agent coalitions game. Journal of Management Information Systems **12** (Winter 1995-96) 97–125
21. Kimbrough, S.O., Dworman, G.O., Laing, J.D.: Bargaining by artificial agents in two coalition games: A study in genetic programming for electronic commerce. In Koza, J.R., Goldberg, D.E., Fogel, D.B., Riolo, R.L., eds.: Genetic Programming 1996: Proceedings of the First Annual Genetic Programming Conference, July 28-31, 1996, Stanford University. The MIT Press (1996) 54–62
22. Wu, D.J.: Discovering near-optimal pricing strategies for the deregulated power marketplace using genetic algorithms. Decision Support Systms **27** (1999) 25–45
23. Wu, D.J.: Artificial agents for discovering business strategies for network industries. International Journal of Electronic Commerce **5** (Fall 2000)
24. Papadimitriou, C.H.: Algorithms, games, and the internet. In: Proceedings of the 33rd annual ACM Symposium on Theory of Computing: STOC'01, July 6–8, 2001, Hersonissos, Crete, Greece, ACM Press, New York (2001) 749–753
25. Forst, C.V.: Molecular evolution. Journal of Biotechnology **276** (1997) 546–547

PNA-mediated Whiplash PCR

John A. Rose[1], Russell J. Deaton[2], Masami Hagiya[3], and Akira Suyama[4]

[1] Institute of Physics, The University of Tokyo
johnrose@genta.c.u-tokyo.ac.jp
[2] Department of Computer Science and Computer Engineering,
The University of Arkansas
rdeaton@uark.edu
[3] Department of Computer Science, The University of Tokyo
hagiya@is.s.u-tokyo.ac.jp
[4] Institute of Physics, The University of Tokyo
suyama@dna.c.u-tokyo.ac.jp

Abstract. In Whiplash PCR (WPCR), autonomous molecular computation is achieved by the recursive, self-directed polymerase extension of a mixture of DNA hairpins. A barrier confronting efficient implementation, however, is a systematic tendency for encoded molecules towards backhybridization, a simple form of self-inhibition. In order to examine this effect, the length distribution of extended strands over the course of the reaction is examined by modeling the process of recursive extension as a Markov chain. The extension efficiency per polymerase encounter of WPCR is then discussed within the framework of a statistical thermodynamic model. The efficiency predicted by this model is consistent with the premature halting of computation reported in a recent *in vitro* WPCR implementation. The predicted scaling behavior also indicates that completion times are long enough to render WPCR-based massive parallelism infeasible. A modified architecture, PNA-mediated WPCR (PWPCR) is then proposed in which the formation of backhybridized structures is inhibited by targeted PNA_2/DNA triplex formation. The efficiency of PWPCR is discussed, using a modified form of the model developed for WPCR. Application of PWPCR is predicted to result in an increase in computational efficiency sufficient to allow the implementation of autonomous molecular computation on a massive scale.

1 Introduction

In Whiplash PCR (WPCR), autonomous computation is implemented by the recursive polymerase extension of a mixture of DNA hairpins [1]. Although the basic feasiblity of WPCR has been experimentally demonstrated [1,2,3], a barrier which confronts efficient implementation is a tendency for single-stranded (ss) DNAs to participate in a form of self-inhibition known as *backhybridization* [1,2]. To illustrate, consider the WPCR implementation of the 3 step path, $0 \rightarrow 1 \rightarrow 2 \rightarrow 3$, shown in Fig. 1. Computational states are represented by unique DNA words of length, l bases. Each strand is composed of 3 regions. The *transition rule*

N. Jonoska and N.C. Seeman (Eds.): DNA7, LNCS 2340, pp. 104–116, 2002.
© Springer-Verlag Berlin Heidelberg 2002

region encodes the computation's transition rules (in Fig. 1, $0 \to 1$, $1 \to 2$, and $2 \to 3$). The *head region* contains a record of the strand's computation, where the 5'-most and 3'-most code words encode for the strand's initial and current state, respectively (in Fig. 1, 0 and 1). The *spacer region* guarantees adequate spacing for hybridization. A single round of computation is achieved by the hybridization of the 3' head with a matching code word in the transition rule region, followed by extension by DNA polymerase. Extension is terminated by a short poly-Adenine *stop* sequence, combined with the absence of free dTTP in the buffer. In Fig. 1 (top structure) this process has appended codeword 1 to the strand's 3' end, implementing the transition, $0 \to 1$. Although the second extension requires the formation of hairpin (a), this process is complicated by the ability of the strand to form *backhybridized* hairpin (b), which is much more energetically favorable than hairpin (a). The number of alternative, backhybridized configurations increases with each extension. For a ssDNA undergoing the rth extension, a total of r alternative hairpin structures will be accessible, only one of which is extendable by DNA polymerase. Occupancy of the $r - 1$ backhybridized structures reduces the concentration of ssDNAs available for the computation.

WPCR Molecule after 1 Successful Extension:

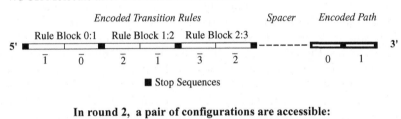

In round 2, a pair of configurations are accessible:

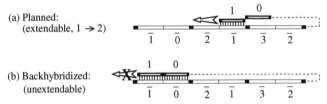

Fig. 1. Backhybridization. After the first extension process (top structure), two hairpins are accessible to the extended molecule. Occupancy of hairpin (b) reduces the concentration of extendable structures (a), and inhibits further computation. A total of $r - 1$ backhybridized structures will be accessible during extension process, r.

In Sec. 2, the length distribution of extended strands, as a function of the reaction temperature and the number of polymerase encounters per strand, is examined by modeling the recursive extension of each strand as a Markov chain. The extension efficiency per polymerase-strand encounter is then discussed using a statistical thermodynamic model of DNA hybridization. Model predictions are shown to be consistent with the premature halting of computation observed in a recent *in vitro* WPCR implementation [3]. Based on the scaling behavior of the model, completion times are predicted to be long enough to render WPCR-based massive parallelism infeasible. In Sec. 3, a modified architecture, PNA-mediated WPCR (PWPCR) is proposed in which the formation of backhybridized structures is inhibited by targeted PNA_2/DNA triplex formation. The efficiency of PWPCR is then discussed by application of the statistical thermodynamic model developed for WPCR, combined with a simplified all-or-none model of iterative extension. Targeted triplex formation is predicted be accompanied by a large increase in efficiency, which is sufficient to support the implementation of autonomous molecular computation on a massive scale.

2   The Efficiency of Whiplash PCR

The appeal of WPCR lies in the potential for the parallel implementation of a massive number of distinct computational paths. For this purpose, a distinct DNA species must be included in the initial reaction mixture for each acyclic path in the instance graph. Although a general analysis of hairpin extension efficiency would require an assessment of strand-strand interaction, in WPCR the DNA molecules are anchored to a solid support. As a result, the impact of intermolecular interaction may be neglected, allowing the recursive extension of each WPCR species to be modeled independently. The fundamental details of WPCR efficiency are therefore contained in an analysis of the single-path case.

The process of recursive extension for each DNA strand may be modeled as a Markov chain [4]. For a q-step WPCR implementation, let the *extension state*, r of each strand be defined to equal the number of times the molecule has been successfully extended plus 1. Note that a strand's extension state is distinct from a strand's *computational state*. During the course of the reaction, extending strands may occupy a total of $q + 1$ extension states, ranging from $r = 1$ (completely unextended) to $r = q + 1$ (fully extended). Let ϵ_r denote the probability that a polymerase encounter with a DNA strand in extension state r observes the strand in an extendable configuration. With each polymerase encounter, a DNA strand will increment its extension state by either 0 or 1, with probabilities $1 - \epsilon_r$, and ϵ_r, respectively. For molecules which reach the final absorbing state, $q + 1$, no further extension is possible (*i.e.*, $\epsilon_{q+1} = 0$). The state occupancies resulting from N_e polymerase encounters/strand at temperature T_{rx} are given by the product of the N_e-step transition matrix, $\mathbf{P}(T_{rx}, N_e)$ and the initial state occupancy vector, $[N_o\ 0\ \dots\ 0]$, where N_o is the total strand number. $\mathbf{P}(T_{rx}, N_e)$ is given by the Chapman-Kolmogorov eq. [4],

$$\mathbf{P}(T_{rx}, N_e) = \begin{bmatrix} 1 - \epsilon_1 & \epsilon_1 & \dots & 0 & 0 \\ 0 & 1 - \epsilon_2 & \dots & 0 & 0 \\ \vdots & \vdots & \ddots & \vdots & \vdots \\ 0 & 0 & \dots & 1 - \epsilon_q & \epsilon_q \\ 0 & 0 & \dots & 0 & 1 \end{bmatrix}^{N_e}. \tag{1}$$

The estimation of N_e and ϵ_r is discussed in Sec. 2.1 and Sec. 2.2, respectively. The resulting state occupancies estimate the length distribution, in terms of number of extensions, among all N_o strands, for particular values of T_{rx} and N_e. Accounting for a more complicated thermal program is straightforward. For a thermal cycle which consists of several polymerization periods of diverse duration and temperature, the process of extension is modeled by (1) estimating an N_e value for each subcycle, (2) constructing a transition matrix for each subcycle according to the T_{rx} employed, and (3) applying the resulting set of matrices iteratively to the initial state occupancy vector.

2.1 The Efficiency per Polymerase-DNA Encounter

The quantity ϵ_r may be discussed within the framework of a statistical thermo-dynamic model. Consider an ensemble, S_r of identical WPCR molecules, each of which has been extended r-1 times. Assuming an all-or-none model of duplex formation, members of S_r will be distributed amongst $r+1$ configurations: an unfolded ssDNA species, an extendable hairpin species, and a set of r-1 unextendable hairpin species, each of which is a backhybridized artifact from a previous round of extension. The statistical weight of a simple hairpin configuration, which consists of an end loop of n unpaired bases and a lone duplex of length j paired bases is estimated by $K = \sigma Z_j(n+1)^{-1.5}$, where Z_j is the statistical weight of stacking and σ is the cooperativity parameter [5].

   In order to ensure the uniformity of the various extension reactions of an implementation, WPCR code words are typically selected to have uniform GC content [2]. This procedure results in an approximately equal Gibbs free energy of stacking for each codeword with its Watson-Crick complement [3]. The statistical weight of stacking for a length j duplex is then estimated by $Z_j = s^{j-1}$ [6], where s is the statistical weight for the average base pair doublet of the implementation. The equilibrium fraction of extendable ensemble members, ϵ_r is estimated by the ratio of the statistical weight of the extendable hairpin to the sum of the statistical weights of all structures. Constructing this ratio with the particular values, $j = l$ and $j = 2l$ for the single planned, and $r - 1$ backhybridized hairpin configurations, respectively yields,

$$\epsilon_r = \left[1 + \gamma_r s^l + \frac{(n_r + 1)^{1.5}}{\sigma s^{l-1}}\right]^{-1}, \qquad (2)$$

for the extension efficiency per polymerase-DNA encounter of the single-path WPCR implementation. Here, $\gamma_r \approx \sum_{i=1}^{r-1}(n_r/n_i)^{1.5}$ expresses the impact of variations in loop length between competing hairpin strucures, n_r is the terminal loop length of the extendable configuration, and each n_i is the loop length of the hairpin structure extended during previous round i.

   The single path case may be generalized to apply to parallel WPCR if variations in ϵ_r due to differences in the specific ordering of transition rule blocks within the rule region are neglected. It is straightforward to demonstrate that the values, $\overline{\gamma}_r \approx 1.66r$ and $\overline{n}_r \approx (q+r)l$ are those characteristic of an implementation with mean loop lengths in all rounds, where the average is taken over all transition rule orderings. Combining these mean values with Eq. 2 yields,

$$\overline{\epsilon}_r \approx \left\{1 + 1.66rs^l + \frac{[l(q+r)]^{1.5}}{\sigma s^{l-1}}\right\}^{-1} \tag{3}$$

for the mean efficiency of a parallel, q-step WPCR implementation with parameters l and s. This expression may also be used to estimate the efficiency of the mean q-step single-path implementation. In the following text, estimates which have been obtained using $\overline{\epsilon}_r$ will be distinquished by an overscore.

2.2   The Mean Polymerase/DNA Encounter Rate

The mean number of polymerase encounters per strand, during a polymerization period of length Δt_p may be estimated as follows. Let N_u denote the number of units of *Taq* DNA polymerase utilized, where 1 unit corresponds to the synthesis of 10 nmol of added bases in 30 minutes, using an excess of activated salmon sperm DNA as substrate [7]. Let v_t denote the number of distinct extensions/second by 1 unit of polymerase under optimal conditions, using excess substrate (target and primer), and in the absence of unextendable substrate. *Taq* DNA polymerase is fast and highly processive [7]. It is therefore assumed that (1) the mean polymerase-DNA dissociation time is much larger than both the time required for oligo-length extension and the mean time between polymerase-DNA encounters, and (2) each encounter results in the all-or-none (oligonucleotide length) extension of the encountered molecule. In this case, the total number of enzyme-substrate encounters in time Δt_p is invariant to the DNA substrate extendability, and may be estimated by the product $N_{enc} = N_u v_t \Delta t_p$. Assuming that encounters are distributed uniformly over all N_o strands, the number of encounters/strand which occur in time Δt_p is estimated by,

$$N_e = \frac{N_{enc}}{N_o} = \frac{N_u v_t \Delta t_p}{N_o}. \tag{4}$$

2.3   Comparison with Experiment

The WPCR implementation of an 8 step path was recently reported [3]. The experimental protocol in [3] was as follows. An estimated total of $N_o \approx 1.2 \times 10^{13}$ immobilized strands was utilized, with 5 units of *Taq* DNA polymerase, in a total volume of 400 μL. Constant conditions of pH = 7.0 and I = 0.205 M ([K$^+$] = 0.05 M, [Mg^{++}] = 1.5 mM) were maintained. The first extension process of each strand was implemented separately, by "input PCR". The remaining 7 extensions were implemented by the application of 15 thermal cycles, each of which consisted of (1) 30 s at 337 K, (2) a rapid increase to 353 K in 60 s, (3) 300 s at 353 K, and (4) a decrease to 337 K in 120 s. The success of each extension was evaluated in all-or-none fashion, by means of a novel "output PCR" technique.

Success of the output phase was evaluated using gel electrophoresis. Bright bands were reported at the mobilities characteristic of the fully extended product for each of the first 5 extensions (including the extension implemented by input PCR). This result was taken to indicate the success of the first five extensions. Very faint bands reported at various other mobilities are assumed to indicate error extension during WPCR and output PCR.

In [3], it was maintained that problems due to backhybridization had been overcome by the applied thermal program, and that the observed poor performance was due to other factors. The validity of this view may be tested theoretically by a comparison of the observations reported in [3] with the predictions of the Markov chain model. For this purpose, the free energies of the code word set in [3] were estimated using the nearest-neighbor model of [8]. Computed values were verified to approximately satisfy the assumption of code word energetic uniformity. For instance, the mean code word standard enthalpy and entropy of stacking for each $l = 15$ base DNA code word was estimated at 114 ± 2.04 kcal/mol and 303 ± 5.62 cal/mol K, respectively, at 1.0 M [Na$^+$]. Values were then adjusted to account for the reported experimental K$^+$ and Mg^{++} concentrations, using the methodology described in [9], The statistical weight of the mean single stacked doublet in [3] was then estimated from the Gibbs free energy of stacking, $\langle \Delta G^\circ \rangle$ by the Gibbs factor, $s_{nn} = -\langle \Delta G^\circ \rangle / RT_{rx}$, where R is the ideal gas constant. The consensus value of the cooperativity parameter, $\sigma = 4.5 \times 10^{-5}$ was assumed [6]. The temperature dependence of $\bar{\epsilon}_r$ was estimated for the implementation in [3] using Eq. 3. A maximal extension efficiency per encounter of roughly 3×10^{-5} is predicted at 350 K. This predicted optimal T_{rx} is in good agreement with the experimentally determined optimum of 353 K.

In addition to the parameters discussed above, an estimation of overall efficiency requires an estimate of v_t. The estimate, $v_t \approx 6.70 \times 10^{10}$ encounters/unit/s, was obtained by taking the ratio of the rate of nucleotide addition defined to equal 1 unit of enzyme, and the mean number of bases added per polymerase-DNA encounter. Based on the manufacturer's estimate, a mean processivity of 50 bases/encounter was assumed [11]. The present Markov chain model of recursive extension, has been used to estimate the number of strands, $\overline{N}_r$ in [3] having undergone each of from 1 $(r = 2)$ to 8 $(r = 9)$ extensions, as a function of thermal cycle. Results are illustrated in Fig. 2(a). The implementation of the first extension by input PCR was modeled by assigning an efficiency of unity for the first extension. As shown, the production of fractions of molecules which have successfully undergone from 1 $(r = 2)$ to 4 $(r = 5)$ extensions is predicted during the first thermal cycle. The production of longer strands, however, is delayed until the 11th cycle, when the appearance of 5-fold extended $(r = 6)$ molecules is predicted. The production of 6 to 8-fold extended $(r > 6)$ molecules is not predicted to occur during the course of the experiment. These predictions are in agreement with the experimental behavior reported in [3], which reported the production of strands with up to 5 extensions. This agreement between model predictions and experimentally observed behavior lends strong support to the theory that backhybridization was responsible for the premature failure observed in [3], and calls into question the success of the isothermal protocol in eliminating problems stemming from backhybridization.

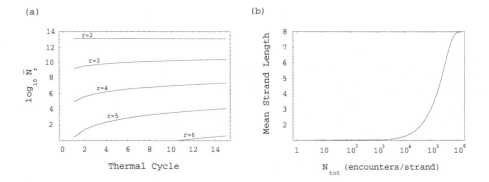

Fig. 2. The Efficiency of WPCR. (a) The mean number of strands, $\overline{N}_r$ predicted to undergo a total of from 1 extension ($r = 2$) to 5 extensions ($r = 6$), as a function of thermal cycle, for the WPCR implementation in [3]. The total strand number was roughly 1.2×10^{13}. (b) Mean strand length, in terms of extension number, as a function of the total number of polymerase encounters/strand, N_{tot}.

Continued application of a large number of thermal cycles must eventually result in completion. However, this process is predicted to require unrealistic reaction time. As shown in Fig. 2(b), the WPCR implementation in [3] (adjusted to the optimal $T_{rx} = 350$ K) is predicted to require $\approx 5 \times 10^4$ polymerase encounters/strand to exceed a mean efficiency of 2 extensions/strand. At the estimated rate of 8.4 encounters/strand/5 minute round, this corresponds to a total time of ≈ 500 hours. Furthermore, 4.0×10^5 encounters/strand are required to reach a mean of 7 encounters/strand (165 days). Mean completion is reached at roughly 10^6 encounters/strand (1.1 years). The linear scaling of encounter number predicted with N_u (*cf.*, Eq 4) also indicates that an attempt to reduce reaction time by using excess polymerase will encounter limited success. For instance, if $N_u = 54$ units of polymerase are used (90.7 encounters/round), the completion time for the 8-step path in [3] is reduced to 38 days.

3   PNA-mediated WPCR

3.1   Inhibiting Backhybridization

WPCR may be redesigned to enable the specific inhibition of backhybridized structures by targeted PNA_2/DNA triplex formation. The ability of peptide nucleic acid strands (PNAs) to bind to complementary ssDNA with extremely high affinity and sequence-specificity is well characterized [12]. For a pair of homopyrimidine PNA strands, binding to a complementary ssDNA target sequence occurs with stoichiometry 2 PNA:1 DNA, indicating the formation of a PNA_2/DNA triplex. Under appropriate reaction conditions, rapid, irreversible formation of the triplex structure occurs, even if the target sequence is embedded in a dsDNA duplex. This strand invasion results in the extrusion of the target-complementary DNA strand, formating a "P-loop" [13].

The rule block structure of WPCR may be modified to enable directed triplex formation. In particular, separation of each source/target codeword pair by the sequence, $T_4CT_2CT_2$ results in the separation of state-encoding sequences in the head region by $A_2GA_2GA_4$, the target sequence of the highly efficient cationic bis-PNA molecule reported in [14]. This is shown in Fig. 3(a). Exposure of the reaction mixture, after each polymerization round to a low $[Na^+]$, excess [bis-PNA] wash then results in a high saturation of target sequences with bis-PNA (Fig. 3, panel b). For the reported first-order rate constant of 2.33 min^{-1} at 1.0 μM bis-PNA, 20 mM $[Na^+]$ [14], a fractional saturation of 0.999 is achieved within 3 min. Cytosine-bearing, cationic bis-PNAs of length 10 bases have been reported to melt from complexed ssDNA at $\approx 85°$ C (in 0.1 M $[Na^+]$), with a very narrow melting transition [15]. The maintenance of PNA_2/DNA triplexes formed during the bis-PNA wash, during subsequent polymerization can therefore be assured by the selection of a polymerization temperature substantially less than 80° C. In each round, the presence of a PNA_2/DNA triplex immediately 5' to the new head region will not inhibit planned hybridization, due to the extreme compactness of the P-loop. The stability of the extended backhybridized configuration (shown in Fig. 3, structure c1), however will be diminished due to the separation of the duplex islands by a PNA_2/DNA triplex. This modified protocol will be referred to as PNA-mediated WPCR (PWPCR).

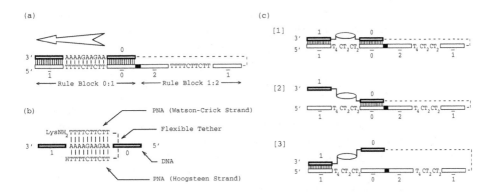

Fig. 3. PNA-mediated WPCR. [a] A "target" sequence, $A_2GA_2GA_4$ is produced between codewords during each extension. [b] Addition of bis-PNA results in the formation of a PNA_2/DNA triplex at the target sequence. Triplex represented by an oval in subsequent structures. [c] Accessible backhybridized structures have decreased stability relative to those in WPCR.

3.2   The Efficiency of PNA Mediated WPCR

The effect of the presence of the PNA_2/DNA triplex on the stability of hybridized structures, and on the per encounter polymerization extension efficiency may be estimated by means of a statistical thermodynamic model. Due to the experimentally reported compactness of the P-loop, the presence of a triplex region immediately adjacent to the head sequence is assumed to have a negligible impact on the ability of the head to hybridize with a complementary sequence in the transition rule region. Each successful extension may facilitate the later formation of three distinct backhybridized structures (see Fig. 3): (1) an extended structure, composed of a pair length l duplex islands punctuated by a P-loop (structure C1), or (2,3) two shorter structures, each of which is generated by formation of one of the duplex islands of the extended structure (structures C2, C3). Like the planned configuration, backhybridized hairpins C2 and C3 each have the form of a simple hairpin structure, with a statistical weight given by $K = \sigma Z_{l-1}(n+1)^{-1.5}$, where n is the terminal loop length of the particular structure, as discussed in Sec. 2.1. Here, the longer (n_i'') and shorter (n_i') of the associated terminal loop lengths are related by $n_i'' = n_i' + 10\,l/3$. The statistical weight of the extended backhybridized configuration, C1 has the form $Z_i = Z_p\sigma^2 s^{2l-2}(1 + n_i')^{-1.5}$, where Z_p is the post-triplex formation statistical weight of the P-loop. As discussed earlier, given the use of a polymerization temperature substantially less than 80° C, the presence of the triplex may be assumed (statistical weight of 1). Z_d then reduces to the statistical weight of interaction between the P-loop components (*i.e.*, the established triplex and the extruded single strand). The P-loop's distinctive eye structure [13] suggests the absence of stabilizing interactions between the extruded single-stranded target-complementary strand and the PNA_2/DNA triplex. Z_p is therefore assumed to be entirely entropic in origin, and is modeled as a Gaussian chain with excluded volume. For a target region of length $\frac{2}{3}\,l$, the loop region is assigned a statistical weight of $Z_p = (2 + 4l/3)^{-1.7}$. Taking the ratio of the statistical weight of the expected configuration to that of all configurations, and assuming the mean transition rule ordering, yields

$$\bar{\epsilon}_r' \approx \left\{1 + 1.94r\left[2 + \frac{\sigma s^{l-1}}{(2 + 4l/3)^{1.7}}\right] + \frac{[l(2r + 4q/3)]^{1.5}}{\sigma s^{l-1}}\right\}^{-1}, \qquad (5)$$

for the extension efficiency/polymerase-substrate encounter for a strand undergoing the rth extension process, in the mean-path PWPCR implementation with static characteristics l, q, and s. A comparison of expressions 3 and 5 indicates that the primary effect of targeted PNA_2/DNA triplex formation on the per encounter extension efficiency is the destabilization of the full length backhybridized configuration by a factor of σ.

3.3   The Overall Extension Efficiency

The Markov chain model of extension used to discuss WPCR may also be applied to PWPCR. This procedure, however is complicated by the need to separate the PNA treatment from each extension process. In particular, the two processes may not be performed concurrently, because of the very low ionic strength required for high efficiency PNA_2/DNA triplex formation. As a result, application of a Markov chain model requires the definition of an additional set of intermediate states, and the use of a second transition matrix, to model the formation of triplexes during each PNA treatment. A simpler stochastic model of performance, however may be constructed by modeling the extension process for each strand, during the polymerization period of each PWPCR cycle as a single-step, all-or-none transition. This approximate treatment, which has the advantage of yielding a closed form estimate of completion efficiency, is motivated by the extremely low efficiency per polymerase encounter predicted for a P-WPCR molecule which has been extended but not treated with bis-PNA, due to the increased length of the non-PNA treated backhybridized structure.

Consider the observation of a ssDNA which has been successfully extended in each of a total of of $c - 1$ PWPCR cycles. The probability that all of the N_e polymerase encounters with this molecule that occur in the polymerization period of cycle c will result in extension failure is equal to $(1 - \bar{\epsilon}'_c)^{N_e}$. The probability of successful extension is then estimated by, $p_{ext}(c) = 1 - (1 - \bar{\epsilon}'_c)^{N_e} \approx N_e \bar{\epsilon}'_c$. If $\langle N_{c-1} \rangle$ denotes the mean number of fully extended DNA strands present in a WPCR mixture at the end of cycle $c - 1$, then the mean number of fully extended structures present after cycle c can be written as $\langle N_c \rangle = \langle N_{c-1} \rangle p_{ext}(c)$. This relationship may be applied $c - 1$ times to yield the estimate,

$$\chi(c) \equiv \frac{\langle N_c \rangle}{N_o} \approx \frac{N_e^{c-1}}{N_o} \prod_{i=2}^{c} \epsilon'_i, \qquad (6)$$

for the fraction of c-fold extended strands produced after cycle c. Here, the first extension process for each strand in the first cycle has been assumed to proceed with an efficiency of unity, due to the absense of backhybridization.

The impact of PNA_2/DNA triplex formation on the overall efficiency of computation may be illustrated by concrete application. For this purpose, the efficiency of a PWPCR implementation of the 8-step computational path described in [3], in terms of the log of the number of fully extended substrate molecules, where $\langle \overline{N}_c \rangle = N_o \overline{\chi}'(c)$, was estimated using Eqs. 5 and 6, and is illustrated in Fig. 4(a). For consistency, a codeword set energetically equivalent to the set presented in [3] was assumed. Buffer conditions and total polymerization time were also assumed to be identical to [3]. A comparison of Fig. 2(a) and Fig. 4(a) indicates that the triplex-induced inhibition of backhybridization results in a substantial increase in predicted overall efficiency of computation. If each extension is performed at the predicted optimal reaction temperature of 60° C, roughly 1.3×10^9 of the initial 1.2×10^{13} encoded strands are predicted to be fully extended after the completion of all rounds.

3.4   The Parallelization of PWPCR

The ultimate aim of both WPCR and PWPCR is to effect the parallel, *in vitro* simulation of a massive number of distinct paths. Consider the parallelization of a PWPCR implementation, in which the set of N_o strands has been parsed into P distinct species, each of which represents a different computational path, and is present with equal copy number, $N_{copy} = N_o/P$. The maximum parallelism obtainable by this implementation is equal to $P = N_o/N_{copy}$. However, this is practically obtained only when N_{copy} is sufficiently large to ensure the full extension of at least one copy per path. Given the absence of bimolecular interaction, it is straightforward to demostrate that the threshold of completion for a parallel PWPCR implementation is reached when N_{copy} is chosen such that,

$$[1 - \overline{\chi}'(q)]^{N_{copy}} = \frac{N_{copy}}{N_o}. \tag{7}$$

For the 8-step PWPCR implementation under discussion, $N_o = 1.2 \times 10^{13}$ and $\overline{\chi}'(8) = 1.1 \times 10^{-4}$. According to Eq. 7, maximum parallelism for this implementation is achieved when $N_{copy} \approx 1.6 \times 10^5$. This copy number yields a maximum parallelism of $P \approx 7.5 \times 10^7$, which corresponds to the implementation of roughly $N_{ops} = qP \approx 1.5 \times 10^9$ distinct computational operations. A substantial improve-

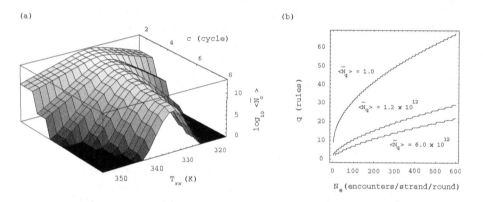

(a)   (b)

Fig. 4. The Efficiency of PWPCR (a) An estimate of the number of c-fold extended strands, $\langle \overline{N}_c \rangle$ as a function of T_{rx}, c PWPCR cycles, for the mean implementation of an 8 step computational path. (b) The contours which define the line of failure, 10% efficiency, and 50% efficiency after all q cycles ($\langle \overline{N}_q \rangle = 1.0$, 1.2×10^{12}, and 6.0×10^{12}, respectively), vs. N_e and q, for PWPCR implementations of length $q = 2 - 65$ rules. Accompanying surface and z-axis omitted for clarity.

ment may be obtained by modest modification of the protocol. Fig. 4 illustrates the contours which are predicted to define the lines of 0%, 10%, and 50% completion efficiency per strand for PWPCR implementations of length $q = 2-65$, as a function of N_e and q. The use of a codeword set energetically equivalent to that in [3] was assumed. Under this protocol, the application of a realistic set of extension reaction conditions ($N_u = 54$ units, $\Delta t_p = 30$ min, yielding $N_e \approx 542.7$)

allows the implementation of computational paths of length $q = 20$, with a per strand efficiency at completion of $\overline{\chi}'(20) \approx 0.5$. According to Eq. 7, the maximum parallelism under these conditions, $P \approx 3.1 \times 10^{11}$ paths, is achieved when $N_{copy} \approx 40$ copies per path. This corresponds to the implementation of $N_{ops} \approx 6.2 \times 10^{12}$ distinct operations.

4   Conclusion

In this work, the impact of backhybridization on WPCR efficiency was investigated by modeling the extension of each hairpin as an independent Markov chain, and estimating the associated state transition probabilities using the statistical thermodynamic theory of DNA melting. This model was shown to predict that the poor performance of WPCR observed in [3] was due to backhybridization. This is significant, because in [3], it was maintained that problems due to backhybridization had been overcome by the applied thermal program, and that the observed poor performance was due to other factors. The scaling behavior of the model also predicts that mean completion times are sufficiently long to render WPCR impractical for massive parallelism. In an effort to enhance computational efficiency by reducing the impact of backhybridization, a modified architecture, PWPCR was then introduced, which enables the specific inhibition of backhybridized structures through targeted PNA_2/DNA triplex formation. Application of this protocol is predicted to result in an efficiency increase which is sufficient to allow the realistic implementation of massive parallelism.

Acknowledgements. The authors are grateful to A. Nishikawa of the Osaka Electro-Communication Jr. College and M. Arita of the Tokyo Electrotechnical Laboratory for critical reviews, and to the referee for helpful comments. Financial support provided by the JSPS "Research for the Future" program (JSPS-RFTF 96I00101), and a JSPS Postdoctoral Fellowship and grant-in-aid (J.A.R.).

References

1. M. Hagiya *et al.*, in [16], p. 57.
2. K. Sakamoto *et al.*, *Biosystems* **52**, 81 (1999).
3. K. Komiya *et al.* in *DNA Computing*, edited by A. Condon and G. Rozenberg, (Springer-Verlag, Berlin, 2001), p. 19.
4. S. Ross, *Intro. to Probability Models*, 7th Ed., (Academic Press, San Diego, 2000).
5. D. Poland and H. Scheraga, *Theory of Helix-Coil Transitions in Biopolymers* (Academic Press, New York, 1970).
6. R. Wartell and A. Benight, *Physics Reports (Review Section of PRL)* **126**, 67 (1985).
7. M. Innis *et al.*, *Proc. Natl. Acad. Sci.* **85**, 9436 (1988).
8. J. SantaLucia, Jr., *Proc. Natl. Acad. Sci.* **95**, 1460 (1998).
9. J. Wetmur, in [16], p. 1.
10. S. Kozyavkin, S. Mirkin, and B. Amirikyan, *J. Biomol. Struct. Dyn.* **5**, 119 (1987).
11. http://www.pebio.com/pc/catalog/pg9.html

12. A. Lomakin and M. Frank-Kamenetskii, *J. Mol. Biol.* **276**, 57 (1998).

13. D. Cherny *et al.*, *Proc. Natl. Acad. Sci.* **90**, 1667 (1993).

14. H. Kuhn *et al.*, *Nuc. Acids Res.* **26**, 582 (1998).

15. M. Griffith *et al.*, *J. Am. Chem. Soc.* **117**, 831 (1995).

16. H. Rubin and D. Wood (Eds.), *DNA Based Computers III*, (American Mathematical Society, Providence, RI, 1999).

Biomolecular Computation in Virtual Test Tubes

Max H. Garzon[1]* and Chris Oehmen[2]

[1] Computer Science Division
The University of Memphis
Memphis, TN 38152-3240, U.S.A.
mgarzon@memphis.edu
http://www.cs.memphis.edu/~garzonm
[2] Biomedical Engineering
The University of Memphis
Memphis, TN 38152, U.S.A.
coehmen@memphis.edu

Abstract. Biomolecular computing (BMC) aims to capture the innumerable advantages that biological molecules have gained in the course of millions of years for computational purposes. While biomolecules have resolved fundamental problems as a parallel computer system that we are just beginning to decipher, BMC still suffers from our inability to harness these properties to bring biomolecular computations to levels of reliability, efficiency and scalability that are now taken for granted with solid-state based computers. In the same way that evolutionary algorithms capture, in silico, the key properties of natural evolution, we explore an alternative approach to exploiting these properties by building virtual test tubes in electronics that would capture the best of both worlds. We describe a distributed implementation of a virtual tube, EdnaCo , on a cluster of PCs that aims to capture the massive asynchronous parallelism of BMC. We report several experimental results, such as solutions to the Hamiltonian Path problem (HPP) for large families of graphs than has been possible on a single processor or has been actually carried out in wet labs. The results show that the paradigm of molecular computing can be implemented much more efficiently (in terms of time, cost, and probability of success) in silico than the corresponding wet experiments, at least in the range where eDNA can be practically run. Consequently, we pinpoint the appropriate range of problem sizes and properties where wet biomolecular solutions would offer superior solutions.

1 Introduction

Biomolecular computing (BMC) is now a fairly known field in computer science and biology. Like genetic algorithms a few decades earlier, it aims to capture the advantages that biological molecules (DNA, RNA and the like) have gained in the course of millions of years of evolution to perform computation out of reach through conventional electronic computers. Several conferences [22,23,4,2,5] have

---

* Max Garzon is the corresponding author.

N. Jonoska and N.C. Seeman (Eds.): DNA7, LNCS 2340, pp. 117–128, 2002.
© Springer-Verlag Berlin Heidelberg 2002

established the potential of the field to achieve some of this goal, either through new experimental biomolecular protocols using state-of-the-art biotechnology, or through theoretical results, such as universality and complexity [26], comparing it with the standard computational models such as Turing machines.

It is becoming increasingly clear, however, that the realization of this potential cannot be achieved without addressing the fact that biomolecular protocols in use are relatively unreliable, inefficient, unscalable, and expensive compared to conventional computing standards. Current efforts in the field aim at overcoming these problems by tapping on a number of properties that biomolecules must exercise to accomplish their evolved goal in natural organisms [6,16]. A good example of such is *in vivo* molecular computing [6], where the protocols are transferred from test tubes to living organisms. While this strategy may produce very interesting results, it presents some shortcomings. For example, our understanding of the information processing ability of biomolecules will not increase substantially, just as cloning an organism does not shed any scientific understanding of the complexity of morphogenesis from a biological point of view. The critical issues of reliability, efficiency and scalability remain unanswered for computation *in vitro*.

Biological phenomena have inspired a number of other computational methodologies. A recent example is genetic algorithms and evolutionary computation, inspired by natural selection. They attempt to capture in-silico the fundamental features of cross-over and mutation present in natural organisms to help program computers with a good deal of success and without losing basic advantages of electronic computing. An alternative approach to shed light on biomolecular computations is thus to introduce an analog of biomolecules and their interactions in electronics that capture their biological counterparts. Preliminary attempts have used evolutionary algorithms to find good encodings [8,17,28,14], abstract bases [13] to gauge the reliability of BMC protocols, numerical simulation of reactions [15,13], and electronic DNA [11,12] to estimate the performance of a BMC protocol before it actually unfolds in the tube. These analogs can serve at least three related but rather different purposes. First, they can be used as a cost-effective tool to pre-process and gauge computational protocols before they are carried out in test tubes. Second, because electronics allows better programmability and control, they allow more controlled experimentation than is possible in wet tubes, with an increased yield in experimental payoffs. Third, like evolutionary algorithms, the biomolecular analogy may become useful as a computational methodology in its own right, if adapted and exploited for computational purposes in conventional computing, perhaps independently of how un/faithfully it may capture the biological processes themselves.

In this paper we explore the concept of virtual test tubes as a general technique to address these three goals, although we focus on the third and partially the second. Previously, we have introduced the notion of a virtual test tube [11] and reported preliminary results on a small-scale proof-of-concept, called **Edna**, that runs on a single processor [12]. These results indicated that, properly scaled, a virtual test tube might effectively solve problems by manipulating electronic analogs of DNA molecules in-silico. We have thus scaled this concept up to a virtual test tube in a distributed environment running on a cluster of PCs. We report several experimental results on solutions of the Hamiltonian Path problem and word set quality criteria for molecular computing. We show that an electronic version of DNA (eDNA) is indeed capable of executing *in practice* Adleman's experiment *in silico* for large families of graphs, including graphs of larger size than was possible on Edna or has been actually carried out in wet labs. We also give further experimental evidence of the soundness of proposed encoding strategies for biomolecular computation. The results show that, within the limits of parallelism attainable on a distributed environment, virtual test tubes can not only capture many phenomena present in real biomolecules, but also permit much higher reliability and efficiency than has been possible with a comparable number of actual biomolecules.

The layout of the paper is as follows. In section 2 we discuss virtual test tubes. In section 3 we present some results that allow us to estimate the reliability of biomolecular computations in silico. We argue that this is an upper bound on the reliability of biomolecular computation in-vitro, at least for a relatively large range of problem size. Finally, in section 4, we discuss some conclusions of these results. A point of debate are the properties of DNA that cannot be attained in eDNA, and more specifically, whether virtual test tubes can compete, in actual experiments, with the computational scales that can be realistically tackled in the wet lab.

2   Virtual Test Tubes

As mentioned above, several approaches have been used to attempt to understand the basic advantages of molecules for computation, including a high-level model of molecular reactions inspired in biochemical reaction models [15,13]. Molecules are represented by concentrations and their interactions are modeled by differential equations reflecting the kinetics of the reactions. The model consists in solving the equations (usually a numerical solution) and thus predicting the concentration of potential solution molecules. Here we focus on a fairly different type of model. A *virtual test tube* is any model of biomolecular reactions in electronic media that captures fairly closely the environment and kinetics of the molecular interactions, while making minimal assumptions about the global behavior of molecular populations. It can be a simulation in software, or a computer chip that serves as a container for the same purpose. We will thus assume a strong form of the Church-Turing Thesis, barring the existence of elemental computational steps achievable only through nucleic acids *in vivo*. That is, we will assume that the competitive advantage of biomolecules resides exclusively in their massive parallelism that could be, in principle, achieved on conventional media (solid-state electronics). We will restrict our attention to conventional solid-state devices, simply because they have matured to satisfactory levels of reliability and implementation.

On these assumptions, what are then the fundamental advantages of biomolecules to be captured in a virtual test tube? Some natural candidates emerge when comparing the behavior of a biomolecular ensemble in a test tube to a parallel computer. The type of computation is *asynchronous, massively parallel* and determined by both local and global properties of the processor molecules. Biomolecules seem to have solved very efficiently key problems such as communication, load balancing, and decentralized control. This seems to be achieved through highly localized base-to-base interactions, template matching, and enzymatic reactions, mediated by randomized motion in both transport and reactions. These characteristics are reminiscent of cellular automata, although their sites are rather asynchronous, randomized, and not localized in space but in mobile molecules. The unfolding of molecular computations is also reminiscent of genetic algorithms driven by a natural fitness function defined by the Gibbs free-energy of hybridization. It is therefore not surprising that molecular computing faces a common basic problem with cellular automata, namely, a *programming methodology* that would unleash their power to solve problems now considered too difficult for conventional machines [25]. This line of reasoning suggests that an appropriate analog may thus capture some of these properties in a suitable electronic computational medium.

2.1   Edna's Architecture

We have introduced the concept of a virtual test tube in [11]. The architecture of a small prototype, Edna, has been described in detail in [12]. In short, Edna is a piece of software that simulates the reactions that might actually happen in a test tube as closely as possible. The tube consists of a regular array of cells similar to the space of a cellular automaton [9]. The cells represent quanta of 2D or 3D space that may be empty or occupied by nucleotides, molecules, or other reactants. Each cell can also be characterized by associated parameters that render the tube conditions in a realistic way, such as temperature, salinity, covalent bonds, etc. DNA nucleotides are represented by states of the nodes where they are located. They move randomly from node to node in an attempt to capture the random, brownian motion present in wet test tubes, although flow simulations of aqueous solutions are possible but have not been implemented since eDNA may be just as effective with the simpler model. Longer DNA molecules are represented by linked lists. In order to preserve their structural integrity as they move in the tube, the header the list makes its random moves and the remaining nucleotides must follow. The molecular interactions are governed by local rules in order to reflect closely the kinetics in wet tubes. When two nucleotides belonging to two different strands come in close physical proximity, the local rule determines whether the strands should hybridize or not. If so, a new duplex is formed in the most stable energetic frame-shift possible and placed anew in the tube. This frameshift is purely determined by well established facts about the kinetics of DNA hybridization based on the Gibbs free energy [27]. Currently, two models of hybridization are implemented, based on the well-known nearest-neighbor and staggered ziper models [3,24]. If the stacking energy of the best hybrid is below the threshold, (as determined by the user), a new duplex is created and placed at random free location in the tube; otherwise, hybridization does not take place and the single strands are simply allowed to continue on their random motion. Copies of the original strands may or may not be programmed to remain in the tube (more below). Other rules of interaction can be easily substituted, such as a more combinatorial H-distance [11], or perhaps stochastic rules based on melting temperatures [27]. The cellular space can also be set up to boundary conditions that reflect tube walls. Edna has an ergonomic interface to choose among these local rules for hybridization, to place and remove strands in the tube, and in general, to set up the various reactions conditions in the desired combination.

The programming of the tube stops here, however. Once the tube contents is initialized with a population of strands, reactions unfold thereafter, unprogrammed. The interaction of the molecules and their random wanderings in the tube is compounded over space and time, in their multiple attempts and eventual success or failure at hybridizing into double stranded molecules. Eventually, the virtual tube will come to equilibrium under the local rule of interaction specified, showing a possible outcome of the experiment. Edna thus serves as a virtual test tube where full chains of complex interactions of electronic structures (eDNA) emerge, although admittedly probably not at the same scale of realism of wet tubes. Since the interactions reflect so closely well known kinetics of the hybridization process, it is reasonable to expect that this type of model bears a significant resemblance to the analogous experiment if implemented in a wet tube. This expectation has been confirmed by the preliminary experimental results in [12] and by further results discussed below. Virtual tubes, on the other hand, offer several advantages of their own. First, Edna already shows evidence that electronics can capture key properties of biomolecules while keeping obvious advantages of electronics. Important facts are ready programmability, robustness, and a higher degree of reliability and control. Also, once programmed, the cost of a run in very low compared to wet tube runs. One can now easily play the what-if game. Once can try various hybridization rules, under various temperatures and reaction conditions and for a variety of purposes, as illustrated below. The major disadvantage, namely the relatively limited scalability compared to wet tubes, can be addressed to a large extent, as discussed in Section 4 below.

2.2   EdnaCo's Architecture

EdnaCo is a distributed implementation of a scaled-up version of Edna in a distributed computing environment. The computational framework of EdnaCo is a complex of interacting data structures distributed over several processing nodes which are joined, transparently to the user, in order to produce a single test tube in each run. The entire tube is distributed over a cluster of processors (currently up to 8) in such a way that each local processor holds an entire tube segment. A segment is itself a copy of Edna, so that one can check the contents of the tube and manage deletions (when a strand leaves the local tube segment) and additions (incoming strands, strand additions at the outset of the simulation, and hybridization events). No nucleotide or strand is split between two different nodes, but their brownian motion may include migration to a any other different node. The migration strategy is programmable. If a strand tries to leave a node, it is sent via message passing to an "adjacent" process, and randomly placed in that tube segment. The tube segments are conceptually strung together to produce a coherent tube structure. This is crucial to the overall functioning of EdnaCo since partial solution strands must be able to potentially hybridize with *all* other strands, many of which may reside outside the local node. The computer interface of EdnaCo is for now a primitive two dimensional boolean (text) array. The columns indicate strand length. Each row is a time slice consisting of a "gel" that summarizes the tube content by strand length and multiplicity. The content of EdnaCo, at any given time, can actually be saved if a more realistic visualization on Edna of the entire distributed test tube is desired.

3   Experimental Results

Now we present a summary of the numerous experimental runs obtained by implementing several biological protocols in EdnaCo. Edna has been protoyped on Adleman's solution [1] of HPP, the Hamiltonian Path Problem. The molecules represent graph vertices and edges, and the hybridization logic carries the brunt of the computational process to produce longer molecules representing paths in the graph. Once the chemistry reaches equilibrium, it is a matter of searching the products of the reaction to determine whether a molecule exists representing the witness Hamiltonian path. Adaptions for other problems are readily made and experiments for other problems (such as MAX-CLIQUE) are being conducted that will be reported elsewhere.

3.1   Scaling Up Adleman's Experiment

The first run was to reproduce and scale up the original experiment performed by Adleman in a wet tube with real DNA molecules [1]. On EdnaCo, we were able to systematically reproduce Adleman's result with electronic versions of his original vertex encodings (or word sets) and several other encodings that were deemed good acoording to two proposed measures, the computational incoherence and the H-distance, described below. We easily succeded with very sparse graphs (such as paths and cyclic graphs). EdnaCo established (non)Hamiltonicity without a problem with a 100% reliability for cycles up to 20 vertices. The result appeared perfectly scalable to any cycle size.

Next, we scaled up the number of edges in the problem instance in order to run in silico an experiment that no one in the literature has reported, in order to test, in the small range allowed by the small cluster (8 processors), the potentially enormous scalability of Adleman's approach that makes DNA-based computing so fascinating. We were successful in scaling the results systematically on EdnaCo up to about 15 vertices on sparse graphs (up to about 25 edges). A sample of the results is shown in Table 1(right) for graphs with 5 vertices. The "mode" refers to the type of hybridization rule (E = nearest neighbor model - threshold in Gibbs free energy; H = H-distance mode - threshold in H-distance). "Path" refers to the witness path actually formed. Numbers in ()'s indicate an edge to or from the vertex number indicated, but no vertex encoding on the hybrid. "C" refers to a cyclic graph, "K" refers to a complete graph and "G" refers to the graph with 5 vertices $0, 1, 2, 3, 4, 5$ and directed edges $0 \to 1, 0 \to 3, 1 \to 2, 1 \to 4, 2 \to 3, 3 \to 2$ and $3 \to 4$. For graphs with more edges, numerous partial paths were produced that swamped the tube, but EdnaCo failed to show the formation of a Hamiltonian path. We are currently refining the implementation of a method of subtractive hybridization so that paths are more systematically groomed to prevent false negative partial paths from swamping the tube, by one of several proposed strategies [22]. Despite several attempts to solve instances of sparse graphs of size 50 vertices, we have been so far unable to report success on the current cluster (where the maximum run can only last 36 hours), but optimizing the efficiency of the eDNA makes it only a matter of time. For example, we have implemented a genetic algorithm in the hybridization rule that prunes molecules with low fitness (such as those revisiting a previous vertex); we are also optimizing the concentration of individual eDNAs with respect to tube size. Too many molecules hinder brownian motion; too few also prevent the molecules from finding each other and the appropriate reactions from taking place. A ratio 1 : 2 of occupied to free cells in the tube seems to be an appropriate ratio. We can maintain a constant concentration or a variable regime of individual molecules, as the reactions progress and the problems demands. We believe these techniques will be sufficient to close the gap. Based on this data, we estimate that the probability of success of an HPP run is about $15/16 \equiv 94\%$. Computational with more runs show similar figures. We estimate that, on larger state-of-the-art clusters running the same implementation of EdnaCo, we will realistically solve random instances of sparse graphs with over 200 vertices with a comparable degree of reliability and within reasonable times. Note that the sparse region of HPP is the range of interest since sufficiently dense graphs are guaranteed to be Hamiltonian by any of well known sufficient conditions (say Dirac's test on the number of edges).

3.2   Evaluation of CI Encodings

The experiments also permit an analysis of the encoding used in the experiments. A systematic comparison of encoding quality similar to that reported in [12] for a graph with 5 vertices was done with larger families of graphs. The first measure for encoding goodness tested was the computational incoherence, ξ, based on statistical mechanics (see [21] for a precise definition). The measure is based on the average probability, at equilibrium, that a randomly observed hybridization within the annealed mixture is in an error configuration with respect to the computation. Using a standard genetic algorithm with $-\log_{10}\xi$ as the applied measure of fitness, sets of DNA words of different lengths were evolved for a Hamiltonian Path problem. Use of this fitness produced encodings with various (small, medium, large) average probability of error. The values of ξ for the different combinations produced is shown in Table 2(left). The specific

Quality	Length	ξ
Good ξ	12	2.34×10^{-7}
Medium ξ	12	8.56×10^{-4}
Bad ξ	12	1.0
Good ξ	20	1.15×10^{-10}
Medium ξ	20	1.24×10^{-6}
Bad ξ	20	1.0
Good ξ	28	3.94×10^{-16}
Medium ξ	28	6.03×10^{-6}
Bad ξ	28	1.0

Encoding	Path found	Run Time
28g1/E/C	$0 > 1 > 2 > 3 > 4$	55 min.
28g2/E/C	$0 > 1 > 2 > 3 > 4$	11 min.
28g1/H/C	$0 > 1 > 2 > 3 > 4$	50 min.
28g2/H/C	$0 > 1 > 2 > 3 > 4$	11 min.
28g1/E/K	cyclic	29 min.
28g2/E/K	$(4) > 1 > 2 > 4 > 3 > (4)$	22 min.
28g1/H/K	$(3) > 0 > 4 > 1 > 2 > (4)$	82 min.
28g2/H/K	$(4) > 3 > 1 > 4 > 0 > (3)$	26 min.
28g1/E/G	$0 > 1 > 2 > 3 > 4$	< 4 hrs.
28g2/E/G	$0 > 1 > 2 > 3 > 4$	41 min.
28g1/H/G	$0 > 1 > 2 > 3 > 4$	< 4 hrs.
28g2/H/G	cyclic	42 min.

Table 1. Encoding quality by CI fitness and EdnaCo solution to HPP on 5 vertices.

encodings are not shown due to space limitations. These encodings were used for both virtual tube simulations of ξ as a measure of encoding goodness using the nearest neighbor model's free energies and the H-metric (discussed below) as hybridization rules. The results are shown in Table 2.

In the EdnaCo runs on the CI encoding sets, long molecules generally formed faster with the free-energy hybridization rule than the H-metric rule. With the energy condition, the good encodings for all lengths produced Hamiltonian paths. This occurred probably fairly quickly from the start of the run, which lasted 4 hours (in the current setup, we have no way to time more precisely). Very few error hybridizations were observed. The medium quality encodings primarily produced molecules that were in the proper hybridization frame, although not long enough. More mishybridizations, however, were observed with the medium than with the good quality encodings, and it is unlikely that longer runs might have produced a Hamiltonian path. The bad quality encodings produced many mishybridizations, and very few hybridizations in the proper frame. After a four (4) hour simulation, it was evident from the molecules formed that no Hamiltonian path was possible. With the H-metric hybridization condition, the CI encodings did much worse. No Hamiltonian paths were formed for any of the encodings. Most of the oligonucleotides hybridized in improper frames. In addition, for hybridization thresholds less than half the encoding length, no hybridizations formed. Once a threshold equal to half the length was reached, hybridization occurred quickly, but in error modes.

3.3   Evaluation of H-metric Encodings

A more computational measure of hybridization likelihood, the H-distance, has been introduced in [10]. Hamming distance between DNA strands is defined as the difference between the number of WC matching pairs from the shorter length

Coding	H-rule	Thr.	Space	HP	Hrs.
CI12g	E	4	40*40	Y	4
CI12g	H	6	40*40	N	4
CI12m	E	4	40*40	N	4
CI12m	H	6	40*40	N	4
CI12b	E	4	40*40	N	4
CI12b	H	6	40*40	N	4
CI20g	E	4	50*50	Y	12
CI20g	H	10	50*50	N	12
CI20m	E	4	50*50	N	12
CI20m	H	10	50*50	N	12
CI20b	E	4	50*50	N	12
CI20b	H	10	50*50	N	12
CI28g	E	4	60*60	Y	36
CI28g	H	14	60*60	N	36
CI28m	E	4	60*60	N	36
CI28m	H	14	60*60	N	36
CI28b	E	4	60*60	N	36
CI28b	H	14	60*60	N	36

Coding	H-rule	Thr.	Space	HP	Hrs.
H12g	E	4	40*40	Y	4
H12g	H	4	40*40	Y	4
H12m	E	4	40*40	N	4
H12m	H	4	40*40	N	4
H12b	E	1	40*40	N	4
H12b	H	1	40*40	N	4
H20g	E	4	50*50	Y	12
H20g	H	4	50*50	Y	12
H20m	E	4	50*50	N	12
H20m	H	4	50*50	N	12
H20b	E	4	50*50	N	12
H20b	H	4	50*50	N	12
H28g	E	4	60*60	N	36
H28g	H	4	60*60	N	36
H28m	E	4	60*60	N	36
H28m	H	4	60*60	N	36
H28b	E	4	60*60	N	36
H28b	H	4	60*60	N	36

Table 2. EdnaCo runs on CI- and H-encoding sets with nearest neighbor rules.

of lined up strands, but it is not appropriate enough since it ignores likely frame-shifts in the tube. The H-distance between two oligos x and y is defined as the minimum of all Hamming distances obtained by successively shifting and lining up y and its WC-complement against x. A small H-distance indicates that the two oligos are likely to stick to each other one way or another; a large measure indicates that *under whatever physico-chemical conditions* y finds itself in the proximity of x, they are far from containing many WC complementary pairs (let alone segments), and are therefore less likely to hybridize. In other words, they are more likely to avoid an unwanted hybridization. The maximum length of these segments is controlled by a threshold parameter τ, that is a fairly coarse expression of the reaction conditions.

We ran a similar test of encodings evolved using the H-metric as fitness function (see [10] for a precise definition). The results are also shown in Table 2. The encodings were evolved used EdnaCo's on-line genetic facility as well. The results were similar to those obtained for the CI-based word sets. In case no Hamiltonian path was obtained, the time indicates how long they were run before giving up. As expected again, the good encoding produced the desired path fairly quickly regardles of the hybridization rule used, while bad encoding produced no results. In summary, the results also show a fairly high correlation between the two criteria for encoding quality. A more careful quantitative comparison is underway and will be reported elsewhere.

4   Conclusions

We have discussed the concept of a virtual test tube and its extension to a distributed version, EdnaCo, that approaches more closely reaction conditions in a wet tube. We have also reproduced and extended *in silico* a number of experiments, some of which have been previously performed in wet tubes. The results show strong evidence that electronic DNA (eDNA) is capable of running instances for fairly large instances of HPP *in practice* and with a high probability of success. We have also used the distributed test tube to run comparisons of two encoding strategies with experimental data. No substantial difference in performance could be discerned between the two criteria in the runs on EdnaCo. Our results give experimental evidence of the quality of previously proposed encoding strategies that have been devised based on theoretical analyses.

On a larger scale, several advantages to virtual test tubes emerge. First, it is clear from their design and comparison of our experimental results, that the outcomes of the virtual test tube bear a very significant resemblance to the outcomes of analogous experiments in a wet tube. Second, it is equally clear that the savings in cost and perhaps even time, at least in the range of feasibility of eDNA, are enormous compared to the equivalent biochemical protocols to solve the same instances, most of which, to our knowledge, no one has attempted. Third, eDNA inherits the customary efficiency, reliability and control now standard in electronic computing, hitherto only dreamed of in wet tube computations. The physics and chemistry is now much more programmable, perhaps at the expense of biochemical realism, but at a level that would be desirable, if it was possible with actual molecules and achievable with virtual molecules.

We are thus led once again to the fundamental question. What are the true advantages of biomolecules such as DNA for computation? It might appear at first sight that virtual tubes lack critical advantages, say because eDNA may never be able to achieve the same degree of massive parallelism or physical realism of wet DNA. Upon reflection, this question must remain open for several reasons. First, real molecules may be packed in picomols to a micrometer, but it also takes thousands, perhaps millions, of them to be successful enough in a reaction to be witnessed in a gel, whereas just *one* success suffices in a virtual test tube. Assuming that one can effectively use every bit in the 1Gb RAM of a PC as a nucleotide (not unreasonable), a virtual test tube idly running EdnaCo on a 1000-node cluster overnight seems in hindsight as powerful, and perhaps more realistic, an approach than a picomol of molecules for solving a 1000 vertex instance of HPP, a challenge to current biotechnology. Second, the problems of scalability posed by conventional electronics itself may be solved by finding different physical implementations of electronic DNA. For example, it is now conceivable from the results in this paper that the duality exhibited by many fundamental particles in the physical universe (e.g., muons and gluons at the quantum level) may afford interesting computational media in the form of some sort of quantum molecular computer.

Finally, our simulations suggest that massive parallelism is not the only true source of power in biomolecular computations. Two more candidates emerge, *randomness* and *bounded resources*. With hindsight, this is not surprising. Randomness appears to be a generating force in biological evolution and development. In biomolecular-based computations, randomness in the reactions is inherent. Also inherent is the bounded volume in which the reactions must take place. As is already evident in artificial neural networks and limited individual life spans in genetic algorithms and natural selection, bounded resources force devices to improve efficiency. This appears to be the case in biomolecular computation as well. In that sense, biomolecules seem unbeatable by electronics in their ability to pack enormous amounts of information in tiny regions of space and to perform their computations with very high thermodynamical efficiency.

5 Acknowledgements

The authors would like to thank JICS, The Joint Institute for Computational Science of the U. of Tennessee-Knoxville, for use of the PC cluster, to John Rose at The U. of Tokyo for some of the encoding sets, and to Russell Deaton at The University of Arkansas for stimulating conversations.

References

1. L. M. Adleman, Molecular Computation of Solutions to Combinatorial Problems. Science, **266**, 1021 (1994).
2. W. Banzhaf, J. Daida, A.E. Eiben, M.H. Garzon, V. Hanovar, M. Jakiela, R.E. Smith, (eds.), *Proc. of The Genetic and Evolutionary Computation Conference GECCO*, Orlando, FL, July 1999, Morgan Kaufmann.
3. C. R. Cantor, P. R. Schimmel, *Biophysical Chemistry, Part III: The Behavior of Biological Macromolecules* Freeman, New York, 1980.
4. A. Condon, G. Rozenberg (eds.), *DNA Computing* (Revised papers), Proc. of the 6th International Workshop on DNA-based Computers. Leiden University, The Netherlands, 2000. Springer-Verlag Lecture Notes in Computer Science **2054** (2000), 247–258, Heidelberg.
5. D. Whitley, D. Goldebrg, E. Cantu-Paz, L. Spector, I. Parmee, H.G. Beyer (eds.), *Proc. of The Genetic and Evolutionary Computation Conference GECCO-00*, Las Vegas, 2000, Morgan Kaufmann.
6. T.L. Eng, "On Solving 3CNF-satisfiability with an in-vivo algorithm." In [22], 135–141.
7. R. Deaton, M. Garzon, R. E. Murphy, J. A. Rose, D. R. Franceschetti, S.E. Stevens, Jr. The Reliability and Efficiency of a DNA Computation. Phys. Rev. Lett. **80** (1998), 417.
8. R. Deaton, R. E. Murphy, J. A. Rose, Max Garzon, D. R. Franceschetti, S.E. Stevens, Jr. A DNA based Implementation of an Evolutionary Search for Good Encodings for DNA Computation. Proc. IEEE Conference on Evolutionary Computation ICEC (1997), 267–271.
9. M. Garzon, *Models of Massive Parallelism* (Analysis of Cellular Automata and Neural Networks). Springer-Verlag, Berlin, 1995.
10. M. Garzon, P. Neathery, R. Deaton, R.C. Murphy, D.R. Franceschetti, S.E. Stevens, Jr. A New Metric for DNA Computing. In [18], 472–478.

11. M. Garzon, R. Deaton, J.A. Rose, D.R. Franceschetti, Soft Molecular Computing. Proc. of the 4th workshop, Princeton University, 1998. In [23], 89–98.
12. M. Garzon, E. Drumwright, R.J. Deaton, D. Renault, Virtual Test Tubes: a New Methodology for Computing. Proc. 7th Int. Symposium on String Processing and Information Retrieval. A Coruña, Spain. IEEE Computer Society Press, 2000, 116–121.
13. A. Nishikawa and M. Hajiya.Towards a System for Simulating DNA Computing with Whiplash PCR, Proc. of the Congress on Evolutionary Computation CEC-99.
14. A. J. Hartemink, D. K. Gifford, Thermodynamic Simulation of Deoxyoligonucleotide Hybridization of DNA Computation. In [22], 25–38.
15. A. J. Hartemink, T. Mikkelsen, D. K. Gifford, Simulating Biological Reactions: A Modular Approach. In [23], 109–120.
16. S. Ji, "The Cell as the smallest DNA-based Molecular Computer." In [22], 123–133.
17. J. Khodor, D.K. Gifford, A. Hartemink (1998), Design and Implementation of Computational Systems Based on Programmed mutagenesis. In [22], 93–97.
18. J.R. Koza, K. Deb, M. Dorigo, D.B. Fogel, M. Garzon, H. Iba, R.L. Riolo, eds. (1997). *Proc. 2nd Annual Genetic Programming Conference*, Morgan Kaufmann.
19. J.R. Koza, K. Deb, M. Dorigo, D.B. Fogel, M. Garzon, H. Iba, R.L. Riolo, eds. (1998). *Proc. 3rd Annual Genetic Programming Conference*, Morgan Kaufmann.
20. L.F. Landweber, E.B. Baum (eds.), *DNA Based Computers II*, Proc. of the 2nd workshop, Princeton University, 1996. DIMACS series of the American Mathematical Society **44** (1999), 247–258, Providence RI.
21. J. A. Rose, R. Deaton, D. R. Franceschetti, M. Garzon, S. E. Stevens, Jr., A Statistical Mechanical Treatment of Error in the Annealing Biostep of DNA Computation. In [2], 1829–1834.
22. L. Kari, H. Rubin. D. Wood (eds.), 4th DIMACS workshop on DNA Computers. Special Issue of *Biosystems* (*J. of Biological and Information Processing Sciences*) **53**:1–3. Elsevier.
23. L. Kari, E. Winfree and D. Gifford (eds.), Proc. 5th workshop on DNA Computers, MIT, Cambridge, MA, 1999. DIMACS series of the American Mathematical Society **54** (1999), 247–258, Providence RI.
24. J. SantaLucia, Jr., A unified view of polymer, dumbbell, and oligonucleotideDNA nearest-neighbor thermodynamics. *Proc. Natl. Acad. Sci.* **95** (1998), 1460.
25. M. Sipper, *Evolution of Parallel Cellular Machines* (The Cellular Programming Approach). Springer-Verlag, Berlin.
26. W. D. Smith, *DNA Based Computers, Princeton University, 1996*, DIMACS Proc. Series (American Mathematical Society, Providence, RI, 1996).
27. J. G. Wetmur, *Preliminary Proceedings of the Third Annual Meeting on DNA Based Computers, University of Pennsylvania, 1997*, DIMACS Proc. Series (American Mathematical Society, Providence, RI, 1997).
28. B-T Zhang, S-Y Shin, Molecular Algorithms for Efficient and Reliable DNA Computing. In [19], 735–742.

Developing Support System for Sequence Design in DNA Computing

Fumiaki Tanaka[1], Masashi Nakatsugawa[1], Masahito Yamamoto[1], Toshikazu Shiba[2], and Azuma Ohuchi[1]

[1] Division of Systems and Information Eng.,
Graduate School of Eng., Hokkaido University
[2] Division of Molecular Chemistry Eng.,
Graduate School of Eng., Hokkaido University
Nishi 8, Kita 13, Kita-ku, Sapporo, Hokkaido, 060-8628, JAPAN
Phone: +81-11-716-2111 (ext. 6498), Fax: +81-11-706-7834
{fumiaki,masashi,masahito,shiba,ohuchi}@dna-comp.org
http://ses3.complex.eng.hokudai.ac.jp/

Abstract. Sequence design is the important factor which governs the reaction of DNA. In related researches, the method to minimize (or maxmize) the evaluation function based on knowledge of sequence design has been used. In this paper, we develop support system for sequence design in DNA computing, which minimizes the evaluation function calculated as the linear sum of the plural evaluation terms. Our system not only searches for good sequences but also presents contribution ratio of each evaluation term to the evaluation function and can reduce the number of combination of evaluation terms by reduction of the evaluation function. It helps us to find a good criteria for sequence design in DNA computing.

1 Introduction

Currently, DNA computing is expected to be applied to various fields of study, massive parallel computing, nanotechnology, genome informatics, and so on. However the realizations of these applications are influenced by the reliability to control the reaction of DNA. In particular, since the first stages of the study of DNA computing, the importance to design good sequences has been pointed out. Indeed knowledge of sequence design is obtained experientially, however the guide for the design of sequences is unknown. When we try to utilize the knowledge acquired empirically, it is effective approach to minimize evaluation function, which consists of the plural evaluation terms satisfying constraints based on knowledge of sequence design.

Garzon et al. proposed the function, which is called H-measure [2]. This function calculates the minimum value of hamming distance between a DNA strand and a complementary DNA strand. H-measure is well known as an evaluation term which two DNA strands hybridize or not. He demonstrates that solving the optimization problem of sequence design is a NP-hard problem itself. Therefore

N. Jonoska and N.C. Seeman (Eds.): DNA7, LNCS 2340, pp. 129–137, 2002.
© Springer-Verlag Berlin Heidelberg 2002

this approach needs probabilistic search method (e.g. genetic algorithm). Deaton et al. also considered hamming distance as criteria of mishybridization [3]. He demonstrated the validity of it by a chemical experiment though it was the case where they did not consider the shift of DNA strand. Arita et al. considered evaluation function as the sum of some evaluation terms [1]. They proposed two different sequence generators. One is genetic algorithm, the other is random generate-and-test algorithm.

In sequence design, many constraints which should be satisfied exist, and the function which satisfies each constraint can be made innumerably. Additionally, because in case n evaluation items exist, the number of all combination is 2^n, so it is laborious to find the optimum combination set out of many evaluation terms. Therefore in combining plural evaluation terms, we must consider the effectiveness of each evaluation item to the evaluation function because if we can find the evaluation term with low effectiveness, it is possible to reduce the number of combination of evaluation terms by deleting it.

In this paper, we introduce the contribution ratio as an index of such effectiveness and propose reduction of evaluation function as the calculation technique which eliminates the evaluation term with low contribution ratio from evaluation function.

2   Support System for Sequence Design

In this section, we propose a support system for sequence design. This system optimizes evaluation function which is calculated as the linear sum of the plural evaluation terms with weights by SA. However, we do not know an evaluation term to measure the goodness of sequences appropriately. Therefore it is necessary to integrate the evaluation terms by trial and error. In our system, users can integrate the evaluation terms by selecting them from the set of evaluation terms provided by the system and add the new evaluation terms proposed by users. The system provides the following evaluation terms for users.

2.1   Evaluation Terms

In the following discussion, suppose $x_i (1 \leq i \leq m)$ be DNA sequence. And suppose m and n be the number and the length of DNA sequence x_i respectively.

H-measure   Garzon et al. proposed the H-measure as follows [2].

$$|x_i, x_j| := \min_{-n < k < n} H(x_i, \sigma^k(\overline{x_j})) \tag{1}$$

where $H(*, *)$ denotes the Hamming distance, σ^k denotes the right (left) shift in case of $k > 0$ $(k < 0)$, k denotes the number of the shift, and $\overline{y}$ denotes the Watson-Crick complementary pair.

Here, since we formulate the evaluation function as a minimization problem, we use the following evaluation term based on H-measure.

$$f_H = \max_{i,j,i<j} \max_{-n<k<n} \{n - H(x_i, \sigma^k(\overline{x_j}))\} \tag{2}$$

Self-complementary This is the case where $i = j$ in formula 2 .

$$f_{self} = \max_{i} \max_{-n<k<n} \{n - H(x_i, \sigma^k(\overline{x_i}))\} \tag{3}$$

The reason why we distinguish between above two evaluation terms is that because of spatial nearness DNA sequence hybridizes itself frequently. Therefore, this evaluation term also means the frequency of secondary structure.

GC Content It is important to arrange the GC content for keeping the chemical character uniform. Thus we adopt the following evaluation term proposed by Arita et al. [1].

$$f_{GC} = \sum_{i=1}^{m} (GC^{(i)} - GC^{(i)}_{user\_defined})^2 \tag{4}$$

where $GC^{(i)}_{user\_defined}$ is the target value of GC content of DNA sequence x_i which user can set to the value with the range $[0,100]$.

Similarity It is better for mutual sequences not to have a common portion, since sequences will mishybridize frequently if the base of sequences is similar too well. Thus we propose the following evaluation term.

$$f_{Sim} = \max_{i,j,i<j} \max_{-n<k<n} \{n - H(x_i, \sigma^k(x_j))\} \tag{5}$$

Continuity If the same base appears continuously, a reaction is not well controllable since the structure of DNA will become unstable. Thus we propose the following evaluation term.

$$f_{Con} = \sum_{i=1}^{m} \sum_{j=1}^{n} (j-1) N_j^{(i)} \tag{6}$$

where $N_j^{(i)}$ denotes the number of times to which the same base appears j-times continuously in DNA sequence x_i.

Tm Melting temperature is important factor for efficiency of the reaction of DNA. So we propose the following evaluation term for uniform melting temperature.

$$f_{Tm} = \sum_{i=1}^{m}(Tm^{(i)} - Tm^{(i)}_{user\_defined})^2 \tag{7}$$

where $Tm^{(i)}_{user\_defined}$ is the target value of Tm of DNA sequence x_i which user can set to the value with the range [0,100].

We calculate melting temperature by using the approximate expression known as Nearest Neighbor method [4].

Completely Complementary at 3'-end If there exist the point where some bases at 3'-end of each DNA sequence is completely complementary, the unexpected extension would occur. So we propose the following evaluation term for preventing unexpected extension.

$$f_{3end} = \sum_{i=1}^{m}\sum_{j=i}^{m}CN(x_i, x_j^{(k)}) \tag{8}$$

where $CN(x_i, x_j^{(k)})$ is the number of completely complementary site between sequence x_i and k-base sequence from 3'-end of sequence x_j. k is defined by users.

Evaluation Function Users must weight each evaluation term depending on its level of importance. Thus we get the evaluation function as follows.

$$F = \sum_{i=1}^{n} w_i f_i \tag{9}$$

where n is the number of evaluation terms, f_i is each evaluation term chosen by users (e.g. f_H), and w_i is the weight of each evaluation term.

2.2   Composition of a System

This system consists of two modules. One is the reduction module of evaluation function. This module reduces evaluation function (i.e. delete the evaluation term which seems to be irrelevant and distribute the weight of the deleted evaluation term to other evaluation terms). The other is the sequence-search module. This module searches for good sequences according to the evaluation function obtained by reduction.

Reduction Module of Evaluation Function The purpose of this module is to show the user the contribution ratio and reduce the number of combination of evaluation terms by reduction of evaluation function in case the contribution ratio of the evaluation term is low. The procedure of this module is as follows:

1. User inputs the evaluation terms f_i $(i = 1, 2, \cdots, n)$ and their weights w_i into the system. The system arrays the evaluation terms in descending order of the weights.
2. The system generates the sample sequences, randomly. In this paper, the number of sample sequences is denoted as $k=1000$. The term value x_{ij} $(j = 1, 2, \cdots, k)$ of each evaluation term f_i is calculated from these sequences. These evaluation terms, weights and n sets of k term values are shown in Table.1
3. The system calculates the average μ_i and the standard deviation σ_i of the term value x_{ij} in each evaluation term.

$$\mu_i = \frac{1}{k} \sum_{j=1}^{k} x_{ij} \tag{10}$$

$$\sigma_i = \sqrt{\frac{1}{k} \sum_{j=1}^{k} (x_{ij} - \mu_i)^2} \tag{11}$$

4. The system calculates the normalized value X_{ij} of the term value x_{ij} by the average μ_i and the standard deviation σ_i.

$$X_{ij} = \frac{x_{ij} - \mu_i}{\sigma_i} \tag{12}$$

5. The system calculates the orthogonalized value z_{ij} of the normalized value X_{ij} by Schmit's orthogonalization.

$$z_{ij} = X_{ij} - \sum_{l=1}^{i-1} b_{il} z_{lj} \tag{13}$$

Table 1. Evaluation terms

evaluation term	weight	term value
f_1	w_1	$x_{11}, x_{12}, \cdots, x_{1k}$
f_2	w_2	$x_{21}, x_{22}, \cdots, x_{2k}$
$\vdots$	$\vdots$	$\vdots$
f_n	w_n	$x_{n1}, x_{n2}, \cdots, x_{nk}$

$$w_1 \geq w_2 \geq \cdots \geq w_n$$

Where,

$$b_{il} = \frac{1}{kV_l} \sum_{j=1}^{k} X_{ij} z_{lj} \tag{14}$$

$$V_l = \frac{1}{k} \sum_{j=1}^{k} z_{lj}^2 \tag{15}$$

6. Generally, it is wasteful to take the evaluation term with the high correlation with other terms into consideration. The correlation can be eliminated by Schmit's orthogonalization. If the correlation is high, the variance of the orthogonalized terms z_{ij} becomes smaller. Therefore, the variance V_i can be considered as the usefulness of each evaluation term. This is the view of contribution ratio. The system calculates contribution ratio ρ_i of each evaluation term to evaluation function as follows.

$$\rho_i = \frac{V_i^2}{\sum_{p=1}^{n} V_p^2} \tag{16}$$

7. The system deletes the evaluation term f_i whose contribution ratio ρ_i to evaluation function is subthreshold. If there is no such an evaluation term, the system skips from this module to the next module.
8. The system distributes the weight w_i of the deleted evaluation term f_i to other evaluation terms f_q ($q = 1, 2, \ldots, i-1$). The modified weight w_q' of w_q is calculated as follows.

$$w_q' = w_q + \frac{r_{iq}}{\sum_{l=1}^{i-1} r_{il}} (1 - \rho_i) w_i \tag{17}$$

9. The system re-arrays the evaluation terms f_i ($i = 1, 2, \ldots, n-1$), and turns back to 5.

Sequence-Search Module The purpose of this module is to search for good sequences minimizing the evaluation function by SA. The evaluation function optimized by this module consists of the plural evaluation terms obtained by processing in the reduction module.

SA minimizes the evaluation function as follows.

1. choose an initial feasible solution x (i.e. x is a sequence set)
2. while the temperature $T > \epsilon$ (ϵ is a small number), the following is performed
3. choose a feasible solution y from neighborhood randomly
4. if $x < y$, y is substituted for x with probability $\exp \frac{-(f(y)-f(x))}{KT}$
5. otherwise, y is substituted for x
6. T is lowered according to cooling schedule c

In SA, we define the neighborhood as a flip of one base from the DNA set, because we can calculate the score of the DNA set easily by calculating only about the relation with the flip point. And initial temperature and cooling schedule about SA are set to 1000 and 0.9998, respectively. If the search stops at $T < 0.000001$, we can search 103606 candidate solutions.

3   Experiments

In order to evaluate our system, we carried out the following experiments.

First we performed an experiment for investigating about the relation between the reduction of the evaluation function and the contribution ratio of each evaluation term to the evaluation function. Here we introduce the following evaluation term about AT content.

$$f_{AT} = \sum_{i=1}^{m}(AT^{(i)} - AT^{(i)}_{user\_defined})^2 \tag{18}$$

where $AT^{(i)}_{user\_defined}$ is the target value of AT content of DNA sequence x_i which user can set to the value with the range [0,100].

AT content is the rate for which A and T account in sequences. Therefore, the sum of AT content and GC content always becomes 100%. In this experiment, we build the evaluation function as the sum of two evaluation terms, GC content and AT content. Both of these weights are set to 1. In other words, the evaluation function is as follows.

$$F = f_{GC} + f_{AT} \tag{19}$$

In this experiment, we set both $GC_{user\_defined}$ and $AT_{user\_defined}$ at 50% first and then raise only the $GC_{user\_defined}$ by 1%. Then the correlation coefficient between f_{GC} and f_{AT} are plotted in Fig. 1.

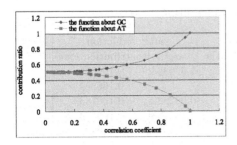   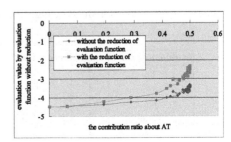

Fig. 1. The correlation coefficient   **Fig. 2.** The effect of the reduction

We performed the search for sequences by using both evaluation function without reduction and one with reduction. In case of evaluation function with reduction, we deleted the f_{AT} regardless of the contribution ratio. The target problem is the design for sequences under the condition that the number of DNA is 10 and the length of each DNA (i.e. the number of base of each DNA) is 100. The result is shown in Fig. 2.

The figure shows that the lower the contribution ratio of the evaluation term to evaluation function is, the more effective the reduction of evaluation function is. So it seems that it is appropriate to reduce an evaluation function on the basis of the contribution ratio.

Next, we applied our system to the evaluation function which consisted of seven evaluation terms mentioned above f_H, f_{self}, f_{GC}, f_{Sim}, f_{Con}, f_{Tm}, f_{3end}. We set $GC_{user\_defined}$ at 50%, $Tm_{user\_defined}$ at 350K and k at 5. The target problem is the design for sequences under the condition that the number of DNA is 7 and the length of each DNA (i.e. the number of base of each DNA) is 20 (this is similar to Adleman's experiment).

Table 2 shows the correlation coefficient between evaluation terms. Table 3 shows the weight and contribution ratio of each evaluation term in the case of both before reduction and after reduction.

Table 2. Correlation coefficient

	f_H	f_{self}	f_{GC}	f_{Sim}	f_{Con}	f_{Tm}	f_{3end}
f_H	1	-0.006804	-0.02707	0.003681	-0.015259	-0.011736	0.030687
f_{self}	-0.006804	1	0.092293	-0.008876	-0.029155	0.065794	0.055172
f_{GC}	-0.02707	0.092293	1	-0.011328	0.202055	0.740378	0.034056
f_{Sim}	0.003681	-0.008876	-0.011328	1	-0.009967	-0.02112	-0.011076
f_{Con}	-0.015259	-0.029155	0.202055	-0.009967	1	-0.013646	0.01596
f_{Tm}	-0.011736	0.065794	0.740378	-0.02112	-0.013646	1	0.020442
f_{3end}	0.030687	0.055172	0.034056	-0.011076	0.01596	0.020442	1

Table 3. weight and contribution ratio

		before reduction		after reduction
evaluation term	weight	contribution ratio	weight	contribution ratio
f_{Tm}	1	0.156424	1.672854	0.166975
f_{Sim}	1	0.156304	1	0.166854
f_H	1	0.156244	0.982822	0.166783
f_{Con}	1	0.15618	1.194388	0.166714
f_{3end}	1	0.155906	1.012692	0.166422
f_{self}	1	0.155747	1.039692	0.166252
f_{GC}	1	0.063195	-	-
correlation coefficient of evaluation function				0.825748

From the result of Table 3, we can find that the contribution ratio of the evaluation term f_{GC} is very low. It is because the correlation coefficient between f_{GC} and f_{Tm} is extremely high (0.740378) and one between f_{GC} and f_{Con} is relatively high (0.202055) (see Table 2). Although correlation coefficient gives the relation between evaluation terms, it is difficult to grasp all the relation of them. Therefore, in order to know the effectiveness of evaluation term itself, contribution ratio serves as a good measure. Here, we can understand the evaluation term f_{GC} is the most ineffective.

4   Concluding Remarks

In this paper we developed the sequence design system which searches for the sequences satisfying the plural constraints and provides contribution ratio of the each evaluation term to the evaluation function for users. Additionally in case there exists an evaluation term with low contribution ratio, users can reduce the number of combination of evaluation terms by reduction of evaluation function.

When we try to design good sequences by optimizing the evaluation function consisting of plural evaluation terms, we should consider the validity of each evaluation term and how we should combine them. In this paper we discussed the second issue mainly. The validity of each evaluation term should be checked by laboratory experiments.

In our system users can know the contribution ratio of each evaluation term to evaluation function and reduce the number of combination of evaluation terms by reduction of evaluation function. This assists users in integrating the plural terms into an evaluation function. We believe that our system is useful when we try to search for not only good sequences satisfying severe constraints but also the evaluation terms for sequence design in DNA Computing.

References

1. Masanori Arita, Akio Nishikawa, Masami Hagiya, Ken Komiya, Hidetaka Gouzu and Kensaku Sakamoto: "Improving Sequence Design for DNA Computing," Proceedings of GECCO'00 (Genetic and Evolutionary Computation Conference), pp. 875-882 (2000)
2. M. Garzon, R. Deaton, L.F. Nino, Ed Stevens: Encoding Genomes for DNA Computing, Proc. of the Third Annual Genetic Programming Conf., pp.684-690, 1998
3. R. Deaton, R. C. Murphy, M. Garzon, D. R. Franceschetti and S. E. Stevens, Jr.: "Good Encodings for DNA-Based Solutions to Combinatorial Problems," DNA Based Computers II, DIMACS Series in Discrete Mathematics and Theoretical Computer Science, Vol. 44, pp. 247-258 (1999)
4. JAMES G. WETMUR, "Physical Chemistry of Nucleic Acid Hybridization," DNA Based Computers III, DIMACS Series in Discrete Mathematics and Theoretical Computer Science, Vol. 48, pp. 1-23 (1999)

The Fidelity of the Tag-Antitag System

John A. Rose[1], Russell J. Deaton[2], Masami Hagiya[3], and Akira Suyama[4]

[1] Institute of Physics, The University of Tokyo
johnrose@genta.c.u-tokyo.ac.jp
[2] Department of Computer Science and Computer Engineering,
The University of Arkansas
rdeaton@uark.edu
[3] Department of Computer Science, The University of Tokyo
hagiya@is.s.u-tokyo.ac.jp
[4] Institute of Physics, The University of Tokyo
suyama@dna.c.u-tokyo.ac.jp

Abstract. In the universal DNA chip method, target RNAs are mapped onto a set of DNA tags. Parallel hybridization of these tags with an indexed, complementary antitag array then provides an estimate of the relative RNA concentrations in the original solution. Although both error estimation and error reduction are important to process application, a physical model of hybridization fidelity for the TAT system has yet to be proposed. In this work, an equilibrium chemistry model of TAT hybridation is used to estimate the error probability per hybridized tag (ϵ). The temperature dependence of ϵ is then discussed in detail, and compared with the predictions of the stringency picture. In combination with a modified statistical zipper model of duplex formation, implemented by the *Mjolnir* software package, ϵ is applied to investigate the error behavior of small to moderate sized TAT sets. In the first simulation, the fidelities of (1) 10^5 random encodings, (2) a recently reported Hamming encoding, and (3) an ϵ-based, evolved encoding of a 32-strand, length-16 TAT system are estimated, and discussed in detail. In the second simulation, the scaling behavior of the mean error rate of random TAT encodings is investigated. Results are used to discuss the ability of a random strategy to generate high fidelity TAT sets, as a function of set size and encoding length.

1 Introduction

DNA chips are indexed arrays of DNA probes which are immobilized on a solid substrate. When exposed to a set of unbound, target DNA or RNA molecules, the DNA chip essentially performs an exhaustive parallel search for complementary sequences between the immobilized probes and the target species. DNA chips have been successfully applied to simultaneous gene expression profiling (GEP) and genotyping on a genome scale [1]. Although DNA chips provide a powerful tool for gaining insight into overall genome function, the target-dependent nature of current DNA chips poses difficulties for both chip production and validation. Recently, an alternative approach to the use of custom DNA chips for

N. Jonoska and N.C. Seeman (Eds.): DNA7, LNCS 2340, pp. 138–149, 2002.
© Springer-Verlag Berlin Heidelberg 2002

genome-scale GEP has been reported, which is based on the concept of a *universal* DNA chip [2]. This protocol, which is based on DNA computing, consists of 3 steps. In the first step, a 1-to-1 mapping is defined, such that each species of RNA which may be present in an input mixture is associated with a unique, oligonucleotide-length DNA sequence, or DNA *tag*. In the second step, an RNA input mixture of interest is mapped, using an *in vitro* molecular algorithm, onto a corresponding set of DNA tags. Following PCR amplification, combined with fluorescence-labeling, the resulting tags are removed and washed across a universal DNA chip. This chip is composed of an indexed array of DNA *antitags*, each of which corresponds to the Watson-Crick reverse complement of one of the tag species. If both the mapping and amplification are accomplished in a uniform manner, the relative fluorescence intensities of the hybridized tags at equilibrium will approximate the relative concentrations of the RNA species present in the input mixture. The set of tags and complementary antitags, taken together, has been referred to as a *Tag-Antitag System* [3].

Because the use of a molecular mapping obviates the need for organism-specific chips, the DNA code words used to implement a Tag-Antitag (TAT) system may be determined arbitrarily. In order to minimize errors in decoding, however, DNA words should be designed to have minimal potential for mishybridization. Although several heuristic methods for generating TAT sets appropriate for implementing universal DNA chips have been proposed [3,4,5,6,7], no physically principled model of TAT system fidelity has yet been reported.

The principles of equilibrium chemistry have been used previously to investigate the fidelity of DNA-protein interactions [8], nucleic acid-based antisense agents [9], and the annealing [10] and annealing-ligation biosteps [11]. In Sec. 2, the fidelity behavior of the TAT system is discussed in terms of an equilibrium treatment of DNA hybridization. Sec. 2.1 describes the derivation of an expression which estimates the equilibrium probability of error per hybridized tag, ϵ (Eq. 2). The temperature dependence of ϵ is then discussed in Sec. 2.2. Model predictions indicate that the applicability of the stringency picture of DNA hybridization fidelity is limited to the special case in which the total enthalpy of unplanned TAT interactions dominates that of planned interactions. Sec. 2.3 addresses the dependence of ϵ on system input (*i.e.*, tag concentrations), and establishes a set of input-independent expressions for estimating the mean (ϵ_w), maximum (ϵ_+), and minimum (ϵ_-) fidelities over the set of all inputs.

In Sec. 3, the practical application of the equilibrium model of TAT fidelity developed in Sec. 2 is undertaken. Following a brief review of current approaches to modeling DNA duplex formation (and related software), Sec. 3.1 describes a modified version of the statistical zipper model of duplex formation, which is implemented by the new *Mjolnir* software package. Sec. 3.2 and Sec. 3.3 then report the results of a pair of simulation experiments performed using *Mjolnir*, in combination with the model of duplex formation outlined in Sec. 3.1. In Simulation A, the predicted fidelity performance of the Hamming-based TAT set described in [5,6] is examined relative to that of 10^5 randomly generated encodings. In order to establish the evolvability of TAT encodings with improved performance, an encoding produced using a standard genetic algorithm is also discussed. Simulation B addresses the scaling behavior of the mean fidelity of a randomly generated encoding, in the range of small to moderate sized sets. Sec. 4 outlines the overall predictions of the model, and discusses current work, which is focused on the experimental validation of the model.

2   TAT Fidelity: Equilibrium Model

2.1   The Error Probability per Hybridized Tag, ϵ

Consider a universal DNA chip array, composed of n anchored antitag ssDNA species, in equilibrium with a mixture of n unbound tag species, each of which is the Watson-Crick reverse complement of the corresponding antitag. Let C_i° and C_i refer to the initial and equilibrium concentrations of the tag member of TAT pair $\{i, i^*\}$, respectively. Similarly, let $C_{i^*}^\circ$ and C_{i^*} denote the initial and equilibrium concentrations of the antitag member of TAT pair $\{i, i^*\}$, respectively. Furthermore, let K_{ij^*} denote the total equilibrium constant of bimolecular duplex formation between tag i and antitag j^*. At equilibrium, the mean error probability per antitag-hybridized tag is estimated by the ratio,

$$\epsilon = \frac{\sum_i \sum_{j^*} C_i C_{j^*} K_{ij^*}^e}{\sum_i \sum_{j^*} C_i C_{j^*} K_{ij^*}} \equiv \frac{1}{SNR},\tag{1}$$

where $K_{ij^*}^e$ is the sum of the error equilibrium constants of formation between species i and j^*, and SNR corresponds to the conventional, experimentally observed measure of hybridization error, the *signal to noise ratio*. This expression may be rendered into a tractable form by the successive application of a set of approximations. First, antitags are assumed to be present in equal, excess concentration C_a, relative to each tag, so that $C_{j^*} \approx C_a(1+K_{j^*}^{hp})^{-1}, \forall j^*$. In analogy with [8], TAT encodings are then assumed to be sufficiently well encoded so that the total equilibrium constant of error interaction for each tag-tag and each TAT pair is small relative to that of the full-length, planned TAT interaction. Finally, it is noted that the equilibrium constants of hairpin formation for an antitag and it matching tag are roughly equal. Eq. 1 then reduces to the form,

$$\epsilon \approx \frac{\sum_i C_i^\circ \left[\frac{1+K_i^{hp}}{(1+K_i^{hp})^2 + C_a K_{ii^*}} \sum_{j^*} \frac{K_{ij^*}^e}{1+K_{j^*}^{hp}} \right]}{\sum_i C_i^\circ \left[\frac{K_{ii^*}}{(1+K_i^{hp})^2 + C_a K_{ii^*}} \right]}.\tag{2}$$

As the set $\{C_i^\circ\}$ may be estimated from fluorescence measurements using the expression, $C_i^\circ \approx C_{ii^*}[1+(C_a K_{ii^*})^{-1}(1+K_i^{hp})^2]$, a post-experiment, experiment-specific estimation of ϵ may be straightforwardly obtained.

2.2   The Temperature Dependence of ϵ

Let $K_{i,e}$ and $\Delta H_{i,e}^\circ$ denote the sum of the statistical weights and the enthalpies of formation of all error configurations involving tag i, respectively. Similarly, let ΔH_{ii*} denote the sum of the enthalpies of all duplex configurations between TAT pair $\{i, i*\}$. For the typical hybridization experiment, that is performed at a temperature beneath the melting transition of planned TAT pairs, the temperature dependence of ϵ is given by,

$$\frac{d\epsilon}{dT} \approx \left[\frac{\sum_i \frac{C_i^\circ K_{i,e}}{1+C_a K_{ii*}} (\Delta H_{i,e}^\circ - \Delta H_{ii*}^\circ)}{\sum_i \frac{C_i^\circ K_{i,e}}{1+C_a K_{ii*}}} \right] \frac{\epsilon}{RT^2}, \qquad (3)$$

where the impact of hairpin formation has been neglected. The simplest application of Eq. 3 is to a single tag species in equilibrium with a a number of antitag species (i.e., $C_i^\circ = C_t$, $C_j^\circ = 0, \forall j \neq i$). In this case, $\frac{d\epsilon}{dT}$ is a monotonically varying function of T, whose sign is determined completely by the sign of the quantity, $\Delta\Delta H^\circ \equiv \Delta H_{i,e}^\circ - \Delta H_{ii*}^\circ$. There are two cases of interest.

For the case in which the sum of the enthalpies of formation of planned interaction dominates the sum of the enthalpies of the set of error interactions (Case I), $\Delta\Delta H^\circ > 0$, and maximal fidelity is predicted to be obtained by application of the minimum practical T_{rx}. Although this result appears to conflict with the well-established *stringency* picture of DNA primer hybridization fidelity [3], which predicts that the maximal fidelity is achieved at a T_{rx} both (1) greater than the melting temperatures (T_m) of all error hybrids, and (2) lower than the T_m of each planned hybrid, this conflict is only apparent. In actuality, this temperature behavior applies only to the special case of a DNA mixture which encodes for a low number of error interactions.

For the case in which the sum of the enthalpies of formation of error interaction dominates (Case II), $\Delta\Delta H^\circ < 0$, and maximum fidelity is predicted at the highest practical T_{rx}. This behavior applies to the DNA mixture which encodes for both a target structure and a large number of suboptimal, but marginally stable mismatched structures, as in the PCR amplification of a target sequence from genomic DNA. The optimal T_{rx} is then predicted to be that temperature which co-optimizes the signal to noise ratio and the total yield of planned hybrids, a goal which is achieved at a T_{rx} marginally beneath the T_m of the full length planned hybrid, as predicted by the stringency picture of fidelity.

For the universal DNA chip, which is composed of a number of TAT pairs, cases I and II correspond to the temperature behavior for the extreme cases in which each tag encoding is relatively error-free ($\Delta H_{ii*}^\circ < \Delta H_{i,e}^\circ, \forall i$), or relatively error-prone ($\Delta H_{i,e}^\circ < \Delta H_{ii*}^\circ, \forall i$) respectively. In addition to these cases, a third distinctive type of temperature behavior (Case III) is predicted when the encoding of interest consists of a mixture of error-free and error-prone tag encodings. For encoding sets in which some tag encodings satisfy the condition $\Delta H_{ii*}^\circ < \Delta H_{i,e}^\circ$ (relatively error-free), while others satisfy the condition, $\Delta H_{i,e}^\circ < \Delta H_{ii*}^\circ$ (relatively error-prone), Eq. 3 will contain both positive and negative terms. The overall error rate of such an encoding is then predicted to assume a more complicated, multiphasic temperature dependence. Let TAT encodings which exhibit these three types of temperature behavior be classified as Type I, Type II, and Type III encodings, respectively.

2.3   The Input Dependence of ϵ

For purposes of standardization, and to enable rational design, it would be desirable to associate a single, input-independent measure of fidelity to a given DNA chip. According to Eq. 2, however, the experimentally observed error rate will in general be dependent upon the relative concentrations of tag species present in the experimental wash, in addition to the details of chip design. Let the variation of ϵ over the set of all dilute inputs be termed the *error response* of a DNA chip. Although the weakly nonlinear nature of ϵ complicates the derivation of an exact expression for the mean error response, an approximate estimate is provided by ϵ_w, the error rate in response to an input distributed uniformly over all tag species (*i.e.* a *white* input). For this case, Eq. 2 reduces to the form,

$$\epsilon_w \approx \frac{\sum_i \left[\frac{1+K_i^{hp}}{(1+K_i^{hp})^2 + C_a K_{ii*}} \sum_{j*} \frac{K_{ij*}^e}{1+K_{j*}^{hp}} \right]}{\sum_i \left[\frac{K_{ii*}}{(1+K_i^{hp})^2 + C_a K_{ii*}} \right]}. \tag{4}$$

The extreme values of the error response are also determined straightforwardly. From Eq. 2, the error rate in response to an impulse-like input composed of a single tag species, i is given by the concentration independent expression,

$$\epsilon_i = \frac{1 + K_i^{hp}}{K_{ii*}} \sum_{j*} \frac{K_{ij*}^e}{1 + K_{j*}^{hp}}. \tag{5}$$

Let the complete set of error-impulse responses, $S_i \equiv \{\epsilon_i\}$ be termed the *error spectrum* of the chip. The extrema of the error response for a DNA chip then correspond to the extreme members of the error spectrum, $\epsilon_- = \sup S_i$ and $\epsilon_+ = \inf S_i$. For convenience, the *error spectral width*, which is a measure of the input-related uncertainty attendant upon use of the chip, is defined in terms of the logarithms of the extreme values:

$$w = \log_{10} \epsilon_- - \log_{10} \epsilon_+. \tag{6}$$

3   Simulations of TAT Fidelity

3.1   Estimation of the Equilibrium Constants

The practical application of Eqs. 2-6 requires the estimatation of an overall equilibrium constant of secondary structure formation (K_{eq}) for each tag and each TAT pair. Although estimating K_{eq} for any configuration requires $O(l)$ TIME resource, where l is strand length, the exponential scaling of the number of accessible configurations poses a significant barrier to calculation [12]. Various approaches for circumventing this difficulty have appeared in the literature. In the Dynamic Programming (DP) algorithm, the required time is reduced to $O(l^2)$ for either a single, folding strand [12] or a pair of hybridizing strands [13], by restricting attention to the identification of the most stable accessible structure.

The DP algorithm has been implemented by Zuker [14] in the form of the *m-fold* software package, to support optimal secondary structure prediction for single RNA and DNA molecules. An undesirable feature of the simplest form of DP is that it neglects the occupancy of a large number of suboptimal, but significant structures. Although the set of structures with free energy above some threshold value (*i.e.*, within 10% of that of the optimal structure) can also be generated by DP, in practice the size of this set may still grow exponentially, an occurance which cannot, in general, be determined prior to runtime [12].

The alternative approach is to restrict attention to some tractable subset of configuration space whose occupancy has been determined experimentally to dominate the process under study. For instance, the use of an all-or-none, perfectly aligned model, which considers only the occupancy of the completely melted and unmelted configurations, is satisfactory for modeling the melting of short oligonucleotides and has been used to establish sets of nearest-neighbor parameters [15]. At the other extreme, the melting of long, quasirandom DNAs is well modeled by a perfectly aligned model which retains configurations with internal loops, but neglects configurations in staggered alignments. This model is the physical basis for the Poland algorithm, which requires a computational time of $O(l^2)$, or $O(6l)$, when combined with the Fixman-Friere approximation [16]. Software packages based on this algorithm include Steger's POLAND package [17] and the *MELTSIM* package of Blake and SantaLucia [18].

The analysis of renaturation-based processes requires an assessment of the potential for nucleation in multiple alignments, particularly for strands longer than short oligonucleotides. The hybridization of a short ssDNA primer with a longer template has been modeled using a modified all-or-none model, in which the full-length configuration in each alignment is retained (including all mismatches), and base-pair mismatches are modeled as virtual stacks [19]. This algorithm is implemented by Hartemink's *BIND* [19] and *SCAN* [20] software packages, although a stringency picture of hybridization fidelity is ultimately applied. While the attainment of an $O(l)$ scaling behavior is attractive computationally, the unrealistically small penalty obtained by modeling internal loops in terms of virtual stacks limits model applicability to short oligonucleotides.

A statistical zipper model (SZM), which retains configurations in all alignments, but discards those which require multiple nucleation events, is appropriate for modeling the thermal denaturation and renaturation of quasirandom DNAs shorter than about 100 bases [21], although deviations are expected for DNAs with repeating GC-rich regions longer than about 25 bases [22]. Theoretically, adequacy of an SZM is justified by the large multiplicative penalty ($\sigma \approx 4.5 \times 10^{-5}$) assigned to configurations with internal loops. A standard, Watson-Crick SZM, which requires $0(l^2)$ TIME resource, has been applied previously to the fidelity of both the annealing [10] and annealing-ligation [11] biosteps of DNA computing, and is implemented by the *NucleicPark* software package. As pointed out in [11], however, a standard SZM may substantially underestimate the overall occupancy of error configurations, due to the assumption of broad negligibility for configurations with single mismatches, tandem

GAs, and single base bulges. For this reason, in the current work a new Java-based software package, *Mjolnir* has been developed, which adopts a modified SZM that includes these configurations. For each configuration, the free energy of duplex formation is estimated using the nearest-neighbor parameter set of SantaLucia [15], which includes the impact of single internal mismatches [23]. The following modifications have also been adopted: an internal tandem GA is modeled as a normal nearest neighbor doublet [24]; a single base bulge is treated as a destabilizing energetic perturbation [25]; each dangling end is modeled as a small additive energetic correction [26]; the impact of antitag anchorage is accounted for as suggested in [27]. For hairpins, terminal loop statistical weights are estimated as described in [16]. This modified SZM requires $O(l^3)$ TIME resource per tag or TAT pair. Copies of *Mjolnir* are freely available, by personal request.

CTTGGGCCCGCGTATG	TGCCCTTCCTGGGGAG	GCTGTGCATCACGGGC	CGGCCTCTCCCCTCAT
TGCTACTACTCGTCGG	CCGTGACCGTTCTTCA	CGTTTACAGGGTGCGT	CGCGCGAGCTTGTAAG
CCCAGACTTGCCGATG	GACGACACCCCCACTT	CGCTCCAACAACCTAT	CGGTTGAGAGTCAGCA
GCGGCAGTGATTTGCG	GCTGGATGGAACGGAC	AACCACGCCTGTTCGC	TCGCAAAAGCCAAGGC

Fig. 1. Encoding of an $N = 16$, $L = 16$ TAT system (antitags), evolved using a standard genetic algorithm. Encodings are shown in a 5' to 3' orientation.

3.2   Simulation A - The Fidelity of the Small TAT System

Recently, a SAT instance was solved *in vitro* using a TAT system composed of 16 TAT pairs ($N = 16$), each 16 bases in length ($L = 16$) [6]. The TAT set employed was designed to prevent mishybridization by means of a perfectly aligned, Hamming-based strategy [5]. In order to investigate the effectiveness of Hamming encoding, the error response of this set has been estimated using *Mjolnir*. The parameters $C_a = 10^{-6}$ M and $[Na^+] = 1$ M were assumed. For standardization, the error responses of 10^5 random encodings of a $N = 16$, $L = 16$ TAT system were also examined. As fidelity characteristics were distributed lognormally, population means and standard deviations were computed in terms of the logarithms of the error responses. A third set, shown in Fig. 1 was evolved using a standard genetic algorithm. A complete discussion of TAT encoding evolvability is beyond the scope of this work. However, the performance of the encoding listed in Fig. 1 establishes the availability of good encodings, at equilibrium. The applied genetic algorithm, which was implemented by *Mjolnir*, evolved a TAT population of size 100 for 150 generations, using the fitness measure ϵ_w^{-2}, with a crossover rate of 0.7 and a mutation rate of 0.01.

Simulation results for all encoding sets are shown in Fig. 2. The validity of the approximation set used to derive ϵ was verified for each encoding, at runtime. All

	10^5 Random Encodings		Hamming Encoding		Evolved Encoding	
T_{rx} Response	293 K	303 K	293 K	303 K	293 K	303 K
$\log_{10}\epsilon_+$	-14.4 ± 0.9	-12.5 ± 0.8	-11.9	-9.9	-15.1	-12.9
$\log_{10}\epsilon_w$	-10.0 ± 0.9	-8.6 ± 0.8	-9.8	-8.4	-14.4	-12.4
$\log_{10}\epsilon_-$	-9.0 ± 1.0	-7.6 ± 0.9	-9.0	-7.5	-14.0	-12.0

Fig. 2. A simulated set of error responses for the following $N = 16$, $L = 16$ TAT encodings: (1) the Hamming encoding reported in [3], (2) 10^5 random encodings, and (3) an encoding generated by a standard genetic algorithm using the fitness, ϵ_w^{-2} .

fidelity responses followed a Type I temperature behavior, as predicted for a set of high affinity encodings in a relatively error-free background (*cf.* Section 2.2). This result supports the conclusion that a stringency-based methodology [3] is not universally well motivated for TAT design. Based on the fidelity responses predicted for the encoding reported in [6], enhanced fidelity is not guaranteed by application of the simplest form of a Hamming-based encoding strategy. In particular, the predicted performance of this set is approximately equivalent to that of a random set. The most obvious explanation for this result is that a perfectly-aligned strategy makes no attempt to account for hybridization in shorter alignments, which generally accounts for the majority of hybridization error. A more fundamental problem, however involves the soundness of the underlying assumption of a Hamming-based strategy for DNA encoding: that all mismatches are fundamentally destabilizing. Recent experiments indicate that *all* single mismatches confer at least some energetic stability, when located within a Watson-Crick duplex [23]. As a result, the presence of an arbitrary set of mismatches does *not* necessarily enforce the energetic unfavorablity of a structure.

The good performance predicted to accompany the random encoding of a small TAT set has a strong underlying physical basis. At equilibrium, the occupancy of a given tag species amongst accessible structures is determined by the relative magnitudes of the corresponding Gibbs factors, which scale *exponentially* with length. As a result, a very large number of randomly occurring, suboptimal error structures is required to have a sizable impact on the occupancy of the planned, full-length structure. This line of reasoning has led some researchers to conclude that the high occupancy of a planned structure may be guaranteed, in general by encoding for high target affinity [9]. In practice, of course, the ability of a high affinity, but otherwise random encoding to establish a suitably high specificity depends upon the application (i.e., the level of error which is acceptable), as well as the size of the error background. The key issue is an estimation of the set size above which an "affinity guarantees specificity" strategy is likely to fail, on average.

3.3   Simulation B - The Scaling Behavior of Random Encoding Fidelity

In order to estimate the scaling behavior of the fidelity of randomly generated TAT sets, a second simulation experiment was performed. The experimental design was a two-level, three factor, full factorial experiment [28]. The significant factors were N, L, and T_{rx}, and the levels investigated were: $N = \{16, 128\}$; $L = \{14, 24\}$; and $T_{rx} = \{293 \text{ K}, 303 \text{ K}\}$. A single full-factorial set of simulations (a *replicate*) required a total of $2^3 = 8$ simulation runs. Two independent replicates were performed, for a total of 16 runs. An additional pair of runs at the experimental centerpoint ($N = 72$, $L = 20$, $T = 298$ K) were also performed to investigate curvature. Simulations focused on the influence of the factors on the population average values of the error responses, $\log_{10} \epsilon_w$, $\log_{10}, \epsilon_-$, and w. Regression analyses on the resulting data yielded, for each effect a linear fit with negligible curvature, at 95% confidence. In terms of the population means, the resulting linear fits were given by,

$$\langle \log_{10} \epsilon_w \rangle = -45.7 + 0.021\,N - 0.91\,L + 0.17\,T_{rx}, \tag{7}$$

$$\langle \log_{10} \epsilon_- \rangle = -45.3 + 0.028\,N - 0.90\,L + 0.17\,T_{rx}, \tag{8}$$

$$\langle w \rangle = \quad 24.3 + 0.018\,N + 0.23\,L - 0.078\,T_{rx}\,. \tag{9}$$

The mean worst case error response, $\langle \log_{10} \epsilon_- \rangle$ at 25 °C is illustrated in Fig. 3, as a function of N and L. Response values are shown as a series of a discrete contours, $\langle \log_{10} \epsilon_- \rangle = \{-3, -6, -9, -12\}$. The accompanying error response surface and z-axis have been omitted for clarity. Within the experimental range investigated (dashed box), TAT system performance is predicted to scale exponentially with both L and N in a compensatory fashion: a modest increase in L is easily able to compensate for even substantial increases in N.

Current glass slide-based DNA chips contain roughly 6×10^{13} anchored strands of each antitag species (*i.e.*, $C_a \approx 10^{-6}$ M, in 100 μL). Although the total tag concentrations will vary with input, the most numerous species can be expected to be present at $\approx 0.1 C_a - 0.01 C_a$. Assuming that a minimum of $\approx 10^2 - 10^3$ error hybrids are required for detection, an error rate less than roughly 10^{-12} is required for ideal fidelity. For current oligonucleotide-based TAT systems, which generally lie within the experimental box of Fig. 3, the performance of randomly generated encodings is predicted to be excellent (*i.e.*, on average, a worst case error rate of $\log_{10} \epsilon_- = -12$ may be obtained, given appropriate selection of L). The linear model of error response represented by Eqs. 7 - 9 is strictly valid only within the experimental box. This linear model, however, when extrapolated beyond the box provides an approximate, quantitative basis for predicting the relatively rapid onset of failure for encodings produced using a random strategy, as TAT systems scale up in size and sensitivity. If the contour, 10^{-12} is taken to define an approximate line of failure for error-free computation, the error-free application of randomly generated oligonucleotide-based TAT systems is predicted to be feasible for at most $N \approx 180$ TAT pairs.

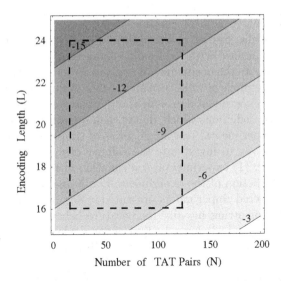

Fig. 3. The worst-input, mean error response of a TAT set as a function of N and L, in terms of the contours $\langle \log_{10} \epsilon_- \rangle = \{-3, -6, -9, -12, -15\}$. Values outside of the experimental box (dashed lines) represent extrapolated behavior.

4 Conclusions and Future Work

In this work, an equilibrium chemistry based model of TAT system fidelity has been developed and explored. The predictions provided by this model support the following general conclusions:

1. The stringency picture of hybridization fidelity is a special case of a more general temperature behavior. In particular, the predicted temperature behavior applies only to the special case of the low-fidelity TAT system.
2. The application of a perfectly aligned Hamming encoding strategy does not guarantee enhanced TAT system fidelity, relative to random encodings. This result is due to the failure of the primary assumption of Hamming encoding, namely that single base pair mismatches are inherently destabilizing.
3. A relatively high specificity may be obtained on average, for the modest-sized TAT system, by using a random encoding strategy combined with the application of a T_{rx} beneath the melting transition of all TAT pairs. The ability of a random strategy to generate high fidelity TAT sets is predicted to be limited to roughly 180 oligonucleotide pairs.
4. The evolvability of TAT encodings requires further study. However, the ability of a genetic algorithm to quickly generate encodings of enhanced fidelity for small TAT sets suggests that this method may also be well-motivated for larger sets.

The validity of the fidelity model presented in Sec. 2 rests principally on the soundness of the applied approximations. As discussed previously, each approximation was verified to hold at runtime for all encodings examined. However, simulation results also rely upon the ability of a nearest-neighbor model of duplex energetics, combined with the modified staggered zipper model of Sec. 3.1, to accurately estimate the occupancy of planned structures. Although physically well-motivated, the range of validity of an SZM for evaluating TAT systems needs to be subjected to experimental investigation. An advantage of the current model is that it provides predictions which, due to their quantitative nature, lend themselves to direct experimental testing. A series of simple DNA chip experiments is currently underway to investigate the validity of model predictions. Of particular interest are: (1) experimental confirmation of the ability of ϵ to provide a quantititive prediction of the experimentally obtained error rate, given use of the modified staggered zipper model presented in Sec. 3.1, (2) experimental confirmation of the non-stringency-like temperature behavior predicted for Type I TAT sets by Eq. 2, and (3) investigation of the deviation of model predictions from experimental error rates, away from equilibrium.

Acknowledgement. Financial support provided by the JSPS "Research for the Future" program (JSPS-RFTF 96I00101), and a JSPS Postdoctoral Fellowship and grant-in-aid (J.A.R.).

References

1. D. Lockhart and E. Winzeler, "Genomics, gene expression, and DNA arrays," *Nature* **405**, 827–836, 2000.
2. A. Suyama, et al., "Gene Expression Analysis by DNA Computing," In S. Miyano, et al., editors, *Currents in Computational Molecular Biology*, (Univeral Academy Press, Tokyo, 2000), 12.
3. A. BenDor, et al., "Universal DNA Tag Systems: A Combinatorial Scheme," *J. Comput. Biol.* **7**, 503 (2000).
4. R. Deaton, et al., "Reliability and Efficiency of a DNA-Based Computation," *Phys. Rev. Lett.* **80**, 417 (1998).
5. Q. Liu, et al., "Progress toward demonstration of a surface based DNA computation: a one word approach to solve a model satisfiability problem," *Biosystems* **52**, 25 (1999).
6. Q. Liu, et al., "DNA Computing on Surfaces," *Nature* **403**, 175 (2000).
7. H. Yoshida and A. Suyama, "Solution to 3-SAT by Breadth-First Search," in *DNA Based Computers V*, E. Winfree and D. Gifford eds., (American Mathematical Society, 2000), 9.
8. P. von Hippel and O. Berg, "On the specificity of DNA-protein interaction," *Proc. Natl. Acad. Sci. USA* **83**, 1608 (1986).
9. B. Eaton, et al., "Let's get specific: the relationship between specificity and affinity," *Chemistry and Biology* **2**, 635 (1995).
10. J. Rose, et al. 1999, "A Statistical Mechanical Treatment of Error in the Annealing Biostep of DNA Computation," in W. Banzhaf, et al., eds., *Proceedings of the Genetic and Evolutionary Computation Conference, Volume 2*, (Morgan Kauffman, San Francisco, 1999), 1829.

11. J. Rose and R. Deaton, "The Fidelity of Annealing-Ligation: A Theoretical Analysis," in *DNA Based Computers*, A. Condon and G. Rozenberg, eds., (Springer, Berlin, 2001), 231.

12. M. Zuker, "The Use of Dynamic Programming Algorithms in RNA Secondary Structure Prediction," in *Mathematical Methods for DNA Sequences*, M. Waterman, ed., (CRC Press, Boca Raton, 1989), 159.

13. A. Suyama, "RNA secondary structure and its relation to biological functions", in *Proceedings of the 10th Taniguchi International Symposium*, A. Wada ed., (Taniguchi Foundation, Kyoto, 1984), 162.

14. M. Zuker, D. Mathews, and D. Turner, "Algorithms and Thermodynamics for RNA Secondary Structure Prediction: A Practical Guide," in *RNA Biochemistry and Biotechnology*, J. Barciszewski and B. Clark, eds., (NATO ASI Series, Klewer Academic Publishers, 1999).

15. J. SantaLucia, Jr., "A unified view of polymer, dumbbell, and oligonucleotide DNA nearest-neighbor thermodynamics," *Biochemistry* **35**, 3555 (1998).

16. R. Wartell and A. Benight, "Thermal Denaturation of DNA Molecules: A Comparison of Theory and Experiment," *PHYSICS REPORTS* (Review Section of Physics Letters) **126**, 67 (1985).

17. G. Steger, "Thermal denaturation of double-stranded nucleic acids: prediction of temperatures critical for gradient gel electrophoresis and polymerase chain reaction," *Nuc. Acids Res.* **22**, 2760 (1994).

18. R. Blake *et al.*, "Statistical Mechanical Simulation of Polymeric DNA melting with MELTSIM," *Bioinformatics* **15**, 370 (1999).

19. A. Hartemink and D. Gifford, "Thermodynamic Simulation of Deoxyribonucleotide Hybridization for DNA Computation," *DNA Based Computers III*, edited by H. Rubin and D. Wood, (American Mathematical Society, Providence, RI, 1999), 25.

20. A. Hartemink, D. Gifford, and J. Khodor, "Automated Constraint-Based Nucleotide Sequence Selection for DNA Computation," *Biosystems* **52**, 227 (1999).

21. A. Benight, R. Wartell, and D. Howell, "Theory Agrees with Experimental Thermal Denaturation of Short DNA Restriction Fragments," *Nature* **289**, 15 (1981).

22. D. Lando and A. Fridman, "Role of Small Loops in DNA Melting," *Biopolymers* **58**, 374 (2001).

23. N. Peyret, et al., "Nearest-Neighbor Thermodynamics and NMR of DNA Sequences with Internal A:A, C:C, G:G, and T:T Mismatches," *Biochemistry* **38**, 3468 (1999).

24. S-H Ke and R. Wartell, "The thermal stability of DNA fragments with tandem mismatches at a d(CXYG)/d(CY'X'G) site," *Nucleic Acids Research* **24**, 707 (1996).

25. J. Zhu and R. Wartell, "Influence of neighboring base pairs on the stability of single base bulges and base pairs in a DNA fragment," *Biochemistry* **38**, 15986 (1999).

26. M. Senior, et al., "Influence of Dangling Thymidine Residues on the Stability and Structure of Two DNA Duplexes," *Biochemistry* **27**, 3879 (1988).

27. A. Fotin, et al., "Parallel thermodynamic analysis of duplexes on oligodeoxyribonucleotide chips," *Nuc. Acids Res.* **26**, 1515 (1998).

28. G. E. P. Box and N. R. Draper, *Empirical Model-Building and Response Surfaces*, (Wiley, New York, 1987).

PUNCH: An Evolutionary Algorithm for Optimizing Bit Set Selection

Adam J. Ruben[1], Stephen J. Freeland[2], and Laura F. Landweber[2]

[1] Department of Molecular Biology, Princeton University,
Princeton, NJ 08544
ajruben@alumni.princeton.edu
[2] Department of Ecology and Evolutionary Biology, Princeton University,
Princeton, NJ 08544
{sfreelan, lfl}@princeton.edu

Abstract. Nearly every nucleotide-based computing problem attempted thus far has involved the prearranged assignment of nucleotide sequences to represent bits. However, no general program is yet available to optimize those bit sequences. Careful selection of bit sequences can promote strong annealing between a bit and its intended complement while at the same time minimizing unintended interactions with other bits. In this paper, we present a program that uses an evolutionary algorithm to generate optimum bit sets using given (changeable) criteria. We also test some properties of the program and discuss future applications.

1 Introduction

An intuitive conclusion to draw about DNA computing, without much knowledge of the field, is that the individual bases (A, C, G, and T) can themselves be used to represent bits. In reality, however, it is immensely difficult to manipulate such small elements-not to mention reading out the results without resorting to time-consuming DNA sequencing. Consequently, even as far back as the first demonstration of DNA computing [1], scientists have resorted to using longer sequences to represent bits. Each bit used in Leonard Adleman's experiment in 1994, for example, consisted of 20 randomly preassigned nucleotides.

A potentially significant problem, though, lies in the possible interaction of these bits, particularly as the library size increases. A randomly selected bit set may experience any number of physical problems during computation: The molecule may adhere to itself, for example, or a bit may adhere too easily to another's complement. Though many papers have proposed the use of combinatorial constraints on the composition of a set of DNA code words [1, 2, 4, 5, 6, 7, 8, 11, 12, 13, 21, 22, 26, 29, 30, 31], currently no method exists for finding the "best" bit set.

In this paper we describe a computer program called PUNCH (Princeton University Nucleotide Computing Heuristic) that implements an evolutionary algorithm to find an optimized bit set given input parameters (e.g. bit length and number of bits).

N. Jonoska and N.C. Seeman (Eds.): DNA7, LNCS 2340, pp. 150–160, 2002.
© Springer-Verlag Berlin Heidelberg 2002

The goal of its creation was twofold: One, the program was designed to be compatible with various current DNA computing methods and should therefore be useful in selecting bit sets for that purpose. And two, using PUNCH, one should be able to test the limits of current methodology (e.g. the maximum number of unique bits that a current protocol can handle) and possibly suggest new techniques for DNA computing that may work better.

In the remainder of this paper, we outline the program's architecture (Section 2), show its relevance to experiments already performed (Sections 3, 5), introduce new scoring functions (Sections 4, 5), and discuss the program's potential future applications (Section 6).

2   Program Architecture

The primary functional unit of PUNCH is a three-dimensional array, the size of which is determined by three defined constants: N (the number of bits in the problem), B (the number of nucleotides in each bit), and V (the number of variations on each bit set, to be explained below). These three constants can be varied as needed. The initialization function begins by selecting random integers between 0 and 3 and fills the array with corresponding nucleotides ($0 = A$, $1 = C$, $2 = G$, $3 = T$)[1]. The randomization function is taken from [24] and passes all statistical tests for randomness until called 108 times. Each of the bit sets, those of dimension B_N, is assigned the maximum possible score, which is equal to

$$2w(1 + BN - w)^2,$$

where w is the window length, to be explained below.

First, the two-dimensional set is converted to a one-dimensional string with each of the bits in order. This is done so that we may consider problems within sequences as well as between them; i.e., one bit may be similar to the ordered combination of the last part of a different bit and the first part of another. Next, a window (of definable length w) fills itself with the first w nucleotides in the string and scans the entire length of the string for that combination of nucleotides. Each time it encounters an identical sequence, it subtracts w from the current score[2].

In addition, if only $w - 1$ nucleotides are identical (but not necessarily in a row), $w - 1$ is subtracted from the current score; it is the same with $w - 2$, $w - 3$, and so on, until one encounters a predefined threshold called sensitivity. The biological rationale behind this scoring function is as follows: Assume $w = 7$.

---

[1] In later versions of this program which use folding energy to score the bit sets, $3 = U$ instead of T. The effect is the same.

[2] This fact explains the derivation of the maximum possible score: A window of length w scanning the entire length of a BN-element string makes $(1+BN-w)$ comparisons, $(1 + BN - w)$ times, each with a possible penalty of w. The coefficient 2 is included because the scan is performed twice with two different w-length windows, explained later in this section.

Having the same 7 nucleotides in a row in multiple locations within the same string of bits, whether within or between bits, would certainly pose problems during the biological implementation of those bits. Thus, 7 is subtracted from the score. Having 6 nucleotides in a row is also problematic, though less so; hence 6 would be subtracted from the score. Note that the 6 nucleotides need not be sequential, but merely within the same window of size 7–that is, if only 1 nucleotide differs somewhere within the center of that sequence, they may still bind to the same complement. This logic also follows for 5 similar nucleotides. However, it may be unrealistic to penalize 1, 2, 3, or even 4 identical nucleotides within the same window of length 7, since these numbers do not reflect any real biological difficulty.

After the window has checked the entire string, it refills itself with the second w nucleotides in the string and checks the string again, using the same criteria and subtracting from the current score as necessary.

In [11] Frutos et al. report that the two most important criteria for bit set selection are "(i) no two 8mers [the bit length, or word size, Frutos et al. chose] in the set should hybridize with each other's complements (i.e., hybridization adsorption should only occur between a word and its perfectly matched complement), and (ii) no two 8mers in the set should hybridize with each other (this could be important, for example, in the process of hybridizing surface-bound words with a combinatorial set of complements)." Roweis et al. [26] report an equivalent set of requirements.

To check for this, PUNCH repeats its assessment using a second window of the same length. This time, however, rather than fill that window with sequences of w bases, it fills that window with the reverse of their complements. It then checks the string in the same manner, penalizing the score accordingly each time.

The final score can be expressed as a percentage by dividing it by the maximum possible score given above. This process normalizes the score and is useful in comparing scores of different-sized bit sets. The score can also be left alone as a very high integer. Both methods are used throughout this paper; the important aspect in each case is not the bit set's absolute score, but rather its score relative to other bit sets.

Now each of the V bit sets has a numeric score. The next set of functions selects the bit set with the highest score, determines which variation contains it, and fills the entire three-dimensional array with that bit set, such that there are V identical copies of the best bit set in the array.

Let m represent a mutation rate between 0 and 1. The next step is the mutation of random bases within the array, so that each bit set resembles the highest-scoring bit set from the previous trial but with minor changes. To accomplish this, $mBN(V-1)$ bases are randomly selected and reassigned. We determined that the best mutation rate to use is

$$m = 1/BN,$$

a result that takes into account the well-known fact that small changes give better optimized results over several generations than large ones [10].

It is also important to note that variation 0 is consistently left unmutated; hence $(V - 1)$ in the equation above. This feature ensures that, if no mutation produces a better score than the previous generation, the best score among the variations will not decrease.

The new bit sets are re-scored, and again the highest-scoring set is isolated, copied throughout the array, and mutated. This process is repeated until the highest-scoring bit set in each generation fails to improve r times, where r is typically a large integer (e.g. 1000).

Later we will describe a test in which PUNCH is asked to optimize bits that do not need to be evaluated in sequence. In other words, for certain nucleotide computing protocols, the bits are treated as one linear string. For others, however, the bits are left independent, making it unnecessary and in fact inaccurate to treat the bits this way. Therefore we created a second version of PUNCH to evaluate independent bits.

Essentially, the difference between this version of PUNCH and the original version is that this version replaces B with $B + 1$ wherever it appears, and the $(B + 1)$st nucleotide becomes an "end-of-bit marker," the digit 4, converted to character "X" (recall that the integers 0–3 represent the four nucleotides). The bits are still placed into a one-dimensional string, but when either the window scanning or the length of string being scanned contains a 4, the penalty is automatically set to 0, thus ignoring inter-bit combinations. This modification also changes the maximum possible score, which now becomes

$$2w((1 + B - w)N)2.$$

3   "Computing on Surfaces" Using PUNCH

Thus far we have demonstrated that PUNCH optimizes bit sets (i.e., it finds a bit set which scores higher than a large number of random sequences). However, this finding does not include support for the validity of the scoring function.

There are two ways to test the usefulness of the scoring function. One is to perform an experiment using the bits it suggests. Another, easier, method is to assess the bits used in an existing, successful experiment.

Frutos, Liu, et al., who suggested the fundamentals of bit set design in [11] and used them to design 108 "ideal" 8-mers, later used 16 of them in an experiment [20]. To summarize, those bits were synthesized within fixed sequences, fluorescently labeled, annealed independently to 16 locations on a DNA chip, digested or left intact as necessary, and the result was read out by fluorescence detection [20].

Initially, PUNCH was designed based on a different procedure, one performed by our laboratory last year [9], that involved the bits lining up sequentially with spacer sequences between them. This meant that the scoring function had to scan the entire length of the string and check not only within bits but for combinations between adjacent ones as well. In [20], however, the bits do not line up end-to-end; rather, they remain independent on the chip. Thus, the scoring function had to be modified as described in Section 2.

After implementing that change, we performed three analyses. One was an assessment of 10,000 randomly generated Monte Carlo bit sets with the input parameters adjusted to match those described in [11] ($B = 8$, $N = 16$, $V = 10$, $w = 4$, sensitivity $= 0$). Another was an assessment of the exact sequences Liu et al. used in [20]. Finally, we performed a PUNCH optimization using the same input parameters. As hoped, the Liu et al. bit set scored higher than most Monte Carlo sequences, and the PUNCH-optimized set scored higher than that of Liu et al.

It should be noted that these results later took one more factor into account: base equality. Scoring for equal amounts of each nucleotide became necessary, as Liu et al. used GC content as a criterion for designing their own bits. Therefore an additional routine counts the number of each base in a given bit set, divides it by the total number of bases in the set, and finds the absolute value of its deviance from 0.25. Next it totals the deviance of all four bases, subtracts the result from 1, and ends up with a coefficient between 0 and 1 by which to multiply the score. Thus, having equal amounts of each base is rewarded.

This method, and a few others to be mentioned later, work better if each aspect scored can be expressed as a floating-point decimal between 0 and 1. Such a conversion is easily done; it simply involves normalizing the score by dividing it by the maximum possible score. The modified scoring routine then averages the normalized score and the equality coefficient.

4   Introducing Folding Scores

Until this point, the scoring function has assessed bit sets based on two criteria: (i) similarity to other bits or inter-bit sequences, and (ii) similarity to the reverse complements of other bits or inter-bit sequences. Since the latter criterion is used to minimize secondary structure problems, it is possible to replace it with an existing program that scores based on single-stranded DNA folding energy.

We used a publicly available C-based program called the Vienna RNA Package [16]. Though this program is designed to find the folding energy of RNA structures, it can be tweaked to determine the folding energy of single-stranded DNA by interchanging between T and U and by disallowing GU base pairs.

Since bit set similarity is normalized to a value between 0 and 1, as is base equality, it would be a good idea as well to convert the folding score into a normalized value-in other words, to divide it by the maximum possible folding energy.

For a single-stranded sequence of length n, if n is even, the sequence with the maximum folding energy is $G_{(n-4)/2}GAAAC_{(n-4)/2}$. If n is odd, the maximum folding energy sequence is $G_{(n-3)/2}GAAAC_{(n-5)/2}$ [14]. Here, n is the length of the continuous string generated when the B-length bits are laid end-to-end—in other words, $n = BN$.

To correctly use the output score from the Vienna RNA Package, one must subtract it from 1, after dividing it by the maximum possible score. This is because the "best" bit set should be the *least* stable—more stable bit sets have branches that bind more strongly to parts of themselves; thus, their secondary structure is problematic for DNA computing.

To test its effectiveness, we ran PUNCH with $B = 15$, $N = 10$, $V = 10$, $w = 7$, and *sensitivity* $= 3$. Two nested binary "for" loops helped output results for three cases: (i) one in which the bit sets were scored using only folding energy and base equality, (ii) one in which the bit sets were scored using only similarity and base equality, and (iii) one in which all three scores averaged to give the final score.

One would expect (i) to give a nontrivial (i.e., four-base) result with few secondary interactions, but which may have similar bits. One would also expect that (ii), while optimized to avoid similarity, would have a more stable secondary structure than (i). Finally, (iii) should be somewhere in between regarding both qualities.

This prediction is certainly true mathematically. Bit set (i) optimizes to a score of 0.94545, bit set (ii) optimizes to 0.98306, and (iii) gives a score in between: 0.94667.

The sequences generated were fed into an RNA secondary structure prediction program [17], a web interface to the RNAfold program [16]. While this program makes predictions for RNA, it can also be set to use single-stranded DNA parameters [28]. An examination of the program's output shows that (i) has fewer and smaller hairpins than either (ii) or (iii), (ii) appears to have the most egregious secondary structure, and (iii) appears to be compromise between (i) and (ii). Clearly the addition of the folding routine [16] gives PUNCH some biological basis.

5   Comparison to Contiguous Bit Sets

With the folding routine firmly in place, it becomes possible to test PUNCH from the inside out again. Last year our laboratory, using bit sets designed by a program called PERMUTE, performed a successful though imperfect DNA computing experiment [9]. Therefore, the bit set generated by PUNCH should score higher than any of the bit sets we used, which should score rather highly themselves relative to a large number of Monte Carlo sequences.

Because we used 5-nt spacers between 15-nt bits, our procedure can be evaluated equivalently in PUNCH using $B = 20$ (since the bits are evaluated end-to-end as a continuous string anyway). Also, PUNCH used $N = 9$, since our experiment only used nine of the ten bits synthesized. The remaining defined constants were V=10, $r = 1000$, $w = 7$, and *sensitivity* $= 3$.

The result is mixed: PUNCH outscores all $10,000$ Monte Carlo sequences, as is expected. However, the bit sets we used score quite low in comparison.

An analysis of the scoring function shows why this is true. As with the evaluation of the Liu et al. bit set [20], the problem lies in the base equality criterion-but this time the problem is the opposite. The bit sets we used in [9] make use of a three-base alphabet (with one exception). The base equality scoring routine penalizes such a choice.

There are two ways to modify the scoring routine so that this will not be a problem: Either omit base equality, or overtly reward a three-base system.

5.1   Omitting Base Equality

Changing the scoring routine to omit base equality is easily done. Instead of averaging the similarity score, the folding score, and the base equality score, the new scoring routine averages only the similarity and folding scores. The results show that, compared to 10, 000 Monte Carlo bit sets, a three-base alphabet makes a big difference. Hence the superiority of the bit sets we used to the Monte Carlo bit sets: Restriction to three bases greatly reduces secondary-structure stability.

Even more interesting, however, is the bit set PUNCH optimizes. Using folding scores only, or using both folding and similarity scores, and without insisting on base equality, PUNCH naturally optimizes its bit set to practically a three-base alphabet. In fact, the sequence scored only for folding achieves a perfect score of 1.00000. When similarity is the only criterion, of course, this is no longer true. Furthermore, as in the bit sets our laboratory chose, the omitted base is G. This makes sense from a molecular perspective, since omitting G destroys the potential for G-C pairs, $G : U$ pairs, and G stacking, three of the most stabilizing forces in RNA secondary structure [27]. Indeed, it is the reason we consciously chose to omit G from our alphabet [9].

5.2   Rewarding a Three-Base System

The other way to guarantee a relatively high rank for the sequences we used is the deliberate restriction, in PUNCH, to a three-base alphabet. For this reason, we introduced a new constant, *bases*, capable of being defined such that $3 \leq bases \leq 8$. This constant represents the number of bases in the desired alphabet.

It becomes possible, with a few adjustments, to set bases = 3-almost exactly the supposition imposed by Faulhammer et al. [9] on their bit set assignment. (The use of a 2-base alphabet—or, rather, of various 2-base combinations of the standard 4-base alphabet—has also been proposed [22]. However, as no specific experiments using this scheme have been set forth, it does not seem a realistic situation for which to prepare PUNCH.)

Another benefit to the addition of the constant bases is the possibility of working with more than four bases. Just as it is possible to expand the amino-acid alphabet through the use of such "unnatural" residues as dipropylglycine and dibutylglycine [3], there exist several nonstandard nucleotide bases [23] that can offer perhaps more versatility to bit set design. These bases are represented by slots 5–8 in *bases*.

The final change that must be made is an adaptation imposed on the function that ensures base equality. Rather than subtract the percent of each base from 0.25, they must be subtracted from $1/bases$. However, in this instance, it may be preferable not to score for equality, since the goal is to show that a three-base system has a head start in terms of folding on its own.

When restricted to a three-base alphabet, PUNCH produced a sequence with absolutely *no* secondary interactions, implying that the choice of a three-base alphabet may be a good one.

6   Future Directions

When used to generate bit sets for a protocol like that of Liu et al. [20], PUNCH seems perfectly adequate in its current structure. It is possible that better scoring routines could be discovered, but the architecture of PUNCH is correctly designed to test, evolve, and output a single bit set to use in an experiment.

When PUNCH is used with the parameters of our experiment [9], however, it may be apparent that PUNCH lacks one key feature: It only generates one bit set, and our protocol calls for $2^9 = 512$ strings. To be more exact, the protocol requires two bit sets, not one, each containing N bits of length B. Each of the two bit sets represents a different bit designation, i.e. 0 or 1, and the 2^N combinations compose the entire combinatorial library of N-bit binary numbers.

Furthermore, PUNCH does not make allowance for the existence of spacer sequences between the bits. In our experiment [9], 5-nt spacers separated the 15-nt bits. The spacers therefore had to be dissimilar to the bits themselves, not cause secondary interactions once in place, and not cause any unforeseen sequences to come into existence when taken in combination with the bits on either side.

Some adjustments can be made to PUNCH, as a future exercise, to account for these discrepancies. Rather than a single three-dimensional (N_B_V) array, one could begin with three such arrays, two of dimension N_B_V and one of $(N-1)\_S\_V$, where S is the length of a spacer sequence. (The adjustment to the dimension of $N-1$ must be made since spacer sequences fit between bits.)

The two similar-sized arrays represent the two binary possibilities, 0 and 1. Rather than evaluate and compare each bit set on a single array, however, one should evaluate and compare the average score of the 2^N sequences generated by the combination of each bit set with the corresponding Vth bit set in the other array and the spacer sequences. In other words, for every N-bit binary number, generate and test a bit set of length $N(B + S - 1)$. It should be easy to evolve the bit sequences and spacer sequences in a similar manner to that already performed.

One key feature of PUNCH is its adaptability to various criteria. Built into the robust architecture is the opportunity to alter, remove, or add functions as necessary. This is especially true of the scoring routine-future users can make very simple modifications to the existing function and plug in their own scoring routines. This would be beneficial if the new scoring routine is either more accurate than existing ones, or at least better suited to the specific DNA computing protocol whose bits it is intended to optimize.

Equally important, in terms of flexibility, is the simplicity with which one can change a defined value. Unlike PERMUTE, the program our laboratory used to design bit sets for the Knight Problem [9], PUNCH does not begin with a hard-coded notion of how many bits it needs to produce, their length, or several other factors. Each of a handful of globally defined constants can be readily altered to meet the needs of the researcher.

Perhaps the most useful future application of PUNCH is one of its original intentions: determining the limits of current methods. Graphing the highest attainable score for various numbers of bits at constant length, for example, could show a trend that would plummet before a certain N. This would uncover flaws in certain current methods, since perhaps it is impossible to design, say, over 108 "unique enough" 8-mers [11], which would force scientists to rethink their approach to that problem.

Critics of DNA computing have said that the field perhaps rewards impractical proof-of-concept experiments before discussing their scalability. PUNCH can help identify which experiments may not be realistic on a larger scale. Or, rather than dismissing unpromising protocols entirely, PUNCH may help point to the combinatorial issues that would otherwise hold them back.

References

[1] Adleman, L.M.: Molecular Computation of Solutions to Combinatorial Problems. Science **266** (1994) 1021–1023
[2] Baum, E.B.: DNA Sequences Useful for Computation. In: Landweber, L.F., Baum, E.B. (eds.): Proceedings of DNA Based Computers II, DIMACS Workshop, June 10–12, 1996. AMS Press, Providence (1999) 235–242
[3] Benner, S.A.: Expanding the Genetic Lexicon-Incorporating Nonstandard Amino-Acids into Proteins by Ribosome-Based Synthesis. Trends in Biotechnology **12** (1994) 158–163
[4] Brenner, S.: Methods for Sorting Polynucleotides Using Oligonucleotide Tags. United States Patent #5,604,097. 18 Feb 1997.
[5] Brenner, S., Lerner, R.A.: Encoded Combinatorial Chemistry. Proceedings of the National Academy of Sciences USA '89 (1992) 5381–5383
[6] Cukras, A.R., Faulhammer, D., Lipton, R.J., Landweber, L.F.: Chess Games: A Model for RNA-Based Computation. Biosystems **52** (1999) 35–45
[7] Deaton, R., Garzon, M., Murphy, R.C., Rose, J.A., Franceschetti, D.R., Stevens, S.E., Jr.: Genetic Search of Reliable Encodings for DNA Based Computation. In: Koza, J.R., Goldberg, D.E., Fogel, D.B., Riolo, R.L. (eds.): Proceedings of First Annual Conference on Genetic Programming
[8] Deaton, R., Murphy, R.C., Garzon, M., Franceschetti, D.R., Stevens, S.E., Jr.: Good Encodings for DNA-Based Solutions to Combinatorial Problems. In: Landweber, L.F., Baum, E.B. (eds.): Proceedings of DNA Based Computers II, DIMACS Workshop, June 10–12, 1996. AMS Press, Providence (1999) 247–258
[9] Faulhammer, D., Cukras, A.R., Lipton, R.J., Landweber, L.F.: Molecular Computation: RNA Solutions to Chess Problems. Proceedings of the National Academy of Sciences USA '97 (2000) 1385–1389
[10] Fisher, R.A.: Darwinian Evolution by Mutations. Eugenics Review **14** (1922) 31–34
[11] Frutos, A.G., Liu, Q., Thiel, A.J., Sanner, A.W., Condon, A.E., Smith, L.M., Corn, R.M.: Demonstration of a Word Design Strategy for DNA Computing on Surfaces. Nucleic Acids Research **25** (1997) 4748–4757

[12] Garzon, M., Deaton, R., Neathery, P., Franceschetti, D.R., Murphy, R.C.: A New Metric for DNA Computing. In: Kaufman, M. (ed.): Proceedings of Second Genetic Programming Conference. (1997) 472–478

[13] Garzon, M., Deaton, R., Nino, L.F., Stevens, S.E., Jr., Wittner, M.: Encoding Genomes for DNA Computing. In: Koza, J.R., Banzhaf, W., Chellapilla, K., Deb, K., Dorigo, M., Fogel, D.B., Garzon, M.H., Goldberg, D.E., Iba, H., Riolo, R. (eds.): Proceedings of Third Genetic Programming Conference, July 1998. AAAI Press, Madison (1998)

[14] Gesteland, R.F., Atkins, J.F.: Thermodynamic Considerations for Evolution by RNA. In: Turner, D.H. and Bevilacqua, P.C. (eds.): The RNA World. Cold Spring Harbor Laboratory Press (1993) 447–464

[15] Goldberg, D.E.: Genetic Algorithms in Search, Optimization, and Machine Learning. Addison-Wesley Publishing Company, Reading, MA (1989)

[16] Hofacker, I., Stadler, P.: Vienna RNA Package. Vienna, Austria, Institut für Theoretische Chemie. http://www.tbi.univie.ac.at/ ivo/RNA/ (2000)

[17] Hofacker, I., Stadler, P.: Vienna RNA Secondary Structure Prediction Program: A Web Interface to the RNAfold Program. Vienna, Austria, Institut für Theoretische Chemie. Internet. http://www.tbi.univie.ac.at/cgi-bin/RNAfold.cgi (2000)

[18] Holland, J.: Adaptation in Natural and Artificial Systems: An Introductory Analysis with Applications to Biology, Control, and Artificial Intelligence. University of Michigan Press, Ann Arbor (1975)

[19] Liu, Q., Frutos, A.G., Wang, L., Thiel, A.J., Gillmor, S.D., Strother, C.T., Condon, A.E., Corn, R.M., Lagally, M.G., Smith, L.M.: Progress Toward Demonstration of a Surface Based DNA Computation: A One Word Approach to Solve a Model Satisfiability Problem. Biosystems **52** (1999) 25–33

[20] Liu, Q., Wang, L., Frutos, A.G., Condon, A.E., Corn, R.M., Smith, L.M.: DNA Computing on Surfaces. Nature **403** (2000) 175–179

[21] Marathe, A., Condon, A.E., Corn, R.M.: On Combinatorial DNA Word Design. In: Winfree, E., Gifford, D.K. (eds.): Proceedings of DNA Based Computers V, DIMACS Workshop, June 14–15, 2000. (2000) 75–89

[22] Mir, K.U.: A Restricted Genetic Alphabet for DNA Computing. In: Landweber, L.F., Baum, E.B. (eds.): Proceedings of DNA Based Computers II, DIMACS Workshop, June 10-12, 1996. AMS Press, Providence (1999) 243–246

[23] Picirilli, J.A., Krauch, T., Moroney, S.E., Benner, S.: Enzymatic Incorporation of a New Base Pair into DNA and RNA Extends the Genetic Alphabet. Nature **343** (1990) 33–37

[24] Press, W.H., Teukolsky, S.A., Vetterling, W.T., Flannery, B.P.: Numerical Recipes in C: The Art of Scientific Computing. Cambridge University Press, Cambridge New York (1992)

[25] Rechenberg, I.: Evolutionsstrategie: Optimierung Technischer Systeme nach Prinzipien der Biologischen Evolution [Evolution Strategies: Optimization of Technical Systems with Principles of Biological Evolution]. Frommann-Holzboog, Stuttgart (1973)

[26] Roweis, S., Winfree, E., Burgoyne, R., Chelyapov, N.V., Goodman, M.F., Rothemund, P.W.K., Adleman, L.M.: A Sticker-Based Model for DNA Comptuation. In: Landweber, L.F., Baum, E.B. (eds.): Proceedings of DNA Based Computers II, DIMACS Workshop, June 10–12, 1996. AMS Press, Providence (1999) 1–29

[27] Salser, W.: Globin Messenger RNA Sequences-Analysis of Base-Pairing and Evolutionary Implications. Cold Spring Harbor Symposium on Quantitative Biology **43** (1977) 985–1002

[28] SantaLucia, J.J.: A Unified View of Polymer, Dumbbell, and Oligonucleotide DNA Nearest-Neighbor Thermodynamics. Proceedings of the National Academy of Sciences USA '95 (1998) 1460–1465

[29] Shoemaker, D.D., Lashkari, D.A., Morris, D., Mittman, M., Davis, R.W.: Quantitative Phenotypic Analysis of Yeast Deletion Mutants Using a Highly Parallel Molecular Bar-Coding Strategy. Nature Genetics **16** (1996) 450–456

[30] Smith, W.D., Schweitzer, A.: DNA Computers in Vitro and in Vivo. NECI Technical Report (1995)

[31] Winfree, E., Liu, F., Wenzler, L.A., Seeman, N.C.: Design and Self-Assembly of Two-Dimensional DNA Crystals. Nature **394** (1998) 539–544

Solving Knapsack Problems in a Sticker Based Model

Mario J. Pérez–Jiménez and Fernando Sancho–Caparrini

Department of Computer Science and Artificial Intelligence,
University of Seville, Spain

Abstract. Our main goal in this paper is to give molecular solutions for two **NP**–complete problems, namely Subset-sum and Knapsack, in a sticker based model for DNA computations. In order to achieve this, we have used a finite set sorting subroutine together with the description of a procedure to formally verify the designed programs through the labeling of test tubes using inductive techniques.

1 Introduction

The *sticker model* was introduced by S. Roweis, E. Winfree et al [3] as an abstract model of molecular computing based on DNA with a random access memory and a new form of encoding the information.

The main goal of this work is the resolution, in this model, of two **NP**–complete problems: the *Subset-Sum* problem and the *Knapsack* problem, in its 0/1 bounded and unbounded versions.

The information is represented in the sticker model in a different way from that used in the Adleman-Lipton paradigm. A (n, k, m)-memory strand, with $n \geq k \cdot m$, is n bases in length subdivided into k non-overlapping substrand each m bases long. The substrands should be significantly different from each other. A sticker associated to a (n, k, m)-memory strand is m bases long and complementary to exactly one of the k substrands in the memory strand. If a sticker is annealed to its matching substrand on a memory strand, then the particular substrand is said to be *on*. If no sticker is annealed to a substrand, then the region is said to be *off*. A (n, k, m)-memory complex is a (n, k, m)-memory strand along with its annealed stickers (if any). In a direct way, (n, k, m)-memory complexes represent bit strings of $\{0, 1\}^k$. For this reason, it is usual to identify them either as binary functions ($\sigma : \{1, \ldots, k\} \longrightarrow \{0, 1\}$, such that $\sigma(i) = 1$ if and only if the i-th substrand is *on*), or as subsets of $\{1, \ldots, k\}$ by means of the characteristic function.

Within the sticker model, a tube is a finite multiset whose elements are memory complexes (that is, a collection of memory complexes where each one can be repeated). The following operations over tubes of the sticker model are used in this paper:

- *Merge* (T_1, T_2): the memory complexes from the tubes T_1, T_2 are combined to form the multiset union of all strings in the two input tubes. We write $Merge(T_1, T_2) = T_1 \cup T_2$ as well.

N. Jonoska and N.C. Seeman (Eds.): DNA7, LNCS 2340, pp. 161–171, 2002.
© Springer-Verlag Berlin Heidelberg 2002

– *Separate* (T, i): Given a tube, T, and an integer, i $(1 \leqslant i \leqslant$ number of substrands that form each complex of T), create two new tubes, $+(T, i)$ and $-(T, i)$, where $+(T, i)$ (resp. $-(T, i)$) contain all strings of T having the i–th substrand set to 1 (resp. set to 0). We write $(T_1, T_2) \leftarrow separate(T, i)$ to indicate that $T_1 = +(T, i)$ and $T_2 = -(T, i)$.

– *Set* (T, i): Given a tube, T, and an integer, i $(1 \leqslant i \leqslant$ number of substrands that form each complex of T), this operation produces a new tube where the i–th substrand of each memory complex in T is set to 1. That is, the sticker for that bit is annealed to i–th region on every memory complex in T.

– *Read* (T): Given a nonempty tube T, this operation reads its content. For that, one memory complex must be isolated from T and its annealed stickers, if any, determined.

A (k, l)-library, with $1 \leq k \leq l$, consists of memory complexes with k substrands, the first l substrands are either *on* or *off*, in all possible ways, whereas the last $k - l$ substrands are *off*.

In Section 2, the problem of sorting the elements of a finite family of finite sets, according to their cardinality, is studied, and for the first time, a program that is able to solve this problem is described within the sicker model. If it is taken into account that just two molecular operations have been used, namely *separate* and *merge*, the program designed can also be considered as a program within the unrestricted model of Adleman [1].

In Section 3 we give a filling subroutine within the sticker model to encode the weight of subsets regarding a given positive function that will be used in following sections.

In Section 4, we give a molecular solution within the sticker model for the Subset–Sum problem, using the sorting by cardinality program and the filling subroutine. Formal verification of the programs designed in Sections 2, 3 and 4 is established through the prior labeling of the distinct tubes which appear in the execution. Then we prove the soundness and completeness of these programs using inductive techniques and analyzing the *history* of every molecule in the initial test–tube along the process.

In Sections 5 and 6, we give molecular solutions, within the sticker model, for Knapsack problem (0/1 bounded and unbounded versions), based in both the sorting by cardinality program given in Section 2, and the filling subroutine given in Section 3.

All designed programs use a linear number of tubes, and the number of molecular operations is, basically, quadratic.

2   Sorting by Cardinality

Problem: *Let* $A = \{1, \ldots, p\}$, $B = \{b_1, \ldots, b_s\} \subseteq A$, $\mathcal{F} = \{D_1, \ldots, D_t\} \subseteq \mathcal{P}(A)$. *Sort the sets of $\mathcal{F}$ according to their relative cardinality to B (that is, according to the number of elements of $B \cap D_i$).*

Next, we will design a molecular program within the sticker model which solves the above problem.

- The input tube, T_0, will contain memory complexes, σ, based on DNA, encoding each set of the family $\mathcal{F}$. For this, each complex of T_0 will be represented through a boolean function, that is, $T_0 = \{\{\sigma \ : \ |\sigma| = p \wedge \exists j \, (\chi_{D_j} = \sigma)\}\}$, where χ_{D_j} is the characteristic function of D_j in A $(\chi_{D_j}(i) = 1$ if $i \in D_j$, and $\chi_{D_j}(i) = 0$ if $i \in A - D_j)$.
- The program consists of a main loop FOR with s steps. In the i-th step, $i + 1$ tubes, $T_0, T_1, \ldots, T_i$, are generated verifying the condition: $\forall \sigma \ (\sigma \in T_j \longrightarrow |\sigma \cap \{b_1, \ldots, b_i\}| = j)$. In order to achieve this, we design the body of the loop by induction. Once the tubes $T_0, T_1, \ldots, T_i$ corresponding to the i-th step have been built, the tubes of the next step $T_0, T_1, \ldots, T_i, T_{i+1}$ are generated in this way:

$$
\begin{cases}
T_0 = -(T_0, b_{i+1}) \\
T_j = +(T_{j-1}, b_{i+1}) \cup -(T_j, b_{i+1}) & (1 \leqslant j \leqslant i) \\
T_{i+1} = +(T_i, b_{i+1})
\end{cases}
$$

The execution of the molecular program can be described starting from a rooted graph that we denominate *labeled merge–binary tree* that is defined by recursion as follows:

- A node with a label is a labeled merge–binary tree of depth 0.
- Let A be a labeled merge–binary tree of depth h. From it, a labeled merge–binary tree, A', of depth $h + 1$ is built in this way:
  - Initially, each leaf of A determines two children.
  - The right child of each leaf and the left one of the next leaf give a node of A' whose label is the composition of the labels of this children by a certain fixed binary operation.

In the description that has been carried out so far, the nodes of the merge–binary tree of execution are labeled by means of tubes. The left and right children of a tube, T, of depth h are labeled starting from the separate operation: $(T_{left}, T_{right}) \longleftarrow separate(T, b_{h+1})$. Finally, the binary operation considered is the molecular merge operation applied to the tubes indicated by the labels of the nodes.

These ideas suggest the following molecular program:

```
Input: (T₀, B)
        for i = 1 to s do
            (T₀, T₁′) ← separate (T₀, bᵢ)
            for j = 0 to i − 1 do
                (Tⱼ″, Tⱼ₊₁′) ← separate (Tⱼ, bᵢ)
                Tⱼ ← Tⱼ′ ∪ Tⱼ″
            end for
            Tᵢ ← Tᵢ′
        end for
Output: T₀, …, Tₛ
```

The procedure described will return $s + 1$ tubes and we will note them as: $\mathtt{Cardinal\_sort}(T_0, B)[j]$, $(0 \le j \le s)$. We have that $|\mathtt{Cardinal\_sort}(T_0, B)[j]| = j$.

This molecular program uses $2s$ tubes and the number of molecular operations is $\frac{s \cdot (s+3)}{2}$.

Let us note that the program we have given to solve the sorting problem is valid in a model without random access memory, like the unrestricted model of Adleman. The simplest way to see this is to adapt the input tube, replacing memory complex for single strands of DNA.

To establish the formal verification of the algorithm program, we will proceed to label the tubes obtained along the execution so that we can individualize them in any moment of the running.

```
Input:  T₀
            T₀,₀ ← T₀;  T₀,₋₁ ← ∅;  T₀,₁ ← ∅
            for i = 1 to s do
                Tᵢ,₋₁ ← ∅;  Tᵢ,ᵢ₊₁ ← ∅
                for j = 0 to i do
                    Tᵢ,ⱼ ← +(Tᵢ₋₁,ⱼ₋₁, bᵢ) ∪ −(Tᵢ₋₁,ⱼ, bᵢ)
                end for
            end for
Output: Tₛ,₀, ..., Tₛ,ₛ
```

By means of convenience, we assume that $T_{0,-1} = T_{0,1} = \emptyset$, and we will note $B_j = \{b_1, \ldots, b_j\}$, and, by definition, we will take $B_0 = \emptyset$.

Proposition 1. $\forall i \, (1 \le i \le s \rightarrow \forall j \le i \, \forall \sigma \, (\sigma \in T_{i,j} \rightarrow |\sigma \cap B_i| = j))$.

Proof. By induction on i. Let us see that $\forall j \le 1 \, \forall \sigma \in T_{1,j} \, (|\sigma \cap B_1| = j)$.

- Let $\sigma \in T_{1,0} = +(T_{0,-1}, b_1) \cup -(T_{0,0}, b_1)$. Since $T_{0,-1} = \emptyset$ and $T_{0,0} = T_0$, it results that $\sigma \in -(T_0, b_1)$, that is, $b_1 \notin \sigma$. Then, $|\sigma \cap B_1| = 0$.
- Let $\sigma \in T_{1,1} = +(T_{0,0}, b_1) \cup -(T_{0,1}, b_1)$. Since $T_{0,1} = \emptyset$, it results that $\sigma \in T_{0,0} = T_0$ and $b_1 \in \sigma$, then $|\sigma \cap B_1| = 1$

Let i $(1 \le i < s)$ be such that $\forall j \le i \, \forall \sigma \in T_{i,j} \, (|\sigma \cap B_i| = j)$. Let us see that the result verifies for $i + 1$. For it, we now proceed by induction on j: $\forall j \le i + 1 \, \forall \sigma \in T_{i+1,j} \, (|\sigma \cap B_{i+1}| = j)$.

- Let $\sigma \in T_{i+1,0} = +(T_{i,-1}, b_{i+1}) \cup -(T_{i,0}, b_{i+1})$. Since $T_{i,-1} = \emptyset$, it results that $\sigma \in T_{i,0}$ and $b_{i+1} \notin \sigma$, by induction hypothesis we deduce that $|\sigma \cap B_i| = 0$. So $|\sigma \cap B_{i+1}| = 0$, since $b_{i+1} \notin \sigma$.
- Let $j > 0$ and $\sigma \in T_{i+1,j} = +(T_{i,j-1}, b_{i+1}) \cup -(T_{i,j}, b_{i+1})$. Then
  - If $\sigma \in T_{i,j-1}$ and $b_{i+1} \in \sigma$, by induction hypothesis we have that $|\sigma \cap B_i| = j - 1$. Since $b_{i+1} \in \sigma$, we conclude that $|\sigma \cap B_{i+1}| = j - 1 + 1 = j$.
  - If $\sigma \in T_{i,j}$ and $b_{i+1} \notin \sigma$, by induction hypothesis we have that $|\sigma \cap B_i| = j$. As $b_{i+1} \notin \sigma$, we have $|\sigma \cap B_{i+1}| = j$.

Proposition 2. $\forall \sigma \in T_0 \ \forall i \ (0 \le i \le s \to \sigma \in T_{i,|\sigma \cap B_i|})$.

Proof. By induction on i. For $i = 0$, the result is trivial.
Assume the result holds for i $(0 \le i < s)$; we will prove it for $i+1$.

- If $b_{i+1} \in \sigma$, we have $|\sigma \cap B_{i+1}| = 1 + |\sigma \cap B_i|$. By induction hypothesis, $\sigma \in T_{i,|\sigma \cap B_i|}$, then $\sigma \in +(T_{i,|\sigma \cap B_i|}, b_{i+1}) \subseteq T_{i+1,|\sigma \cap B_i|+1}$.
- If $b_{i+1} \notin \sigma$, then $|\sigma \cap B_{i+1}| = |\sigma \cap B_i|$. By induction hypothesis, $\sigma \in T_{i,|\sigma \cap B_i|}$, then $\sigma \in -(T_{i,|\sigma \cap B_i|}, b_{i+1}) \subseteq T_{i+1,|\sigma \cap B_i|} = T_{i+1,|\sigma \cap B_{i+1}|}$.

From the preceding propositions it may be concluded, respectively, soundness (every molecule of the output tube provides a correct solution associated to that tube) and completeness (every molecule of the input tube appears in the corresponding output tube, according to its cardinality) of the designed program.

Corollary 1 (Soundness). $\forall j \ \forall \sigma \ (0 \le j \le s \wedge \sigma \in T_{s,j} \to |\sigma \cap B| = j)$.

Corollary 2 (Completeness). *If $\sigma \in T_0$ and $|\sigma \cap B| = j$, then $\sigma \in T_{s,j}$.*

As cases of particular interest, that we will use in other molecular programs, we get the following:

- Cardinal_sort (T_0), when $B = A$.
- Cardinal_sort (T_0, l, k), when $B = \{l, l+1, \ldots, k\}$.

3   A Filling Subroutine

In this section we show a molecular program that will be used as auxiliary subroutine to solve the Subset–Sum problem and the Knapsack problem in the following sections.

Let $A = \{1, \ldots, p\}$, $r \in \mathbb{N}$ and $f : A \longrightarrow \mathbb{N}$ a function. If $B \subseteq A$, we note $f(B) = \sum_{i \in B} f(i)$. For convenience we define $f(0) = 0$. Let $q_f = f(A)$, $A_i = \{0, \ldots, i\}$ $(0 \le i \le p)$ and T_0 a multiset of (n, k, m)-memory complexes, σ, with $k \ge p + r + q_f$.

As it was seen in the previous section, each $\sigma \in T_0$ encodes a subset $B_\sigma \subseteq A$ characterized by the condition $B_\sigma = \{i : 1 \le i \le p \wedge \sigma(i) = 1\}$, and reciprocally, each subset, $B \subseteq A$, can be encoded by a molecule $\sigma_B \in T_0$, characterized by the condition: $\sigma_B(i) = 1$ if and only if $i \in B$.

If $\sigma \in T_0$, we can suppose that it is formed by the following *zones*:

$$(A\sigma) = \sigma(1) \ldots \sigma(p), \qquad (F\sigma) = \sigma(p+r+1) \ldots \sigma(p+r+q_f)$$
$$(L\sigma) = \sigma(p+1) \ldots \sigma(p+r), \quad (R\sigma) = \sigma(p+r+q_f+1) \ldots$$

The subroutine works over T_0, and it modifies their elements making that the molecules of the output tube store in $(F\sigma)$ the weight, regarding f, of the subset of A encoded in $(A\sigma)$ (zones $(R\sigma)$ and $(L\sigma)$ have no effect in the process, but they will be useful for a general use). That is:

$$\sum_{i=1}^{p} \sigma(i)f(i) = \sum_{j=p+r+1}^{p+r+q_f} \sigma(j)$$

The designed program will be noted $\texttt{Parallel\_Fill}(T_0, f, p, r)$:

```
Input:  (T₀, f, p, r)
        for i = 1 to p do
            (T⁺, T⁻) ← separate(T₀, i)
            for j = 1 to f(i) do
                T⁺ ← set(T⁺, p + r + f(A_{i−1}) + j)
            end for
            T₀ ← merge (T⁺, T⁻)
        end for
Output: T₀
```

To establish the formal verification of the algorithm, we will proceed to label the tubes obtained along the execution.

```
Input:  (T₀, f, p, r)
        for i = 1 to p do
            (T⁺_{i,0}, T⁻_i) ← separate(T_{i−1}, i)
            for j = 1 to f(i) do
                T⁺_{i,j} ← set(T⁺_{i,j−1}, p + r + f(A_{i−1}) + j)
            end for
            T_i ← merge (T⁺_{i,f(i)}, T⁻_i)
        end for
Output: T_p
```

For each i $(1 \le i \le p)$ we consider the following *regions*:

$$R_i = \{p + r + f(A_{i-1}) + 1, \ldots, p + r + f(A_i)\}$$

Definition 1. *For each $\sigma \in T_0$ and each k $(1 \le k \le p)$, we will note σ^k the molecule obtained from σ after the execution of the k–th step in the main loop of the program.*

That is, the molecules σ^k provide the *history* of the molecule σ of the input tube, while the program is running. Keeping in mind the syntactic structure of the program, it is straightforward to prove the following results:

Lemma 1.

1. *The initial zone of the molecule (encoding the subset of A) does not change along the execution of the program; that is,*

$$\forall \sigma \in T_0 \; \forall k \; (1 \le k \le p \to (A\sigma) = (A\sigma^k)) \qquad (1)$$

2. *The molecules that are obtained in the k-th step of the main loop are stored in the tube T_k; that is,*

$$\forall \sigma \in T_0 \; \forall k \; (1 \le k \le p \to \sigma^k \in T_k) \qquad (2)$$

3. *Every molecule of the k-th tube comes from some molecule in the initial tube; that is,*

$$\forall k \; (1 \le k \le p \to \forall \tau \in T_k \; \exists \sigma \in T_0 \; (\sigma^k = \tau)) \qquad (3)$$

4. *The execution of a step of the main loop does not modify the regions corresponding to previous steps; that is,*

$$\forall \sigma \in T_0 \; \forall i \; \forall k \; (1 \le i \le k \le p \to \sigma^i_{|R_i} = \sigma^k_{|R_i}) \qquad (4)$$

5. *After the execution of the i-th step of the main loop, the region R_i of σ has been modified to agree with the value of $\sigma(i)$; that is,*

$$\forall \sigma \in T_0 \; \forall i \; \forall k \; (1 \le i \le k \le p \to \sigma^k_{|R_i} \equiv \sigma(i)) \qquad (5)$$

6. *The execution of a step of the main loop does not modify the zones $(L\sigma)$ and $(R\sigma)$; that is,*

$$\forall \sigma \in T_0 \; \forall k \; (1 \le k \le p \to (L\sigma) = (L\sigma^k) \wedge (R\sigma) = (R\sigma^k)) \qquad (6)$$

The following result assures us that the main loop modifies the regions R_i of the molecules to encode the partial weight of the set represented by each one of them.

Proposition 3. *Let $B \subseteq A$ such that $\sigma_B \in T_0$, then for each k $(1 \le k \le p)$ we have that:*

$$f(B \cap \{1, \ldots, k\}) = \sum_{j=p+r+1}^{p+r+f(A_k)} \sigma_B^k(j)$$

Proof. From (1) it follows that

$$f(B \cap \{1, \ldots, k\}) = \sum_{i=1}^{k} f(i) \cdot \sigma_B(i) = \sum_{i=1}^{k} f(i) \cdot \sigma_B^k(i)$$

On the other hand, (1) and (5) assures that

$$f(i) \cdot \sigma_B^k(i) = \sum_{j \in R_i} \sigma_B^k(j)) \; (1 \le i \le k)$$

Corollary 3. *For each $B \subseteq A$ such that $\sigma_B \in T_0$ there exists $\tau \in T_p$ such that $f(B) = \sum_{i=p+r+1}^{p+r+q_f} \tau(i)$.*

Proof. Given $B \subseteq A$, let us consider the associated molecule $\sigma_B \in T_0$. It suffices to consider $\tau = \sigma_B^p$, since $f(B) = f(B \cap \{1, \ldots, p\}) = \sum_{j=p+r+1}^{p+r+q_f} \sigma_B^p(j)$.

4   Subset-Sum Problem

Problem: Let $A = \{1, \ldots, p\}$ and $w : A \longrightarrow \mathbb{N}$ a weight function. Let $k \in \mathbb{N}$ be such that $k \leq w(A) = q_w$. Determine whether there exists a subset $B \subseteq A$ such that the sum of the weights of the elements in B is, exactly, k.

Next we will design a molecular program within the sticker model that solves the Subset–Sum problem. The input tube, T_0, will be a $(p + q_w, p)$-library. In a first stage (filling), each molecule, σ, of the input tube is filled in order to obtain in their last q components the weight of the subset that it encodes; the molecules of the resulting tube of the previous stage are ordered according to their cardinality. Finally, the k-th tube is read: it contains the molecules from the input tube encoding subsets of A of weight k, if any.

These ideas suggest the design of the following molecular program:

```
Subset_Sum(p, w, k)
      q_w ← Σ^p_{i=1} w(i)
      T_0 ← (p + q_w, p)-library
      T_1 ← Parallel_Fill(T_0, w, p, 0)
      T_out ← Cardinal_sort(T_1, p + 1, p + q_w)[k]
      Read(T_out)
```

The number of used tubes (including the subroutines) is $4 + 2q$ and the number of molecular operations is $2p + q + 1 + \frac{q \cdot (q+3)}{2}$.

The following result shows the soundness of the molecular program; that is, every molecule in the output tube encodes a correct solution for the Subset–Sum problem.

Theorem 1 (Soundness). If $T_{out} \neq \emptyset$, then there exists $B \subseteq A$ such that $w(B) = k$.

Proof. Taking $\tau \in T_{out}$, and applying (3) and proposition 3, we obtain $\sigma \in T_0$ verifying the result for B_σ.

Next theorem proves the completeness of the given molecular program; that is, every molecule in the input tube encoding a correct solution of the Subset–Sum problem, is in the output tube.

Theorem 2 (Completeness). Let $\sigma \in T_0$ be such that $w(B_\sigma) = k$. Then $T_{out} \neq \emptyset$

Proof. Let $\sigma \in T_0$ be such that $w(B_\sigma) = k$, from corollary 3, after the execution of the filling subroutine, we have a molecule $\tau = \sigma^p \in T_1$ such that $w(B_\sigma) = \sum_{i=p+1}^{p+q_w} \tau(i)$. Then $\tau \in T_{out}$.

Note. The above program does not only solve the problem of decision, but rather, it returns all the solutions to the problem. If we only wanted to solve the decision problem, once the filling stage has been executed, we could use some appropriate restriction enzymes which remove from each molecule, σ, in the tube T_p, the initial region $(A\sigma)$, and then apply an appropriate procedure Cardinal_sort, so that we may work with molecules of smaller length in the final stage.

5 0/1 Bounded Knapsack Problem

Problem: *Let $A = \{1, \ldots, p\}$ be a non empty finite set, $w : A \longrightarrow \mathbb{N}$ a weight function, and $\rho : A \longrightarrow \mathbb{N}$ a function of values. Let $k, k' \in \mathbb{N}$ be such that $k \leq w(A) = q_w$ and $k' \leq \rho(A) = q_\rho$. Determine whether there exists a subset $B \subseteq A$ such that $w(B) \leq k$ and $\rho(B) \geq k'$.*

Next we will design a molecular program in the sticker model that solves the problem 0/1 bounded Knapsack problem: we begin with a $(p + q_w + q_\rho, p)$-library. In a first stage (filling), we proceed as in the previous program: each molecule of the initial tube is filled appropriately so that it encodes the *weight* of the associate subset; next the molecules, σ, of the resulting tube are ordered according to the cardinal of $w(A\sigma)$, being obtained some tubes, $T_0, \ldots, T_{q_w}$, such that $\forall \sigma$ $(\sigma \in T_j \Rightarrow |w(A\sigma)| = j)$. With the tube $T_0 \cup \cdots \cup T_k$ a second stage of filling is carried out, regarding the function of *values*. Then, the molecules, σ, from the resulting tube are ordered according to the cardinal of $\rho(A\sigma)$, being obtained, again, some tubes, $T_0, \ldots, T_{q_\rho}$, such that $\forall \sigma$ $(\sigma \in T_j \Rightarrow |\rho(A\sigma)| = j)$. Finally, the tube $T_{k'} \cup \cdots \cup T_{q_\rho}$, containing the encoded solutions to the problem, is read.

These ideas suggest the following molecular program:

```
Knapsack(p, w, ρ, k, k')
    qw ← Σᵖᵢ₌₁ w(i);  qρ ← Σᵖᵢ₌₁ ρ(i);  T₀ ← (p + qw + qρ, p)-library
    T₀ ← Parallel_Fill(T₀, w, p, 0)
    Cardinal_sort(T₀, p + 1, p + qw)
    T₁ ← ∅
    for i = 1 to k do
        T₁ ← merge (T₁, Cardinal_sort(T₀, p + 1, p + qw)[i])
    end for
    T₀ ← Parallel_Fill(T₁, ρ, p, qw)
    Cardinal_sort(T₀, p + qw + 1, p + qw + qρ)
    T₁ ← ∅
    for i = k' to qρ do
        T₁ ← merge (T₁, Cardinal_sort(T₀, p + qw + 1, p + qw + qρ)[i])
    end for
    Read(T₁)
```

The number of tubes used by the program (again, including the subroutines) is $5 + 2 \cdot \max\{q_w, q_\rho\}$, and the number of molecular operations carried out in the execution is $4p + k - k' + \frac{q_w \cdot (q_w + 5) + q_\rho \cdot (q_\rho + 7)}{2} + 1$.

The formal verification of this program is similar to the one for the Subset–Sum problem, we omit its proof since it does not show any new ideas for the verification methods for the sticker model.

6   0/1 Unbounded Knapsack Problem

Problem: *Under the same conditions as the Knapsack problem, determine a subset $B \subseteq A$ such that $\rho(B) = \max\{\rho(C) : C \subseteq A \land w(C) \le k\}$.*

A molecular solution is obtained from the solution of the bounded problem, changing the final output stage: after the sorting of the tubes regarding the function of values, we will choose the non empty tube with bigger index.

```
Unbounded_Knapsack (p, w, ρ, k)
    q_w ← Σ^p_{i=1} w(i);  q_ρ ← Σ^p_{i=1} ρ(i);  T_0 ← (p + q_w + q_ρ, p)-library
    T_0 ← Parallel_Fill(T_0, w, p, 0)
    Cardinal_sort (T_0, p + 1, p + q_w)
    T_1 ← ∅
    for i = 0 to k do
        T_1 ← merge (T_1, Cardinal_sort (T_0, p + 1, p + q_w) [i])
    end for
    T_0 ← Parallel_Fill(T_1, ρ, p, q_w)
    i ← q_ρ;  t ← 0
    Cardinal_sort (T_0, p + q_w + 1, p + q_w + q_ρ)
    while i ≥ 1 ∧ t = 0 do
        T' ← Cardinal_sort (T_0, p + q_w + 1, p + q_w + q_ρ) [i]
        if T' ≠ ∅ then: Read(T');  t ← 1
        else: i ← i − 1
    end while
```

The number of tubes used by the program is $5 + 2 \cdot \max\{q_w, q_\rho\}$, and the number of molecular operations carried out in the execution of the program, $4p + k - k' + \frac{q_w \cdot (q_w + 5) + q_\rho \cdot (q_\rho + 9)}{2}$, is easily obtained from the bounded case.

7   Conclusions

In this work, a molecular program that allows us to sort a finite family of finite sets, according to their cardinality, has been presented within the sticker model. In order to achieve this, we have used a new technique which is different from the one used by S. Roweis et al [3] to solve the Minimal Set Cover problem.

Also, we have designed original molecular programs in the aforementioned model that solve the Subset–Sum problem and the Knapsack problem (0/1 bounded and unbounded versions), which are NP–complete problems. As a subroutine, we present a molecular program (**Parallel_Fill**) within the sticker model that performs the computation of the weight of subsets regarding a given function. The molecular solutions presented of Subset Sum and Knapsack problems use a linear number of tubes, and the number of molecular operations is, basically, quadratic; nevertheless, the volume of required DNA (space complexity) to perform the computations may vary by exponential factors.

We present formal verification of the designed programs, proving soundness and completeness of these programs through the labeling of tubes and using techniques of induction.

The study of the formal aspects of molecular programs opens up a new research field for their automatic processing by means of theorem provers (ACL2, PVS, ...). The production of prototypes which are able to be executed regarding molecular computational models within the framework of theorem provers, will allow to automate both the soundness and completeness of molecular programs, which have been designed within these models.

References

1. ADLEMAN, L. On constructing a molecular computer, in *DNA based computers*, R.J. Lipton and E.B. Baum, eds., American Mathematical Society, 1996, 1–22.
2. GAREY M.R.; JOHNSON D.S. *Computers and intractability*, W.H. Freeman and Company, New York, 1979.
3. ROWEIS, S.; WINFREE, E.; BURGOYNE, R.; CHELYAPOV, N.; GOODMAN, M; ROTHEMUND, P. AND ADLEMAN, L. A Sticker–Based Model for DNA Computation, *J. Comp. Biol.* 5, 615–629, 1998

A Clause String DNA Algorithm for SAT

Vincenzo Manca[1] and Claudio Zandron[2]

[1] Università degli Studi di Pisa
Dipartimento di Informatica
Corso Italia 40, 56125 Pisa, Italy
mancav@di.unipi.it
[2] Università degli Studi di Milano – Bicocca
Dipartimento di Informatica, Sistemistica e Comunicazione
Via Bicocca degli Arcimboldi 8, 20126 Milano, Italy
zandron@disco.unimib.it

Abstract. A DNA algorithm for SAT, the satisfiability of propositional formulae, is presented where the number of separation steps is given by the number of clauses of the instance. This represents a computational improvement for DNA algorithms based on Adleman and Lipton's *extraction model*, where the number of separations equates the number of literals of the instance.

1 Introduction

Since seminal Adleman's experimental DNA solution of a Directed Hamiltonian Path Problem [1], many experiments, based on molecular biology methods, were carried on to solve *hard* combinatorial problems. In fact, DNA provides a massive computational parallelism which allows us to attack combinatorial problems that, in terms of conventional computation models, are *intractable* (technically, deterministically solvable in a time that is not polynomial with respect to the *dimension* of the instances [6]). In the area of formal language theory, a great number of theoretical studies [19,16,4], related to mathematical models of DNA recombinant behavior, were inspired by this experimental possibility. Recently, in the context of DNA computing, membrane computing, and aqueous computing the combinatorial NP-*complete* problem SAT, of satisfiability for formulae of propositional logic [6], was considered by several authors, e.g. [12,9,17,3,20,21,24,8,23]. In general, in many different fields – ranging from classical combinatorial analysis to statistical physics – there is a growing interest in using SAT as a practical tool for solving real-world problems [22,7]. In this paper we apply a sort of *duality principle* which transforms the candidate solutions expressed as *Literal Strings* in [21] into *Clause Strings*. This will allow us to reduce in a remarkable way the number of more critical DNA operations necessary to solve the propositional satisfiability, according to the Adleman and Lipton's canonical model of DNA computing [5].

N. Jonoska and N.C. Seeman (Eds.): DNA7, LNCS 2340, pp. 172–181, 2002.
© Springer-Verlag Berlin Heidelberg 2002

2   Propositional Satisfiability

SAT can be formulated in the following way. *Given a propositional formula φ, i.e. an instance for SAT, find if it can be satisfied for some values of its propositional variables (i.e. $\varphi \in SAT$).*

An equivalent formulation, directly derived by the *clause representation* of propositional formulae, can be expressed in terms of solvability of a system of boolean equations. Let us say *literal* any boolean variable or any negation of a boolean variable. Consider a system of equations over the boolean algebra of the truth values $0, 1$ with the unary operation of *negation* ($\neg$) and the binary operation of *disjunction* ($\vee$) such that $\neg 0 = 1, \neg 1 = 0, 0 \vee 0 = 0, 1 \vee 1 = 1, 0 \vee 1 = 1, 1 \vee 0 = 1$. Assume that in every equation the left member is a disjunction of literals, called a *clause*, and the right member is 1. We say *assignment* the values associated to the variables. An instance (expressed as boolean equations) belongs to SAT if there is an assignment that satisfy all equations of the system.

$$
\begin{aligned}
\neg X_3 \vee \quad X_6 \vee \quad X_8 &= 1 \\
X_2 \vee \quad X_4 \vee \quad X_8 &= 1 \\
X_3 \vee \quad X_7 \vee \quad X_{11} &= 1 \\
X_1 \vee \neg X_2 \vee \quad X_5 &= 1 \\
\neg X_5 \vee \quad X_6 \vee \quad X_{10} &= 1 \\
\neg X_3 \vee \neg X_4 \vee \neg X_{10} &= 1 \\
\neg X_4 \vee \neg X_{10} \vee \neg X_{11} &= 1 \\
X_4 \vee \neg X_5 \vee \quad X_{10} &= 1 \\
X_5 \vee \neg X_7 \vee \quad X_{11} &= 1 \\
X_3 \vee \neg X_4 \vee \neg X_9 &= 1 \\
\neg X_1 \vee \quad X_7 \vee \neg X_8 &= 1 \\
X_4 \vee \quad X_8 \vee \quad X_9 &= 1 \\
X_4 \vee \neg X_7 \vee \neg X_{10} &= 1 \\
\neg X_2 \vee \quad X_9 \vee \neg X_{11} &= 1 \\
X_1 \vee \quad X_6 \vee \neg X_8 &= 1 \\
\neg X_6 \vee \neg X_8 \vee \neg X_9 &= 1 \\
\neg X_6 \vee \neg X_8 \vee \neg X_{10} &= 1 \\
X_2 \vee \neg X_5 \vee \quad X_{11} &= 1 \\
X_1 \vee \quad X_7 \vee \quad X_{10} &= 1 \\
\neg X_1 \vee \quad X_3 \vee \neg X_9 &= 1
\end{aligned}
$$

A 3-SAT instance as a system of boolean equations

The instance considered, we call it TAMPA [14], is of type 3-SAT(11, 20) because it has 3 literals per clause, 11 variables, and 20 clauses. It derives from a 3-SAT(11, 22) instance randomly generated, that was slightly modified by deleting the clauses 5 and 12 that resulted to be tautologies (equal to 1 for any value of their variables) and by performing some other minor changes. Experiments on its DNA solvability are currently in progress at *Laboratories of Microbial Biotechnology and Environmental Microbiology* of the *Dipartimento Scientifico e Tecnologico* at the University of Verona [14].

3   The Extract Model

The canonical computation model of DNA computing remains so far the *extract model* of Adleman [1] as generalized by Lipton [12]. This model is based on two basic parts [5]: i) a test tube of DNA strands encoding a set of *candidate solutions* of the problem (usually obtained by annealing and ligation from an initial test tube encoding the data), and ii) a procedure that *extracts* the true solutions (the "good strands") from the non-solutions (the "bad strands"). The extraction procedure can be performed by a minimal set of DNA operations on test tubes: *separate, combine, detect.* The operation "separate" takes as input a test tube T and a sequence S, and produces as output a *yes-tube* containing the strands in T where S occurs as subsequence, and a *no-tube* containing the other strands of T. The "combine" operation takes as input two tubes and produces as output a single tube containing the strands of both the input tubes. The "detect" operation checks if the *final* test tube contains any DNA strands, and in that case it chooses one of them and determines the sequence of their nucleotides. Due to the DNA realization of these operations, the computational cost of a DNA computation can be identified with the number of separation steps.

Other operations can be incorporated in the model that are based on PCR, gel-electrophoresis, restriction enzymes, or more complex biotechnological protocols, but they can be viewed as additional tools for performing separation steps that are the *dominant operations* in a DNA computation.

Therefore, according to the paradigm of the extraction model, two parameters are essential in the evaluation of a DNA algorithm based on this model: i) the size of the *solution space*, that corresponds to the amount of DNA necessary to encode the set of candidate solutions, and ii) the number of separation steps necessary to get the final test tube. This means that in order to improve DNA algorithms based on the extract model there are three possible ways, that might be also integrated: 1) decreasing the size of the solution space [24], 2) finding more efficient DNA implementation of the operation "separate" [3], or 3) decreasing the number of separation steps necessary to get the final test tube [20,21]. The best solution, with respect to the third point, would be some implementation of one big-step separation that could perform all the separations in a time to a great extent independent from the instance size. Papers [20] and [21] represent attempts of separation in a constant time. But in the first case a sophisticated biotechnology is necessary that needs more experimental work in order to be applied to significant examples; in the second case, as it will shown later on, the constant time for separation is obtained by increasing too much the number of candidate solutions; therefore the problem remains of finding more efficient implementations that could balance this solution space amplification.

For the further discussion it will be useful to recall the following resolution schema of the first DNA algorithm introduced by Lipton for solving SAT [12]. Encode all possible assignments by DNA *assignment strands* where, for each propositional variable X, a DNA oligonucleotide which encodes either X or $\neg X$ occurs in the strand. When a complete pool of assignment strands is supposed to be generated, then this pool is filtered by means of separation procedures, one for each clause. For example for the clause $(\neg X_3 \lor X_6 \lor X_8)$ the strands are separated into two tubes A, B. In A are collected the strands where (the encoding of) $\neg X_3$ occurs, while in B are collected those where $\neg X_3$ does not occur. The strands of B are then separated into the tubes C, D. In C are collected the strands where X_6 occurs, while in D there are those where X_6 does not occur. Finally, the strands of D are separated into E and F, in E there are the strands where X_8 occur and in F the remaining strands of D. Then the strands of A, C, E are merged in a test tube where the assignments which satisfy the clause are collected. In general, for any clause, the strands are kept where at least one literal of the clause occurs. This means that for a clause where k literals occur we must apply k separation steps. Therefore, in the case of a 3-SAT(n, m) we need $3m$ separation steps.

4   Contact Formulations of SAT

SAT is equivalent to the *Contact Network Problem* [12]. In fact, associate to every clause C a graph with two nodes, the *source* and the *target* $S(C), T(C)$, and, for every literal in the clause, an edge connecting these nodes labeled by the literal. Let $X_1, X_2, \ldots, X_n$ and $C_1, C_2, \ldots, C_m$ be the variables and the clauses of our instance. Then, connect the target of C_i with the source of C_{i+1}, for $i = 1, \ldots, m - 1$.

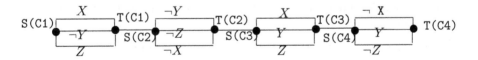

A Contact Network Instance

The following algorithm and an ingenious DNA implementation of it, which uses DNA hairpin structures, was described in [9].

Jonoska et al's algorithm:
Consider many copies of a graph G that is an instance for the Contact Network Problem. Separate these copies into two different almost "equivalent" pools A, B. Remove in all the graphs of A the edges labeled by X_1 and in the graphs of B the edges labeled by $\neg X_1$. Unite the resulting pools into a unique pool and apply the same procedure (separation, edge removing, unification) for $X_2, \ldots, X_n$. The original propositional formula can be satisfied if, at end of this process, a graph remains that connects the source of the first clause with the target of the last one.

In the following, $Lit(C)$ is the set of literals of the clause C, and $Cla(L)$ is the set of clauses where the literal L occurs. Let us call *theory* a set of clauses and *diagram* a *coherent* set of literals (no literal and its negation can occur both in a diagram). Then *Lit* and *Cla* can be extended as functions from the theories to the sets of diagrams, where a natural ordering relation is given by the usual set inclusion. It is easy to verify that in this way *Lit* and *Cla* define a *Galois correspondence* (a pair of functions between ordered sets which reverse the order). This is a special case of a more general correspondence between models and theories in the first order logic, and will be the basis of our formulation of SAT given in the next section (the central role of Galois correspondences in the solvability of algebraic equations is a well known fact which was the basis of modern algebra).

The Contact Network Problem transforms easily in the following *Literal String Problem*, that is, the problem of finding paths (if any) connecting all the clauses by using a *constructive* approach rather than a destructive one, as in Jonoska et al.'s algorithm, followed by a successive *extraction* procedure.

Assume a pair of nodes, source and target $S(C), T(C)$ for any clause C. The following algorithm was described in [21].

Sakamoto et al's algorithm:
Assume a pool of elementary graphs, each one constituted by an edge labeled by a literal L that connects the nodes $S(C), T(C)$ if L belongs to $Lit(C)$. Then start a linking process that constructs a literal string by adding an edge between a target of a clause and the source of the next clause (w.r.t. a prefixed order of the clauses). After that, a *decimation* of the paths is performed by applying a test which controls whether a given path is *incoherent*, that is, whether a literal label occurs in both its positive and negative forms. Any literal string that results to be *incoherent* is then removed. The initial propositional formula is satisfiable if, at end, some *coherent* contact paths remain after the elimination of the incoherent ones.

The following diagram is relative to a SAT(3,4) instance.

$$S(C1) \quad \overline{\underset{\alpha \,\in\, Lit(C1)}{T(C1)}} \quad S(C2) \quad \overline{\underset{\beta \,\in\, Lit(C2)}{T(C2)}} \quad S(C3) \quad \overline{\underset{\gamma \,\in\, Lit(C3)}{T(C3)}} \quad S(C4) \quad \overline{\underset{\delta \,\in\, Lit(C4)}{T(C4)}}$$

Literal Strings

The strings: $(X\ \neg Y\ X\ Z)$, $(\neg Y\ \neg Y\ X\ Z)$, $(Z\ \neg Y\ X\ Z)$ are some coherent literal strings, in the case the clauses are those given in the contact network instance above.

It is easy to realize that for a 3-SAT instance the solution space of this algorithm is 3^m where m is the number of clauses. A DNA solution of this algorithm has been implemented for a 3-SAT(6, 10) instance in [21]. This algorithm uses a more complex separation method that is based on hairpin formation rather than on hybridization-affinity. In this way the separation procedure can be made in a constant time (by using traditional separations, $4n$ steps would be necessary, where n is the number of variables). However, although the method is really interesting, and requires a constant time for separation, in its actual form the size of the solution space is a strong limitation for its scale up. In fact, typically the number of clauses is about 3 or 4 times the number of variables.

5   Clause String Formulation of SAT

Let us consider another *contact* formulation of the propositional satisfiability, introduced in [13], that is related to the Galois correspondence (Lit, Cla) indicated in the previous section.

Consider the pair of nodes $S(X), T(X)$ (source and target) for each variable X. A dual perspective of the literal string formulation of SAT leads to the following algorithm.

A clause string algorithm:
Assume a pool of elementary graphs where $S(X), T(X)$ are connected either with an edge labeled by $Cla(X)$, or with an edge labeled by $Cla(\neg X)$ (the clauses where X occurs and those where $\neg X$ occurs, respectively). Then start a linking process building the clause strings where the target of a variable is linked to the source of the next variable (according to a prefixed order of the variables).

The clause strings coincide with the paths of the graph above that connect one node of the first variable with one of the last variable. A clause string is *complete* if each clause belongs to some set that occurs as a label. Remove the clause strings that are not complete. If some complete clause strings remain, they are solutions of the considered instance of SAT.

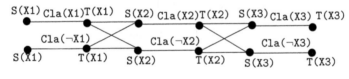

Clause Strings

It is easy to realize that the solution space of the clause string algorithm is 2^n where n is the number of variables. Moreover, as it will be completely clear in the following section, the number of separations that are necessary in order to extract the solutions are m, that is the same number of the clauses of the instance. This fact implies the possibility of a scale up in the DNA solution of SAT, independently from any DNA implementation of the separation steps.

Another advantage of representing candidate solution by means of clause strings is due to the fact that a clause is encoded several times in a clause string, as much as it can belong to $Lit(L)$ for several literals L (3 in the average case, for 3-SAT). This aspect will give for free an increasing in the probability of the expected hybridizations necessary for separations, according to the analysis developed in [2] for *double encoding*.

In order to define our DNA algorithm we use a further formulation of SAT. In fact, we encode the clause string formulation of SAT in terms of the *Bipartite Covering Problem* problem (BCP) that can be stated in the following way.

Given a finite set C and n pairs of subsets of C: $A_1/B_1, \ldots, A_n/B_n$ such that $A_i \cap B_i = \emptyset$ for $i = 1, \ldots, n$, find (or at least, say if it is possible to find) n sets $Y_1, \ldots, Y_n$ such that $C = Y_1 \cup \cdots \cup Y_n$, and either $Y_i = A_i$ or $Y_i = B_i$.

It is a simple exercise to transform any instance of SAT into an equivalent instance of BCP and viceversa. The following is the BCP formulation of the 3-SAT(11,20) instance TAMPA, already given as system of boolean equations. For the sake of brevity, we use the indexes of clauses instead of clauses, following the order of the boolean equations; however, in order to keep the original indexes of the randomly generated instance, numbers 5 and 12 are skipped (for $i = 1, \ldots, 11$, we put $Ai = Cla(Xi)$ and $Bi = Cla(\neg Xi)$).

$$
\begin{array}{rcrcr}
A1/B1 & = & 4,17,21 & / & 13,22 \\
A2/B2 & = & 2,20 & / & 4,16 \\
A3/B3 & = & 3,11,22 & / & 1,7 \\
A4/B4 & = & 2,9,14,15 & / & 8,7,11 \\
A5/B5 & = & 4,10 & / & 6,9,20 \\
A6/B6 & = & 1,6,17 & / & 18,19 \\
A7/B7 & = & 3,13,21 & / & 10,15 \\
A8/B8 & = & 1,2,14 & / & 13,17,18,19 \\
A9/B9 & = & 14,16 & / & 11,18,22 \\
A10/B10 & = & 6,9,21 & / & 7,8,15,19 \\
A11/B11 & = & 3,10,20 & / & 8,16 \\
\end{array}
$$

The BCP Instance

6   The DNA Algorithm

Now we encode A1, B1, ..., A11, B11 with DNA *dominoes*, that is, DNA molecules constituted, for every $i = 1, \ldots, 11$, by two different central parts corresponding to Ai and Bi, but having the same left sticky end Xi, and the same right sticky end –Xi+1. For every $i = 1, \ldots, 11$, Xi and –Xi are complementary DNA single strands. This means that we have, for any variable Xi, two dominoes with sticky ends (Xi, –Xi+1) where in the middle there are encoded either all the clauses satisfied by Xi or all the clauses satisfied by ¬Xi, respectively. For example, the dominoes for A3 and B3 have the following shape:

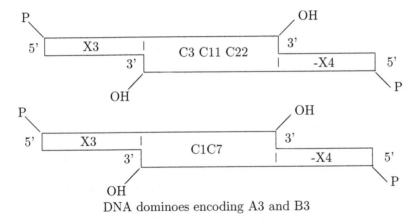

DNA dominoes encoding A3 and B3

The sticky ends may correspond i) either to sites of non-palindromic restriction enzymes obtained by bigger purchased oligos after amplification and digestion with the enzymes relative to the linkers, ii) or to dominoes obtained from single stranded DNA molecules as in the original Adleman's approach (see [25] for a detailed analysis of the second method).

Now, put in a test tube many copies of the dominoes A1, B1, ..., An, Bn. After annealing and ligation of complementary sticky ends, we expect to obtain a pool of linked DNA dominoes (actually, X1 and X12 do not perform any link, but can be useful for amplification by PCR).

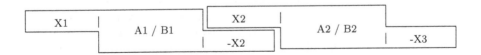

A DNA string with two dominoes

If we indicate by the oligo Li the pairing of the two complementary strands –Xi and Xi, which acts as a linker between two dominoes, then a long string of dominoes where all the linkers occur will have the following shape:

Linked DNA dominoes

Our algorithm is based on the validity of the following proposition that follows from the analysis so far developed.

Proposition
DNA strings where all the clauses are encoded represent solutions of our instance.

7   Conclusions

The proposed algorithm is really different from Lipton's method of generating assignments and then filtering them (which a lot of DNA algorithms for SAT are based on). But, it is also really different from the approach of generating 3^m literal strings (m clauses, where typically m is three or four times the number of variables) and filtering them by 'coherence' [21].

In fact, we have shown that we can generate clause strings and then filter them by 'completeness'. This leaves the dimension of the solution space the same as in Lipton's method, that is 2^n (n variables), but in this case m separations are enough, whereas the number of separations necessary in the Lipton algorithm is given by the number of literals (or the number of binary connectives plus 1). This means that in the case of a 3-SAT instance the number of dominating operations in our case is one third with respect to the algorithms based on Lipton's schema. This analysis is independent from any particular DNA implementation that our algorithm is based on. In other words, any DNA algorithm for SAT that is based on: 1) a ligation protocol of small DNA 'pieces', and 2) a separation protocol for filtering the solutions in the pool of candidate solutions, can be improved if candidate solutions express clause strings rather than assignments or literal strings. In conclusion, in the class of DNA algorithms for SAT based on Adleman and Lipton's extract model the clause strings approach implies a direct scale-up in the size of DNA solvable instances. Other more theoretical aspects of the clause string representation of SAT are related to other formulations of SAT in terms of matrices and membrane systems that are under current investigation [15] and that could turn to be relevant for DNA computing too.

References

1. L. M. Adleman, Molecular Computation of solutions to combinatorial problems, Science, Vol. 266, pp. 1021–1024, November 11, 1994.
2. D. Boneh, C. Dunworth, R. J. Lipton, J. Sgall, Making DNA computers error resistant, in [18] pp. 163–170, 1999.
3. R. S. Brainch, C. Johnson, P.W.K. Rothemund, D. Hwang, N. Chelyapov, L. M. Adleman, Solutions of a Satisfiability Problem on a Gel-Based DNA Computer, Sixth International Meeting on DNA Based Computers, Leiden Center for Natural Computing, A. Condon G. Rozenberg (eds.), Leiden, 2000.
4. C. Calude, Gh. Păun, *Computing with Cells and Atoms*, Taylor and Francis, London, 2000.
5. K. Chen, E. Winfree, Error Correction in DNA Computing: Misclassification and Strand Loss, in [26] pp. 49–63, 2000.

6. M. R. Garey, D. S. Johnson, Computers and Intractability, Freeman, San Francisco, 1979.

7. C. P. Gomes, B. Selman, N. Crator, H. Kautz, Heavy-tailed phenomena in satisfiability and constraint satisfaction problems, J. of Automated Reasoning, Vol. 24 (1/2), pp. 67–100, 1999.

8. T. Head, X. Chen, M. J. Nichols, M. Yamamura, and S. Gal, Aqueous Solutions of Algorithmic Problems: emphasizing knights on a 3×3, in [10] pp. 219–230, 2001.

9. N. Jonoska, S. A. Karl, M. Saito, Three Dimensional DNA Structures in Computing, Biosystems, Vol. 52, 242–245, 1999.

10. N. Jonoska, N. Seeman (eds.), 7th International Meeting on DNA Based Computers, Preliminary Proceedings, Tampa, FL (U.S.A.), 2001.

11. L. Landweber, E. Baum, DNA Based Computers II, DIMACS Series 44, American Math. Society, Providence, RI, 1999.

12. R. Lipton, DNA Solutions of hard computational problems, Science, Vol. 268, 242–245, 1995.

13. V. Manca, Monoidal systems and membrane systems, in Pre-proc. Workshop on Multiset Processing, Curtea de Arges, Romania, TR 140, CDMTCS, Univ. Auckland (New Zealand), pp. 176–190, 2000.

14. V. Manca, and Di Gregorio S., Lizzari D., Vallini G., Zandron C., A DNA Algorithm for 3-SAT(11,20), in [10], pp. 167–178, 2001.

15. V. Manca, Membrane Algorithms for Propositional Satisfiability, in Pre-proc. Workshop on Membrane Computing, Curtea de Arges, Romania, TR 17, GRLMC, Univ. Rovira i Virgili, Tarragona (Spain), pp. 181–192, 2001.

16. Gh. Păun, G. Rozenberg, A. Salomaa, DNA Computing: New Computing Paradigms, Springer-Verlag, Berlin, 1998.

17. Gh. Păun, P Systems with Active Membranes: Attacking NP Complete Problems, J. Automata Languages and Combinatorics, 6, 1, 2001.

18. H. Rubin, D. Wood, DNA Based Computers III, DIMACS Series 48, American Math. Society, Providence, RI, 1999.

19. G. Rozenberg, A. Salomaa, eds., Handbook of Formal Languages, Springer-Verlag, Heidelberg, 1997.

20. K. Sakamoto, H. Gounzu, D. Kiga, K. Komiya, H. Gouzu, S. Yokoyama, T. Yokomori, S. Ikeda, H. Sugiyama, M. Hagiya, State transitions by molecules, Biosystems, Vol. 52, pp. 81–91, 1999.

21. K. Sakamoto, H. Gounzu, K. Komiya, D. Kiga, S. Yokoyama, T. Yokomori, M. Hagiya, Molecular Computation by DNA Hairpin Formation, Science, Vol. 288, pp. 1223–1226, May 19, 2000.

22. K. Selman, H. Kautz, B. Cohen, Local Search Strategies for Satisfiability Testing, in: Cliques, Coloring, and Satisfiability, DIMACS Series in Discrete Mathematics and Theoretical Computer Science, D. S. Johnson, M. A. Trick, eds., vol. 26, AMS, 1996.

23. Y. Takenaka, A. Hashimoto, A proposal of DNA computing on beads and its application to SAT problems, in [10], pp. 331–339, 2001.

24. H. Yoshida and A. Suyama, Solutions to 3-SAT by breadth first search, in [26] pp. 9–22, 2000.

25. M. Yamamoto, J. Yamashita, T. Shiba, T. Hirayama, S. Takiya, K. Suzuki, M; Munekata, and A. Ohuchi, A study of hybridization process in DNA computing, in [26], pp. 101–110, 2000.

26. E. Winfree E., D. K. Gifford, DNA Based Computers V, DIMACS Series 54, American Math. Society, Providence, RI, 2000.

A Proposal of DNA Computing on Beads with Application to SAT Problems

Takenaka Yoichi and Hashimoto Akihiro

Department of Informatics and Mathematical Science, Osaka University,
Machikaneyama 1-3, Toyonaka, Osaka 560-8531, Japan,
{takenaka, hasimoto}@ics.es.osaka-u.ac.jp

Abstract. We propose a strategy using tiny beads for DNA computing. DNA computing is a means of solving intractable computation problems such as NP-complete problems. In our strategy, each bead carries multiple copies of a DNA sequence, and each sequence represents a candidate solution for a given problem. Calculation in our strategy is executed by competitive hybridization of two types of fluorescent sequences on the beads. One type of fluorescent sequences represents a constraint that has not been satisfied, and the other type a constraint that has been satisfied. After competitive hybridization, beads with only the latter type of fluorescent sequences hold "true" solutions. To extract the beads from the test tube, we use fluorescent-activated cell sorter.

We describe the approach to DNA computing on beads through SAT problems. The SAT problem is an NP-complete problem in Boolean logic. Using Megaclone, which allows DNA strands to be attached to beads, we show that DNA computing on beads can solve up to 24 variables.

1 Introduction

Since Adleman [1], and subsequently Lipton [2], demonstrated the possibility of solving NP-complete problems by using DNA [3], biomolecular computation has become a new vista of computation that bridges computer science and biochemistry. They demonstrated an experimental technique for simple examples of the Hamiltonian path and Satisfiability (SAT) problems. Since then, approaches of DNA computing to SAT problems have been investigated. In 1998 and 2000, Smith and Liu proposed a surface-based approach to DNA computation [4,5]. In 2000, Sakamoto et al. proposed an approach using DNA hairpin formation where the computational paradigm differs from the Adleman-Lipton paradigm [8]. Approaches of DNA computing to the other problems have also been investigated [9,10,11].

In this paper, we propose a new DNA computational strategy using tiny DNA beads. A tiny DNA bead is approximately a $5\mu m$ diameter glycidal methacrylate bead which carries multiple copies of a DNA sequence [15]. In typical cases, each bead carries about 10^6 copies of a DNA sequence [15]. In our proposal, we assume each DNA bead represents a candidate solution for a given problem. Calculation in our strategy is executed by competitive hybridization of two types

N. Jonoska and N.C. Seeman (Eds.): DNA7, LNCS 2340, pp. 182–190, 2002.
© Springer-Verlag Berlin Heidelberg 2002

of fluorescent sequences on the beads. One type of fluorescent sequences suggests that at least one of the constraints of the problem is not satisfied, and the other type of sequences suggests that at least one of the constraints is satisfied. After competitive hybridization, beads with only the latter type of fluorescent sequences hold true solutions. To extract the beads from the test tube, we use fluorescent-activated cell sorter.

In the following sections, we first show our approach to DNA computing on beads through a SAT instance. Secondly, we explain the experimental techniques used in realizing our approach.

2   DNA Computing on Beads

In DNA computing on beads, each bead represents a candidate solution for a given problem, and a set of beads include all the candidate solutions. The calculation is executed by competitive hybridization to the DNA strands on the beads. We consider *candidates* to be the DNA strands that encode candidate solutions, and *co-candidates* to be the complementary strands of candidates. The DNA computing on beads requires the following 6 steps.

1. Create beads so that they contain the entire candidate solutions.
2. Synthesize all the *co-candidates* that are labeled with a fluorophore.
3. For each constraint of the problem, synthesize *co-candidates* that don't satisfy the constraint and label them with another fluorophore.
4. Hybridize competitively the *co-candidates* to the beads.
5. Extract beads by the fluorescence-activated cell sorter.
6. Read out the strands.

We describe these 6 steps through a SAT instance. The SAT problem is an NP-complete problem in Boolean logic [3]. An instance of the SAT problem consists of a set of Boolean logic variables separated by the logical OR operation (denoted by $\vee$, where $u \vee v = 0$ if and only if $u = v = 0$) within clauses, and with the clauses separated by the logical AND operation (denoted by $\wedge$, where $u \wedge v = 1$ if and only if $u = v = 1$). The problem is to determine whether there is an assignment of the variables that simultaneously satisfy each clause in a given instance of the problem [3]. We use the following SAT instance for the explanation.

$$(w \vee x \vee y \vee z) \wedge (\neg w \vee \neg y \vee z) \wedge (\neg x \vee y) \wedge (\neg w \vee z). \tag{1}$$

As this instance has four variables, there are a total of 2^4 or 16 candidate solutions.

In the first step, we create 16 kinds of beads. Each bead carries multiple copies of one DNA strand which represents a assignment of the variables, and the whole beads contains the solution space (See Fig. 1). In the Fig. 1, the 4bit numbers (0000 $\sim$ 1111) that are attached to each bead represent the DNA strands that encode each assignment of the variables (wxyz) = (0000 $\sim$ 1111).

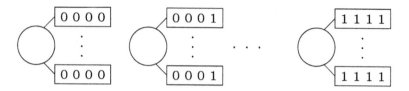

Fig. 1. Beads containing the solution space

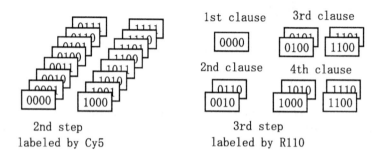

Fig. 2. Complement strands made in the 2nd and 3rd step

In the second step, we synthesize the complementary strands of the candidates, and label them with a fluorophore. With loss of generality, we use Cy5 as the fluorophore. Fig. 2 shows the complementary strands to be synthesized.

Constraints of the SAT problem are clauses to be satisfied. In the third step, for each clause, we synthesize complementary strands of candidates that don't satisfy the clause. Table 1 shows the truth table of the clauses. In the table, "F" shows the assignment of the variable not satisfing the clause. For legibility, we remain blank when the assignment leads the clause true. For the 1st clause, we make complementary strands of (0000). For the 2nd clause, complementary strands of (1010) and (1110) are created. Fig. 2 shows the complementary strands made in this step. The 3rd and 4th clauses are not satisfied by the solution candidate (1100). In this case, the complementary strands of (1100) are synthesized twice. The complementary strands are labeled with another fluorophore. With loss of generality, we use R110 as the fluorophore.

In the fourth step, we competitively hybridize the strands synthesized in the 2nd and 3rd steps to the strands on the beads. After competitive hybridization, the beads with the true assignment hold only Cy5 labeled strands, and the beads with the false assignment hold both Cy5 labeled and R110 labeled strands. The amount of R110 on the bead grows proportionally to the number of non-satisfied clauses. The ideal proportion of Cy5 and R110 on the beads with each assignment is shown in Table 1.

In the fifth step, we regard the beads as cells and extract the beads labeled only with Cy5 using the fluorescence-activated cell sorter. Each extracted bead

Table 1. Truth table of the clauses and the proportion of fluorophores on the beads

variables (wxyz)	1st clause $(w \vee x \vee y \vee z)$	2nd clause $(w \vee \neg y \vee z)$	3rd clause $(\neg x \vee y)$	4th clause $(\neg w \vee z)$	Cy5:R110
(0000)	F				1:1
(0001)					1:0
(0010)		F			1:1
(0011)					1:0
(0100)			F		1:1
(0101)			F		1:1
(0110)		F			1:1
(0111)					1:0
(1000)				F	1:1
(1001)					1:0
(1010)				F	1:1
(1011)					1:0
(1100)			F	F	1:2
(1101)			F		1:1
(1110)				F	1:1
(1111)					1:0

holds a DNA sequence that encodes one of the solutions. After the extraction, we read out the DNA sequences to know what assignments are the solutions in the final step.

In the following section, we introduce two techniques, Megaclone [15] and MPSS (massively parallel signature sequencing) [13]. These techniques have been developed by LYNX therapeutics, Inc. to analyze gene expression. With these techniques, we can implement our DNA computing on beads simply and easily.

3   Techniques to Realize

3.1   Megaclone

Megaclone [15] enables DNA strands to be attached onto the surfaces of 5-μm beads. Each bead carries about 10^6 copies of a specific DNA strand, which are called tag sequences, or simply tags. The tags are composed of eight 4-mer "words" and are synthesized by eight rounds of "Mix and Divide" combinatorial synthesis on glycidal methacrylate microbeads (See Fig. 3). Therefore, the repertoire of tags are up to $8^8 = 2^{24} = 16,777,216$. The eight 4-mer words are TTAC, AATC, TACT, ATCA, ACAT, TCTA, CTTT, and CAAA. Each word uses only three (A, T, and C) of the DNA bases and differs from all of the other words in three of the four bases. Limiting the composition of the four-nucleotide words to three bases eliminates self-complementarity within any sequence made

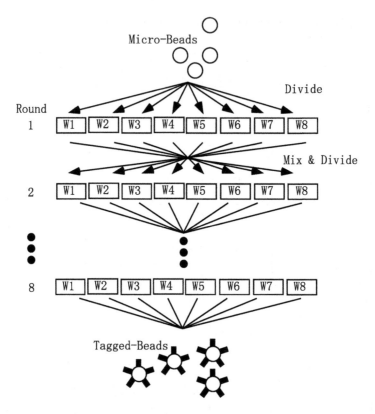

Fig. 3. Tags are synthesized by eight rounds of combinatorial synthesis, where in each word, w1-w8, is added in a separate column of a DNA synthesizer.

up of the 4-mer words. We show an example of one tag and its complement strand below.

$$5' - TACT.TTAC.ACAT.ATCA.CTTT.CTTT.CAAA.AATC - 3'$$
$$3' - ATGA.AATG.TGTA.TAGT.GAAA.GAAA.GTTT.TTAG - 5'$$

3.2   MPSS

MPSS is a particularized method for sequencing DNA whose thousands of copies are attached on a 5μm diameter bead. Brenner et al. [13] assemble a planar array of millions of beads in a flow cell at a density greater than 3×10^6 beads/cm^2. Sequences of the free ends on each bead are simultaneously analyzed using a fluorescence-based signature sequencing method that does not require DNA fragment separation. The signature sequences of 16-20 bases are obtained by repeated cycles of enzymatic cleavage with a type IIs restriction endonuclease, adaptor ligation, and sequence interrogation by encoded hybridization probes. The details of MPSS are shown in [13].

4   A Design of DNA Computing on Beads through SAT

In this section, we design the DNA computing on beads using the Megaclone and the MPSS. We use tags in the Megaclone as the DNA strands that encode candidate solutions of the SAT problem (1).

First, we encode variables to DNA sequences. Because Megaclone has 8 kinds of words, each word can encode three variables. Therefore we use tags made up of two words to compose four-variable candidate solutions. The first word of the tag encodes the assignments of three variables w, x, and y, and the second word of the tag encodes the assignment of a variable z. The words and corresponding assignments of variables are shown in Table 2. For an example, the assignment $(wxyz) = (1010)$ is encoded to $5' - TACT.AATC - 3'$.

We describe 6 steps for DNA computing on beads as follows:

1. Create beads with candidate solution strands by Megaclone. In the Mix & Divide operations, we use eight words (w1 $\sim$ w8) in the first round, and two words (w1,w2) in the second round.
2. Synthesize all the co-candidates. First we synthesize a set of beads with candidate solutions, that are the same as the set of beads created in the first step. Next, we detach the strands from beads and amplify them by PCR to make complementary strands. In PCR amplification, we label the complementary strands by Cy5.
3. Synthesize co-candidates that don't satisfy a clause using Mix & Divide operation. Because the 1st clause $(w \vee x \vee y \vee z)$ isn't satisfied by the assignment (0000), we use only one word "w1" at the first and second round of the Mix & Divide operation. In the case of the 3rd clause, which the assignments (0100), (0101), (1100), and (1101) don't satisfy, we use two words (w3, w7) at the first round and use two words (w1,w2) at the second round. The strands are detached, amplified and labeled by R110.
4. Hybridize competitively the strands sinthesized in 2nd and 3rd steps to the strands on the beads. The protocol of the competitive hybridization on beads is described in [15].
5. Extract the beads labeled only with Cy5 by a fluorescence-activated cell sorter. The extracted beads hold the solutions.
6. Read out the sequences on the extracted beads by MPSS.

The Mix & Divide operation of Megaclone takes $O(N \times M)$ time to execute the first, second, and third steps for N-variable, M-clause SAT problems. The fourth step needs constant time. However, the fifth step needs $O(2^N)$ time with existing techniques on DNA beads.

$8^8 = 2^{24}$ tag variations are available in the Megaclone now. Therefore, DNA computing on beads realized by Megaclone can easily expand to the instances of SAT problems whose number of variables is up to 24.

Table 2. Words and corresponding assignments of variables

word	sequence	1st word (wxy)	2nd word (z)
w1	AATC	(000)	(0)
w2	ACAT	(001)	(1)
w3	ATCA	(010)	-
w4	CAAA	(011)	-
w5	CTTT	(100)	-
w6	TACT	(101)	-
w7	TCTA	(110)	-
w8	TTAC	(111)	-

An experiment for a sensitivity test of competitive hybridization on beads was performed by Brenner[15]. In the experiment, two fluorescent probes in a ratio of 1:1, 1:0 were hybridized competitively to the strands on the beads. The result shows that the beads of 1:1 and 1:0 were always classifiable by a cell sorter. This experiment indicates that our DNA computing on beads can distinguish the beads with the solution from the beads with the assignment which doesn't satisfy only one clause.

5   Conclusion

We have described a new approach to the DNA computing using tiny beads. The features of our DNA computing are as folows:

1. Each bead represents one of the candidate solutions,
2. We use competitive hybridization for the calculation,
3. We extract the solution using a fluorescence-activated cell sorter.

The advantages of our DNA computing on beads are as follows:

First, our DNA computing on beads can solve a 24 variable SAT problems with simple encoded DNA sequence. Only $O(N \times M)$ time is required to synthesize a set of DNA sequences for N-variable and M-clause SAT problems.

Second, extracting the solution and reading out the DNA sequence by a cell sorter and MPSS is simple. In Smith and Liu's DNA computing, the readout operation is composed of PCR and hybridization to an addressed array [4,5]. Their methods may become difficult to read out when the size of the problem becomes large.

The disadvantages of DNA computing on beads are as follows:

First, DNA computing on beads essentially needs larger volume than other DNA computings. We use tiny $5\mu m$ diameter beads, but the beads are tremendously larger than DNA strands. This means the DNA computing on beads is more sensitive to the physical limitation of the volume (or exponential curse) than other DNA computings. Moreover, the use of beads may cause a loss in

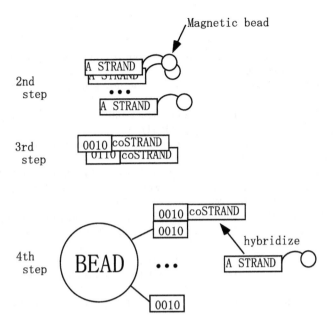

Fig. 4. How to avoid applying all beads to cell sorter

parallelism due to the size of the beads. we have two countermeasures to ease
the exponential curse. One way is to use smaller beads, and the other is to apply
Yoshida's breadth first search algorithm for DNA computing [16] that is based
on dynamic programming. These methods will relieve the disadvantages caused
from the size of beads.

The cell sorter used in the fifth step causes the second disadvantage. The
cell sorter selects the beads sequentially, not in parallel. Therefore, if we apply
all the beads to the cell sorter, no difference is detected from searching all the
solution sequentially. To avoid this situation, we want to use very small (tens
of nm diameter) magnetic beads. The outline of the idea is as follows: (also see
Fig. 4).

1. Determine a specific DNA strand to use.
2. Attach the copies of the specific DNA strand to the magnetic beads.
3. Add the complement strand of the specific DNA strand to the strands made
   in the third step of DNA computing on beads.
4. After fourth step of DNA computing on beads, hybridize the DNA strands
   on the magnetic beads to the DNA strands on the beads.
5. Before using a cell sorter, eliminate the beads with the magnetic beads by
   magnet.

With some advantages and disadvantages, we believe that DNA computing
on beads will become an admirable strategy in the future.

References

1. Adleman, L.: Molecular computation of solutions to combinatorial problems. Science **266(11)**, (1994) 1021–1024
2. Lipton, R.: DNA solutions of hard computation problems. Science **268(11)**, (1995) 542–545
3. Garey, R., Johnson, S.: Computers and Intractability, a guide to the theory of NP-completeness. Freeman and Company (1991)
4. Smith L. M. et al.: A surface-based approach to DNA computation. J. Computational Biology **5**, (1998) 255–267
5. Liu, Q. et al.: DNA computing on surfaces. Nature **403**, (2000) 175–179
6. Frutos, A. G. et al.: Demonstration of a word design strategy for DNA computing on surfaces. Nucl. Acid. Res. **25**, (1997) 4748–4757
7. Frutos, A. G., Smith, L. M., Corn, R. M.: Enzymatic ligation reactions of DNA "words" on surfaces for DNA computing. J. Am. Chem. Soc. **120**, (1998) 10277–10282
8. Sakamoto, K., Gouzu, H., Komiya, K., Kiga, D., Yokoyama, S., Yokomori, T., Hagiya, M.: Molecular computation by DNA Hairpin formation. Science **288**, (2000) 1223–1226
9. Guanirieri, F., Fliss, M., Bancroft, C.: Making DNA add. Science **273**, (1996) 220–223
10. Ouyang, Q., Kaplan, P. D., Liu, S. Libchaber, A.: DNA solution of the maximal clique problem. Science **278**, (1997) 446-449
11. Winfee, E., Liu, F. R., Wenzler, L. A., Seeman, N. C.: Design and self-assembly of two-dimensional DNA crystals. Nature **394**, (1998) 539–544.
12. Paun, G., Rozenberg, G., Salomaa, A.: DNA computing: new computing paradigms. Springer Verlag, (1998)
13. Brenner, S., et al.: Gene expression analysis by massively parallel signature sequencing (MPSS) on microbead arrays. Nature Biotechnology **18**, (2000) 630–634
14. Ausiello, G., Crescenzi, P., Gambosi, G., Kann, V., Marchetti-Spaccamela, A., Protasi, M.: Complexity and Approximation: combinatorial optimization problems and their approximability propoerties. Springer Verlag, (1999)
15. Brenner, S., et al.: In vitro cloning of complex mixtures of DNA on microbeads: physical separation of differentially expressed cDNAs. PNAS **97**, (2000) 1665–1670
16. Yoshida, H., Suyama, A.: Solution to 3-SAT by breadth first search. Fifth International Meeting on DNA Based Computers, (1999) 9–20

Aqueous Solutions of Algorithmic Problems: Emphasizing Knights on a 3 × 3

Tom Head[1][*], Xia Chen[2], Matthew J. Nichols[2], Masayuki Yamamura[3], and Susannah Gal[2]

[1] Department of Mathematical Sciences &
[2] Biological Sciences Department
Binghamton University
Binghamton, New York 13902-6000, USA
[3] Computational Intelligence and System Science
Tokyo Institute of Technology
4259 Nagatsuta, Yokohama, 226 JAPAN

Abstract. A pattern for performing several DNA computations is outlined using the aqueous approach, the essence of which is *writing on molecules dissolved in water*. Four of the indicated computations have been carried out in wet labs in the aqueous style. As an illustration, gel photos will be exhibited that confirm the correctness of a small SAT computation. Emphasis will be placed on the aqueous approach, now in progress, to the problem of producing the set of all patterns in which knights can be placed on a 3 × 3 chessboard with no knight attacking another. Currently the writing technology used is based on molecular biology. In the future we hope that light can replace biochemistry as the writing procedure.

1 Introduction

What we call aqueous computing was developed in the wake of L. Adleman's initiation of DNA computing [A'94]. The aqueous approach has been indicated in a molecular context in [H'00], exposited in a computational architectural context in [H'01c], and surveyed in [H'01a]. The first aqueous computation to be completed was done in Leiden [HRBBLS'00]. Our initial wet lab efforts in Binghamton were reported in [HYG'99] and [YHG'00]. The present article has two major purposes: to indicate how easily aqueous computations can be planned and related to one another (as in Section 4) and to present our laboratory work on the 3 × 3 knight's problem (in Section 5).

In Section 4 we show how a half dozen DNA computations that have been carried out in various laboratories can be done in the aqueous fashion. Moreover, this exposition shows that various DNA computations may be carried out in a manner that allows partial results from one computation to be used in other

---

[*] All correspondence should be sent to the first author, whose e-mail address is: tom@math.binghamton.edu

N. Jonoska and N.C. Seeman (Eds.): DNA7, LNCS 2340, pp. 191–202, 2002.
© Springer-Verlag Berlin Heidelberg 2002

computations. The common parts from several computations are 'factored out' to produce a tree diagram (Fig. 1) for doing these computations with a minimum of repetition. The presentation also indicates the uniformity that aqueous computing provides for treating various problems.

Two of the indicated problems have been solved in the aqueous manner in Leiden. The three-variable SAT problem indicated in [HYG'99] and [YHG'00] has been completed in Binghamton. The confirming gel photos for this SAT will be presented as a demonstration of a reading procedure that is an alternative to sequencing. At this writing the Binghamton lab has made major progress in using the aqueous approach to determine all the patterns in which knights can be placed on a 3 × 3 chessboard. This is the problem treated previously by the Princeton group [FCLL'00]. In Section 5 we present our aqueous program for solving this problem and the laboratory work in progress.

2   The Concept of Aqueous Computing

In aqueous computing we choose a water-soluble molecule on which a set of identifiable locations can be specified at which we can provide a technology for making alterations. We regard this molecule as an analog of a memory register in an ordinary computer. We dissolve a vast number of these molecules in water, and regard the initial state of each of the specified locations on each of the molecules as expressing one of the bits, 0 or 1, whichever we prefer. We compute by altering these bits by the chosen technology. We say that *we write on the molecules* as we alter bits. The first phase of computation consists of a sequence of writing steps. At the end of this phase all molecules would display the same pattern, except for the fact that we subdivide the aqueous solution into separate portions and write at different locations on the molecules in the different portions. The separate portions are then re-united before the next writing step. At the completion of the writing phase it is necessary to determine the result of the computation. In this phase we say we *read* the (appropriate) molecules.

There is no intrinsic connection between biochemistry and the concept of aqueous computing. That aqueous computing has thus far been implemented using DNA molecules and biochemistry is a consequence of its being initiated in the wake of [A'94] and of the previous interest of one of the authors in restriction enzymes [H'87] [HPP'97]. Thus we reiterate that to compute according to the aqueous computing concept requires only the following four mutually dependent choices:

1. a molecule that has a specifiable set of mutually distinguishable local features,
2. water (or other appropriate fluid in which the chosen molecule is soluble),
3. a technology for modifying the specified local features of the molecule, and
4. a technology for detecting the condition of the specified local features of the molecule.

The crucial role of water (or a substitute fluid) is emphasized: It disperses the molecules so that they do not block access to each other when the modification procedures are applied. This also allows the modifications of the molecules to be made in parallel. The dispersal of the molecules allows the assumption that each molecular variety is uniformly distributed throughout the solution. This allows the further assumption that, when the content of a test tube is divided into two or more tubes, each of the resulting tubes contains the same molecular varieties.

3   The Implementation Scheme Chosen Here

As the memory register molecule we chose a small circular cloning plasmid. In Binghamton we have used one of the pBluescript plasmids. For locations at which to write on these plasmids, we choose a set of restriction enzyme sites that lie in the multiple cloning site (MCS) of the plasmid. We compute in water (or appropriate buffers). We write at a specific restriction enzyme site by *locking* the site, i.e., by modifying the site so that the associated enzyme will no longer cut. Probably the easiest way to lock a site would be by methylation, the technique used by living cells [C'75]. We have considered advantages and disadvantages of locking by methylation in [H'01b]. We do not use methylation here. Instead, we have used a more complicated locking process that provides two valuable features not provided by methylation: First, the locking process allows amplification in bacteria of the plasmids such that the locked/unlocked state of each site is preserved. Second, each time a site is locked, the circumference of the plasmid is altered by a fixed amount. For several algorithmic problems, the molecules encoding the solutions will lie in either the longer plasmids or the shorter plasmids found at the end of the writing phase. In such cases gel separations are valuable in the reading phase. Finally, the reading technology may involve the cloning of plasmids in bacteria followed by either sequencing or by determining which sites are locked by applying the associated enzymes. Gel separations can precede cloning when applicable.

The locking process we choose consists of three basic steps. This locking scheme requires that no sites be used for enzymes producing blunt ends. There is additional advantage in using only sites that yield 5'-overhangs all of the same length. Thus far we have used only sites that produce four-base 5'-overhangs. To lock a specific 4-base 5'-overhang site on such plasmids: (1) Linearize the plasmids by cutting with the enzyme associated with the site. (2) Fill-in the single-stranded overhangs to produce fully double-stranded molecules by using a DNA polymerase. (3) Re-circularize the linear molecules by blunt end ligation. Notice that each such locking operation increases the circumference of the plasmids being locked by four base pairs. Such locking operations, affecting the MCS region only, do not interfere with the process of duplication of these plasmids in bacteria.

4   A Uniform Approach to Several Algorithmic Problems

We have participated now in four different wet lab DNA computations: two in Leiden and two in Binghamton. Looking back with the proverbial 20-20 hindsight, we present in this Section a scheme for carrying out these four computations and at least two more in a uniform manner using only materials readily available in college laboratories of molecular biology. We would be delighted if researchers who are entering biomolecular computing would begin by practicing some of the procedures indicated here.

We begin with the choice of a small circular cloning plasmid. There is no further DNA to purchase. Additional plasmids can be produced in bacteria. The number of base pairs in the plasmids we have used is approximately 3000 and the multiple cloning sites (MCS) of these plasmids are approximately 175 base pairs in length. For the computations described here, choose from the MCS eight restriction enzyme sites each of which produces a four-base 5'-overhang when cut with its associated enzyme. Although the plasmids are circular, in order to save space, we indicate the plasmids and the eight chosen enzyme sites using linear diagrams that display only the sites in the MCS. Keep in mind that each of these linear diagrams indicates only about a 175 base pair segment of a 3000 base pair *circular* plasmid. The plasmid in its initial state is represented as:

-o-o-o-o-o-o-o-o-

The lower case o's indicate the eight restriction enzyme sites lying in the MCS that will be used. The plasmid in its initial state can be cut at each of these eight sites. The central writing operation used consists of dividing the content of a test tube T containing plasmids into two equal parts in test tubes T(L) and T(R). Two of the sites in the plasmid are chosen. In T(L) one of the two sites is locked and in T(R) the other is locked. The contents of tubes T(L) and T(R) are then returned to tube T. The result of applying such a pair of locking processes to a tube of molecules in the initial state using the third and fourth sites is represented:

-o-o-o=o-o-o-o-o-

This single line communicates that, after this parallel locking process, the tube T will contain **two distinct plasmid varieties**, one with the third site locked and one with the fourth site locked. One may picture these two molecular varieties as:

-o-o-x-o-o-o-o-o-   and   -o-o-o-x-o-o-o-o- ,

where the symbol 'x' is used to denote a site that has been locked. We may iterate this parallel locking procedure to produce the display, given as Figure 1, that represents the contents of the tube T after various parallel locking procedures have been applied. As a further example, when the parallel locking procedure is now applied to the tube T containing these two varieties, locking the fourth site in T(L) and the fifth site in T(R), there results a tube T the plasmid content of which is represented:

-o-o-o=o=o=o-o-o-o-

This single line communicates that, after this pair of parallel locking processes have been applied, the tube T will contain **four distinct plasmid varieties**, which may be pictured as:

-o-o-x-x-o-o-o-o- , -o-o-x-o-x-o-o-o , -o-o-o-x-o-o-o-o-, and -o-o-o-x-x-o-o-o- .

We may iterate the parallel locking procedure to produce the display, given as Figure 1, that represents the contents of the tube T after various parallel locking procedures have been applied. Six of these represent the contents of tubes T at the completion of the locking phase of the solution of one of the instances of an algorithmic problem previously attacked by members of the DNA computing community. In the remainder of this Section, continuing reference to Fig. 1 is required.

Each of the three problems lying in the horizontal row in Fig. 1 is completed in the following three steps. (1) Cut the MCSs from the plasmids with auxiliary restriction enzymes that cut at the boundaries of the MCS. (2) Gel separate and clone the released MCSs *that have the appropriate length(s)*. (3) 'Read' the results by either sequencing or testing in parallel, with the appropriate restriction enzymes, to determine which sites are locked and which sites remain unlocked. The problem at the extreme right in this row is the subject of Section 5. The '=' sign above the line:

-o=o=o=o=o=o=o=o-

is given to indicate that a parallel locking step has been made for the pair consisting of the leftmost and the rightmost enzyme sites in this row.

Each of the three problems in the vertical column in Fig. 1 is completed in the following four steps. (1) Carry out a sequence of cutting steps (4, 10, 6, respectively) which eliminate all plasmids that do not encode correct solutions. (2) Cut the MCSs from the remaining plasmids with auxiliary restriction enzymes. (3) Gel separate and clone the MCSs (all of which should be of the same length). (4) 'Read' the results by one of the two methods described above.

The writing phase of the SAT problems is merely the formation from the original plasmid of all logically consistent truth assignments. The number of these steps will always be the number of variables of the SAT problem. The cutting phase consists of one parallel step for each clause that must be satisfied. Each such step eliminates all those truth settings that fail to satisfy the clause being treated.

The procedure intended for treating the DHP takes the DHP to be a special case of the following problem: Let R be a subset of the product set A X B, where A and B are sets that have the same number of elements. Problem. Find all bijections of A onto B that are contained in R. We use the notation of [A'94] and embed Adleman's instance of the DHP into the present context by taking A = $\{0, 1, 2, 3, 4, 5\}$; B = $\{1, 2, 3, 4, 5, 6\}$; and R to be the set of 14 directed edges of Adleman's graph. For each element in A, there is a parallel writing step that provides the choices available for the image of that element in B under functions contained in R. The six step writing procedure produces the result indicated by the line

$$-\overline{000}\text{-}o\text{=}o\text{-}o\text{=}o\text{-}o\text{=}o\text{-}o\text{=}o\text{-}\overline{000}\text{-}$$

where the two barred triples of o's at each end require explanation: There are sites for six additional restriction enzymes in the MCS of pBluescript which do not yield blunt ends. The present computation does not use lengths in the reading process; consequently sites that produce overhangs other than 4-base 5'-overhangs can be used. The six extra sites used here are grouped into two groups of three. Each of two vertices in A is the initial vertex of three edges. This requires that for these two vertices the appropriate locking steps must divide tube T into three tubes T(L), T(M), and T(R) in which three different locking operations are performed. The results of these two three-way locking steps are symbolized by the barred triples of o's. The remaining four vertices of the graph are the initial vertices of exactly two edges which allows the use of the four previous parallel locking operations. Thus the line above represents the 144 plasmid varieties that encode the complete collection of the 144 functions from A into B that are contained in R. For each element of B, there is a parallel cut step that eliminates each function for which that element of B is not in the image of the function. Thus, when six appropriate parallel locking steps are followed by six appropriate cutting steps, only plasmids encoding bijections of A onto B remain. In the case of Adleman's graph one can verify that three molecular varieties should remain. Cloning and sequencing should result in the specification of the following three bijections:

$0{\to}6,\ 1{\to}2{\to}3{\to}4{\to}5{\to}1;\ 0{\to}3{\to}4{\to}5{\to}6,\ 1{\to}2{\to}1;$ and $0{\to}1{\to}2{\to}3{\to}4{\to}5{\to}6.$

It should then be noted that the last of these bijections is the unique solution of the original DHP instance. (One may observe that if the surjective feature is assured first then only 72 plasmid varieties arise before cutting is used to insure the function feature of bijections.)

The approach outlined above for finding the bijections contained in a binary relation is treated thoroughly in [H'02], but without laboratory work. It would be a special delight to see this aqueous solution for finding bijections (hence also DHPs) confirmed in a laboratory.

5   Non-attacking Knights on a 3 × 3 Chessboard

Consider a chessboard that has only three rows and three columns. **We ask for a specification of all patterns in which knights can be placed on this board subject to the restriction that no knight attack another.** No limit on the number of knights is made in advance. Such a board is indicated below with its nine squares labeled with digits 0 through 8. It is accompanied by an undirected graph in which the vertices are these same nine digits. Two of these vertices are connected by an edge if knights placed at the squares having these labels would be mutually attacking.

```
0  3  6          0 – 1 – 2 – 3
5  8  1          |           |        8
2  7  4          7 – 6 – 5 – 4
```

Recall that a subset S of the set of vertices of a graph is *independent* if no pair of vertices in S is connected by an edge. It follows that the acceptable patterns of knights on the 3×3 chessboard correspond to the independent subsets of the associated graph. Note that each subset of an independent set is also independent. Consequently, to list all the independent subsets it is sufficient to list those independent subsets having the property that no vertex can be added to the set without destroying independence. Such subsets are said to be *maximal independent* subsets. Since vertex 8 is isolated, it must be contained in every maximal independent subset. Consequently, it is sufficient to list all the maximal independent subsets of the eight-vertex subgraph obtained by deleting vertex 8. To each maximal independent subset of this smaller graph the vertex 8 can be adjoined to obtain a maximal independent subset of the original nine-vertex graph. We make one final elementary observation: no subset having two or fewer vertices can be a maximal independent subset of the smaller graph since at most six of the eight vertices can be occupied by, or be adjacent to, the vertices of such a small subset.

From the previous paragraph we know that to produce the list that solves the problem of placing knights on a 3 × 3 chessboard, it is sufficient to list each maximal independent subset of the eight-element circular graph, each of which will necessarily have three or more vertices. A complete list of all independent subsets of the nine-vertex graph will be obtained by adjoining vertex 8 to each subset and then listing all subsets of the resulting subsets. With these reductions made, the following molecular implementation was carried out.

We chose eight restriction enzyme sites in the multiple cloning site of one of the standard pBluescript circular cloning plasmids and named these enzyme sites by placing them in one-one correspondence with the vertices 0, 1, 2, 3, 4, 5, 6, 7 of the eight-vertex circular graph. Only enzymes producing four-base 5'-overhangs were chosen. The description of our laboratory work follows. Each locking step consists of a sequence of three steps: (1) cut with the appropriate restriction enzyme; (2) fill-in the single stranded four-base 5'-overhangs to full double stranded form by applying a DNA polymerase; (3) re-circularize with a ligase. Processes of purification were interspersed among these three previous steps and the re-circularized molecules were amplified in bacteria. The locking procedure chosen alters site lengths uniformly. This allowed us to take advantage of the fact that we can ignore independent sets having two or fewer vertices, by concentrating, at the reading phase, on MCSs the lengths of which have been increased by at most 5 times 4 (= 20) base pairs – thus leaving an independent set having at least 8 – 5 (= 3) vertices as required for maximality. With these decisions made we proceeded to the laboratory work:

A test tube T was initialized with a vast number of pBluescript plasmids.

The actions stated in the following four lines were then taken in a sequence of eight steps in which N is assigned, in succession, the values 0, 1, 2, 3, 4, 5, 6, 7.

Half of the content of T was poured into each of tubes T(L) and T(R).

In T(L) the enzyme site N was locked.
In T(R) the enzyme site N+1 (mod 8) was locked.

Equal amounts of the plasmid contents of tubes T(L) and T(R) were returned to tube T.

The multiple cloning site (MCS) of pBluescript is bounded by auxiliary restriction enzyme sites that were not used as sites 0 through 7. Using these two auxiliary sites, the plasmids were cut into two linear molecules. One segment consisted of the MCS and the other was the very much longer linear residue from the removal of the MCS. The released MCSs were separated on a gel according to length. The DNA content of the portion of the gel in a region that was adequate to insure that it contained all the MCSs at which at most five of the eight sites had been locked (i.e., at least three were left unlocked) was harvested. This resulting DNA was cloned. Colonies were selected, a few at a time, and sequenced. Each sequence was interpreted as the encoding of a subset of the vertex set of the eight-vertex circular graph. Those subsets that were maximal were added to a list of all maximal independent subsets thus far found. We are continuing to choose colonies and sequence plasmids from selected colonies.

At this writing, of the maximal independent sets, we have obtained the following six, each displayed in its knight pattern interpretation:

```
- K -      - - -      - - -      - - -      - - K      - K K
K - K      K - -      - - -      - - K      - - K      - - K
- K -      K K -      K K K      - K K      - - K      - - -
```

The following four patterns should be obtained, but have not yet been obtained:

```
K - K      K - -      K K -      K K K
- - -      K - -      K - -      - - -
K - K      K - -      - - -      - - -
```

Notice that each of these latter four has a knight in the upper left-hand corner. This corner position is encoded in the molecule as a site for the restriction enzyme Not I. This suggests that at some point in our computational process the ratio of molecules with a locked Not I site to the molecules in which the site was not locked got too large. Perhaps some unintended process locked Not I sites. We are following up these leads now.

At such a time that we recover the four missing molecules, our reduction process, when re-expanded to the complete 3×3 board will produce, after adjoining sub-patterns, the 47 solutions for the outer eight sub-board, and, after adjoining vertex 8, the 94 solutions for the 3×3 board. Through our cloning and sequencing process we have obtained, at this writing, the sequences of plasmids taken from 26 colonies. These 26 include the sequences encoding the six maximal patterns listed above, four sequences that encode non-maximal patterns in which three knights occur, some sequences that encode patterns in which only two knights occur, and repetitions of some of the sequences in each of these categories. Unfortunately, one incorrect solution having a pair of mutually attacking knights has been recovered.

Although the computation described here is not yet complete at this writing, we believe that the clarity and generality of the aqueous approach — as presented here and in Section 4 — shows that the aqueous concept is an attractive and provocative approach to computational problems. The specific biochemical realization of aqueous computing given here may not scale up to provide treatment of large instances — but realizations of the aqueous concept using other modalities may scale up. Several alternate techniques for writing on molecules are possible. The effectiveness of writing on DNA by attaching short PNA strands is currently under investigation [YHM'01].

6   Light in the Future for Aqueous Computing?

We propose to test the feasibility of computing by writing with lasers of controlled frequencies on molecules that would be constructed so that they have local features sensitive to distinct frequencies of light. With no use of biochemistry it may be possible to write on information baring molecules contained in a transparent container with no material being added or removed during the writing process. Perhaps writing steps can be done in this way with extreme rapidity.

Acknowledgements

Partial support is gratefully recognized through DARPA/NSF CCR-9725021, the Leiden Center for Natural Computing, and the Molecular Computing Project of Japan. The concept of aqueous computing developed from a study of [OKLL'97] with encouragement from P. Kaplan. Laboratory work in aqueous computing would not have been initiated without the encouragement and efforts of G. Rozenberg and H. Spaink, which resulted in the first aqueous result [HRBBLS'00].

References

[A'94]   L. Adleman, Molecular computation of solutions of combinatorial problems, *Science*, 266(1994)1021-1024.

[C'75]   G.L. Centoni, Biological methylation: selected aspects, Annual Review of Biochemistry 44 (1975), 435–451.

[FCLL'00]   D. Faulhammer, A.R. Cukras, R.J. Lipton & L.F. Landweber, *Proc. Nat. Acad. Sci.* 97 (2000), 1385–1389.

[GJ'79]   M.R. Garey & D.S. Johnson, *Computers and Intractability — A Guide to the Theory of NP-Completeness*, W.H. Freeman, San Francisco, CA, (1979).

[H'87]   T. Head, Formal language theory and DNA: an analysis of the generative capacity of specific recombinant behaviors, *Bull. Math. Bio.* 49 (1987), 737–759.

[H'00]   T. Head, Circular suggestions for DNA computing, in: A. Carbone, M. Gromov & P. Pruzinkiewicz, Eds., *Pattern Formation in Biology, Vision and Dynamics*, World Scientific, Singapore and London, (2000), 325–335.

[H'01a]   T. Head, Splicing systems, aqueous computing, and beyond, in: I. Antoniou, C.S. Calude & M.J. Dinneen, Eds., *Unconventional Models of Computation, UMC'2K*, Springer-Verlag, London, (2001).

[H'01b]   T. Head, Writing by methylation proposed for aqueous computing, Chapter 31 in: C. Martin-Vide & V. Mitrana, Eds., *Where Mathematics, Computer Science, Linguistics and Biology Meet*, (2001), 353–360.

[H'01c]   T. Head, Biomolecular realizations of a parallel architecture for solving combinatorial problems, (to appear).

[H'02]   T. Head, Finding bijections with DNA, (to appear).

[HPP'97]   T. Head, Gh. Paun & D. Pixton, Language theory and molecular genetics: generative mechanisms suggested by DNA recombination, a chapter in: G. Rozenberg & A. Salomaa, Eds., *Handbook of Formal Languages*, vol. 2, Springer, New York, 1996, pp. 295–360.

[HRBBLS'00] T. Head, G. Rozenberg, R.S. Bladergroen, C.K.D. Breek, P.H.M. Lommerse & H. Spaink, Computing with DNA by operating on plasmids, *BioSystems* 57 (2000), 87–93.

[HYG'99]   T. Head, M. Yamamura & S. Gal, Aqueous computing: writing on molecules, in: *Proceedings of the Congress on Evolutionary Computing*, IEEE Service Center, Piscataway, NJ, (1999), 1006–1010.

[OKLL'97]   Q. Ouyang, P.D. Kaplan, S. Liu & A. Libchaber, DNA solution of the maximal clique problem, *Science* (1997), 446–449.

[PRS'98]   Gh. Paun, G. Rozenberg & A. Salomaa, *DNA Computing — New Computing Paradigms*, Springer Verlag, Berlin (1998).

[R'96]   P.W.K. Rothemund, A DNA and restriction enzyme implementation of Turing machines, in: *DIMACS Series in Discrete Math. & Theor. Comp. Sci.*, vol. 27, Amer. Math. Soc., Providence, RI, (1996).

[YHG'00]   M. Yamamura, T. Head & S. Gal, Aqueous computing – mathematical principles of molecular memory and its biomolecular implementation, Chap. 2 in: Hiroaki Kitano, Ed., *Genetic Algorithms* 4 (2000), 49–73. (In Japanese).

[YHM'01]   M. Yamamura, Y. Hiroto, T. Matoba, Another realization of aqueous computing with peptide nucleic acid, (This Proceedings, 2001).

[YS'00]   H. Yoshida & A. Suyama, Solution to 3SAT by breadth first search, in: *DIMACS Series in Discrete Math. & Theor. Comp. Sci.*, vol. 54, Amer. Math. Soc., Providence, RI, (2000), 9–22.

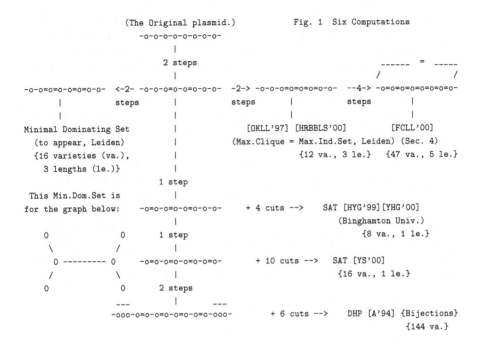

(The Original plasmid.)   Fig. 1 Six Computations

Legend for Fig. 1 This diagram provides a scheme by which instances of five different algorithmic problems can be solved in a uniform systematic manner in which intermediate results of previous computations can be saved and used as the initial steps of later computations. *Reading from left to right across the horizontal row*, the algorithmic problems [GJ'79] treated are:

1. Find the minimal dominating set in a graph.
2. Find the clique(s) of maximum cardinality in a graph, which is immediately equivalent to: Find the independent set(s) of maximum cardinality in a graph.
3. Find those independent set(s) in a graph which cannot be enlarged without destroying independence. The problem of finding all placements of knights on a (3 × 3) chessboard can be viewed as a special case of this problem.

Reading down the column, the algorithmic problems treated are:

4. SAT: satisfiability of a set of four three-variable (disjunctive Boolean) clauses, followed by
5. SAT: satisfiability of a set of ten four-variable clauses.
6. Find all the bijections of A onto B contained in a binary relation R in AXB. The directed Hamiltonian path problem (DHP) can be viewed as a special case of this problem.

Each of the six instances of these five problems has had laboratory treatment previously and several have been reported in the literature: (1.) has been completed in H. Spaink's laboratory in Leiden with an aqueous implementation that differs very slightly from the one described here. (2.) has been treated in [OKLL'97] and in an aqueous form in [HRBBLS'00]. (3.) has been treated in [FCLL'00] for the knights on a 3×3 case concluding with a sample of size 30 from the set all 94 solutions. In Sec. 5 we report our progress toward finding by aqueous computing the 10 'maximal' patterns that are the key to listing all 94 solutions. (4.) has been solved by the aqueous procedure in Binghamton. 'Reading' with restriction enzymes will be illustrated with a gel photo confirming the solution of our three-variable SAT. (5.) has been solved as reported in [YS'00]. (6.), in a DHP form, was solved by Adleman in his paper [A'94] in which DNA computing was initiated. We recommend in Section 4 (without laboratory test) an aqueous solution of this same instance. The pairs of numbers enclosed in braces, for example, the pair {16 va., 3 le.} associated with the Minimal Dominating Set problem, give the maximum number of distinct varieties of plasmids that arise during the aqueous computation (16 in this case) and the number of distinct circumferences of plasmids at the conclusion of the computation (3 in this case). An alternate aqueous approach to the DHP improves {144} to {72}.

Solutions of Shortest Path Problems by Concentration Control

Masahito Yamamoto[1], Nobuo Matsuura[1], Toshikazu Shiba[2], Yumi Kawazoe[2], and Azuma Ohuchi[1]

[1] Division of Systems and Information Engineering,
Graduate School of Engineering,
Hokkaido University,
North 13, West 8, Kita-ku, Sapporo 060-8628, Japan
{matsuura, masahito, ohuchi}@dna-comp.org
http://ses3.complex.eng.hokudai.ac.jp/
[2] Division of Molecular Chemistry, Graduate School of Engineering,
Hokkaido University, Sapporo 060-8628, Japan
{shiba, kawazoe}@dna-comp.org

Abstract. In this paper, we present a concentration control method that may become a new framework of DNA computing. In this method, the concentration of each DNA is used as input and output data. By encoding the numeric data into concentrations of DNAs, a shortest path problem, which is a combinatorial optimization problem, can be solved. The method also enables local search among all candidate solutions instead of a exhaustive search. Furthermore, we can reduce the costs of some experimental operations in detecting process of DNA computing, because we have only to extract and analyze relatively intensive bands. Solutions of a shortest path problem by using a simulator and by laboratory experiments are presented to show the effectiveness of the concentration control method.

1 Introduction

Since Adleman's epochal experiment, various DNA computing algorithms for solving combinatorial decision problems have been presented [1,2,3]. Recently, some remarkable computing algorithms that utilize features of DNA, such as conformation of DNA single strands [4,5] and aqueousness [6], have been proposed as new computing paradigms.

In this paper, we focus on aqueousness of DNA, but in a different way to that reported previously [6], i.e., concentration of DNA. In our computing model, the concentrations of DNA are used as input and output data. Since chemical reactions are controlled by using these concentrations, so we call this method concentration control. As an example of concentration control, a shortest path problem, which is a combinatorial optimization problems, can be solved by encoding the numeric data into concentrations of DNAs. Oliver has shown that calculation of multiplication of Boolean matrix can be performed by the use of

N. Jonoska and N.C. Seeman (Eds.): DNA7, LNCS 2340, pp. 203–212, 2002.
© Springer-Verlag Berlin Heidelberg 2002

concentration of DNA [7]. In this paper, we try to use the concentration of DNA for solving a combinatorial optimization problem.

Our method also enables a local search among all candidate solutions instead of an exhaustive search, since the concentration of hopeless candidate solutions tends to be small. Furthermore, by using concentration control, we can reduce the costs of some experimental operations in detecting process of DNA computing, because we have only to extract and analyze relatively intensive bands. Therefore, this method is expected to become a new framework of DNA computing.

In order to show the effectiveness of the concentration control method, we present results of simulation and then results of laboratory experiments to solve shortest path problems. Based on the results, the effectiveness of our proposed method is discussed.

2   Concentration Control for Shortest Path Problems

2.1   Algorithm

In this paper, we explain how to solve a shortest path problem by using the concentration control method. Fig. 1 (A) shows a directed graph with costs on edges. The shortest path problem for the graph is to find the shortest path (minimizing the total costs including the path) from vertex 0 (start) to vertex 5 (goal).

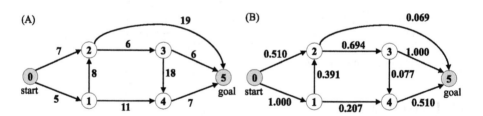

Fig. 1. (A) is a shortest path problem with 6 vertices and 9 edges. This problem has 7 candidate paths (total costs in parentheses): 0-2-5 (26), 0-1-2-5 (32), 0-1-4-5 (23), 0-2-3-5 (18), 0-1-2-3-5 (24), 0-2-3-4-5 (36), and 0-1-2-3-4-5 (44). (B) is the problem encoded with concentration control.

The proposed algorithm for the shortest path problems is as follows.
Step 1: Encoding of the problem in DNAs. Each vertex i in the graph is associated with a designed 20-mer sequence of DNA denoted O_i. For each edge $i \rightarrow j$ in the graph, an oligonucleotide $O_{i \rightarrow j}$ that is 3' 10-mer of O_i followed by 5' 10-mer of O_j is created.
Step 2: Construction of random paths. To construct random paths in the graph shown in Fig. 1 (A), a mixture containing each complementary oligonucleotide

encoding vertices and each oligonucleotide encoding edges is made. The concentrations of complementary oligonucleotides encoding vertices are set to the same values, and the relative concentration D_{ij} of each oligonucleotide encoding an edge $i \rightarrow j$ with cost C_{ij} is calculated by the following formula:

$$D_{ij} = (Min/C_{ij})^{\alpha}, \tag{1}$$

where Min represents the minimum value among costs of all edges in the graph, and α is a parameter value. In this study, we set the value of α to 2 on the basis of results of preliminary experiments. In this step, other formulas, such as exponentially weighed formulas, can be used to translate the costs of edges into the concentrations. The reaction contains oligonucleotides encoding vertices with the same concentrations and edges with different concentrations calculated according to equation (1).

Step 3: Amplification of DNA paths by PCR. Amplification of DNA paths that begin with vertex start and end with vertex goal is performed. Two specific primers that can anneal with vertex start and vertex goal are added to the PCR reaction.

Step 4: Determination of the oligonucleotide encoding the shortest path. Amplified DNA paths are separated by experimental operations such as SSCP (Single Strand Conformation Polymorphism). SSCP is a simple technique for separation and purification of DNA paths according to their size, base sequence and base composition. DGGE and TGGE can be used for this process. As a result of separation, the most intensive band is extracted and analyzed by sequencing.

2.2   Simulation Experiment

In order to verify that the concentration control method is effective, we performed simulation for the problem shown in Fig. 1 (A). The hybridization simulator presented by Yamamoto [8] is a simulation model of the hybridization process based on a concentration dynamics model, and it enables calculation of relative concentrations of resultant DNA paths from initial concentrations of oligonucleotides. We encoded the costs on edges into initial concentrations of DNAs as shown in Fig. 1 (B), and the simulation was then carried out. We compared the results of two cases, i.e., that with and that without concentration control. The results are shown in Fig. 2. In the case without concentration control, the initial concentrations of all complements of vertices and edges are the same. On the other hand, in the case with concentration control, the initial concentrations of edges are calculated according to equation (1).

From these results, the shortest path 0-2-3-5 in the graph had the highest concentration among candidate paths. We verified that concentration control enabled us to efficiently search for the shortest path. Based on the results, we performed laboratory experiments. The results and discussion of the results are presented in the next section.

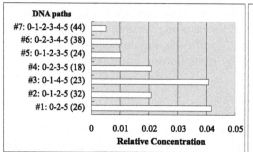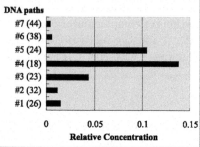

Fig. 2. Comparison of relative concentrations of resultant DNA paths from a hybridization simulator (presented by [8]). The figure on the left shows the results without concentration control. The figure on the right shows the results with concentration control. In both graphs, only paths beginning with vertex 0 and ending with vertex 5 are shown.

3 Laboratory Experiments and Discussions

3.1 Experiments for a Graph

First, we applied our algorithm with concentration control to a graph with 6 vertices, as shown in Fig 1 (A), in order to show the effectiveness of concentration control for the shortest path problem.

Step 1: Encoding of the problem. The encoding process is performed as follows: We encoded the information of vertices and edges into DNA sequences based on Adleman's model [1]. DNA sequences of vertices and edges (DNA vertices and edges) were designed by using the simulated annealing method, which is the encoding method based on the degree of binding between a pair of DNA oligonucleotides [8]. The obtained sequence designs are shown in Table 1. The costs were encoded into initial concentrations of each DNA based on concentration control shown in Fig. 1 (B).

Step 2: Construction of random paths. Random paths (DNA paths) in the problem were generated by the hybridization process. We diluted DNA vertices and edges and mixed them in one tube, according to Fig. 1 (B). The 5' end of DNA vertices and edges was phosphorylated by using 20 units of T4 Polynucleotide Kinase (Takara Co., Japan) in 400 μl of reaction mixture containing 10 × kinase buffer (Takara) and 100 mM ATP. The kination reaction was performed at 37 °C for one hour, and the reaction mixture was heated to about 84 °C and then gradually cooled to make the oligonucleotides anneal. After this process, random DNA paths were generated from DNA vertices and edges. In order to bridge the nicks of DNA paths, 700 units of T4 DNA ligase (Takara), 100 mM ATP and 1 M DTT were added to the kination mixture. The ligation reaction was performed at 16 °C overnight.

Step 3: Amplification of DNA paths by PCR. In order to select the DNA paths that begin with $\overline{O_0}$ (start) and end with $\overline{O_5}$ (goal), DNA amplification

Table 1. Designed sequences of vertices and edges for the graph shown in Fig. 1 (A).

Name	Sequences ($5' \rightarrow 3'$)
$\overline{O_0}$	TGGTTCCGATGTTTAAGCAA
$\overline{O_1}$	AGTCCTATTCATAACGACGC
$\overline{O_2}$	CCACAAGAGGGTATGGGCGG
$\overline{O_3}$	TACGAGCACACGAAGGTAGT
$\overline{O_4}$	GTTCTAGCAGTCTCCAAAAG
$\overline{O_5}$	GATCACTTGACTCAGCCAGA
O_{01}	GAATAGGACTTTGCTTAAAC
O_{02}	CCTCTTGTGGTTGCTTAAAC
O_{12}	CCTCTTGTGGGCGTCGTTAT
O_{14}	CTGCTAGAACGCGTCGTTAT
O_{23}	TGTGCTCGTACCGCCCATAC
O_{25}	TCAAGTGATCCCGCCCATAC
O_{34}	CTGCTAGAACACTACCTTCG
O_{35}	TCAAGTGATCACTACCTTCG
O_{45}	TCAAGTGATCCTTTTGGAGA

was performed by using the polymerase chain reaction (PCR). Before performing PCR, the ligation mixture was diluted to four levels (1, 1/4, 1/16, 1/64, 1/256, 1/1024). PCR was performed in 25 μl of solution containing 2 μl of ligation mixture at each dilution level, 2 mM dNTPs, two primers ($\overline{O_0}$: TGGTTC-CGATGTTTAAGCAA, O_5^R: TCTGGCTGAGTCAAGTGATC), and 1.25 units of KOD Dash DNA Polymerase (TOYOBO Co., Japan). Appropriate cycles of PCR were as follows:

– initial incubation at 94 °C for 20 sec
– 94 °C for 30 sec
– shift down to 50 °C and for 5 sec
– 74 °C for 30 sec

Step 4: Determination of the oligonucleotide encoding the shortest path. We quantified the concentration of DNA paths by performing 6% PAGE and using an image analyzing system, ChemiDoc and Quantity One (BIO-RAD Co.). We performed SSCP to separate each DNA path, but DNA paths were not extracted.

3.2   Result of Gel Analysis

Fig. 3 shows the visualized DNA paths amplified by PCR. In order to detect the DNA path representing the shortest path, we quantified the intensity of DNA bands in lane 4. The results of analysis are shown in Fig. 4.

It can be seen that the results from laboratory experiments and from simulation are well matched. However, in order to show the effectiveness of our method,

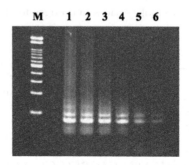

Fig. 3. Confirmation of the formation of DNA paths with 6% PAGE. DNA bands were visualized by ethidium bromide staining. Lane M, DNA size marker (100-bp ladder); lanes 1–6, amplified DNA paths after 16 cycles of PCR. The ligation mixture was diluted as follows: lane 1, 1; lane 2, 1/4; lane 3, 1/16; lane 4, 1/64; lane 5, 1/256; lane 6, 1/1024.

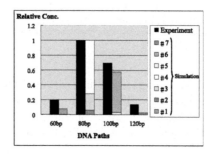

Fig. 4. Comparison of the relative amounts of DNA paths obtained from 6% PAGE experiments and from simulation. We quantified DNA bands shown in Fig. 3, lane 4.

we quantified the relative concentrations of resultant DNA paths. The results are described in the next subsection.

3.3 Quantification and Identification of Generated DNA Paths by Specific Primers

In order to quantify generated DNA paths, a quantitative PCR was performed using specific primers for amplification of specific DNA paths. For example, to amplify DNA path #2 (0-1-2-5), primers P0-1 and P5-2 were used. P0-1 could completely anneal with the complementary strand of $\overline{O_0}$ and the 3' end of it could anneal with 2 bases of the 3' end of the complementary strand of $\overline{O_1}$ but not with that of $\overline{O_2}$ (Table 2). On the other hand, although P0-2 could completely anneal with the complementary strand of $\overline{O_0}$ as well as P0-1, the 3' end of it could anneal with 2 bases of the 3' end of the complementary strand of

$\overline{O_2}$ but not with that of $\overline{O_1}$ (Table 2). Matching of the two bases of the 3' ends of a template and a primer is crucial for a DNA polymerizing reaction catalyzed by DNA polymerase. Following the same principle, we designed 3 primers that can anneal with $\overline{O_5}$ plus 4 bases of the 3' end of $\overline{O_2}$ (P5-2), $\overline{O_5}$ plus 4 bases of the 3' end of $\overline{O_3}$ (P5-3) and $\overline{O_5}$ plus 4 bases of the 3' end of $\overline{O_4}$ (P5-4) (Table 2). These three primers, P5-2, P5-3 and P5-4, can be used for amplification of DNA paths that have direct connections from $\overline{O_2}$ to $\overline{O_5}$, $\overline{O_3}$ to $\overline{O_5}$ and $\overline{O_4}$ to $\overline{O_5}$, respectively.

Table 2. Nucleotide sequences of specific primers. In P0-1 and P0-2, nucleotide sequences that can anneal with O_0 are underlined and those that can anneal with the 3' end of O_1 or O_2 are indicated by bold letters. In P5-2, P5-3 and P5-4, nucleotides that can anneal with the complementary strand of O_5 are underlined and those that can anneal with the 3' end of the complementary strand of O_2, O_3 or O_4 are indicated by bold letters.

Specific primers	Sequences (5'→ 3')
P0-1	TGGTTCCGATGTTTAAGCAA**AG**
P0-2	TGGTTCCGATGTTTAAGCAA**CC**
P5-2	TCTGGCTGAGTCAAGTGATC**CCGC**
P5-3	TCTGGCTGAGTCAAGTGATC**ACTA**
P5-4	TCTGGCTGAGTCAAGTGATC**CTTT**

Using these specific primers, we could easily amplify and quantify each DNA path by PCR separately except for DNA paths #3 and #7 (Table 2). Since both DNA paths #3 and #7 can be amplified by the combinations of primers P0-1 and P5-4, it is difficult to quantify the amount of these DNA paths separately. For this reason, to estimate the relative populations of DNA paths #3 and #7 amplified in the PCR, we also analyzed the amplified DNA paths by gel electrophoresis after 17 cycles of amplification (Fig. 5, lanes 7–9). In addition, gel analysis of DNA paths #2 (Fig. 5, lanes 1–3) and #4 (Fig. 5, lanes 4–6) amplified by 17 cycles of PCR demonstrated that only one DNA path can be amplified using these specific primers.

As a result of quantitative PCR, the relative amount of each DNA path was estimated, and the amounts are shown in Table 3. The relative amounts of DNA paths were calculated from the lowest threshold number of PCR cycles in which amplification of each DNA path was detected by fluorescence (C_T). According to this experimental data, the theoretical order of concentrations of generated DNA paths was consistent with the order of DNA concentrations estimated by quantitative PCR in DNA paths #2, #3, #4, #6 and #7. However, DNA concentrations of #1 and #5 did not correspond to the results of simulation. Since the concentration of 100-bp DNA bands (DNA path #5 and #6) seems to be more intensive than 60-bp DNA bands (DNA path #1) in Fig. 3, the length of DNA paths may have some effect on the quantitative PCR method.

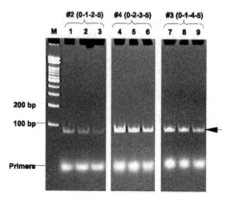

Fig. 5. Species of DNA path amplified by specific primers described in Table 3. Amplified DNA paths after 17 cycles of PCR were analyzed by 6% PAGE and visualized by ethidium bromide staining. DNA templates (ligation mixture) added to the reaction mixtures were diluted as follows: 1/ 20 (lanes 1, 4, 7), 1/ 40 (lanes 2, 5, 8), and 1/ 80 (lanes 3, 6, 9). The arrow indicates 80-bp amplified DNA paths. Lane M represents a size marker (100-bp DNA ladder).

Since only one band whose size corresponded to 80-base pair (bp) was visualized in gel analysis of amplified DNA paths #2 and #4 (Fig. 5), the relative amounts of these DNA paths estimated by quantitative PCR were completely dependent on the amounts of these 80-bp amplified DNAs. In case of PCR with P0-1 and P5-4 primers, DNA path #3 (80-bp) was mainly amplified, and a small amount of DNA path #7 (120-bp) was observed. A comparison of the fluorescent intensities of the DNA bands showed that the amount of DNA path #7 could not be estimated because the intensity was too weak to quantify. This result indicates that the relative amount of amplified DNA estimated by quantitative PCR with P0-1 and P5-4 primers was almost identical to the amount of DNA path #3.

We also performed sequencing analysis to confirm that the amplified DNA paths are really identical with predicted DNA paths supposed to be amplified by the specific primers. Using DNA paths amplified by 40 cycles of quantitative PCR as templates, sequencing reactions were performed using $\overline{O_0}$ and O_5^R primers. It was found that all of the DNA paths amplified by quantitative PCR were identical with the predicted DNA paths even in the 40 cycles of PCR amplification (data not shown).

4   Concuding Remarks

In this paper, we proposed the concentration control method that enables handling of numeric data and implementation of a local search, and we showed that shortest path problems can be solved by using this method. In the experiments,

Table 3. Quantification of each DNA path by quantitative PCR (Smart Cycler System). Reaction mixtures for quantitative PCR include 12.5 ml of 2' SYBR Green Master Mix (PE Biosystems), indicated set of primers (20 pmol each) and 5 ml of 1/100 diluted ligation mixture containing generated DNA paths as a template. Forty cycles of PCR incubation at 94 °C for 15 sec, at 53 °C for 15 sec, and at 72 °C for 20 sec was performed, and amplifications of DNA paths were monitored using a Smart Cycler System (Cepheid). C_T value represents the lowest threshold cycle in which fluorescence of amplified DNA was detected (fluorescent intensity = 2 in Smart Cycler System). N.D.: not determined.

	DNA paths	Size (bp)	Primers	$C_{T\#n}$ value	Relative DNA amount $(2^{-C_{T\#n}}/2^{-C_{T\#4}})$	Relative DNA amount by simulation
#1	0-2-5	60	P0-2, P5-2	12.77	0.28	0.11
#2	0-1-2-5	80	P0-1, P5-2	15.18	0.05	0.08
#3	0-1-4-5	80	P0-1, P5-4	13.36	0.18	0.32
#4	0-2-3-5	80	P0-2, P5-3	10.92	1.00	1.00
#5	0-1-2-3-5	100	P0-1, P5-3	14.00	0.12	0.76
#6	0-2-3-4-5	100	P0-2, P5-4	15.57	0.04	0.04
#7	0-1-2-3-4-5	120	P0-1, P5-4	N.D.	N.D.	0.03

it was easy to detect the optimal (or near-optimal) path by extracting the most intensive band. We believe that the concentration control method will be useful for other computing applications.

In the future, more experiments (both laboratory experiments and simulations) on more scaled-up problems must be performed, in order to verify the effectiveness of concentration control. Theoretical analyses are also needed. In particular, we will have to discuss about the optimality of the path with the highest concentration, in order to investigate other formulas to translate the costs of edges into the relative concentrations.

References

1. L. Adleman: "Molecular Computation of Solutions to Combinatorial Problems," Science, vol. 266, pp. 1021–1024, 1994.
2. R. J. Lipton: "DNA Solution of Hard Computational Problems," Science, vol. 268, pp. 542–545, 1995.
3. Q. Ouyang, P. Kaplan, S. Lir, and A.Libchaber: "DNA Solution of the Maximal Clique Problem," Science, vol. 278, pp. 446–449, 1997.
4. J. Chen, A. Eugene, B. Lemieux, W. Cedeno and D. H. Wood: "In vitro Selection for a Max 1s DNA Genetic Algorithm," Preliminary Proceedings of Fifth International Meeting on DNA Based Computers, pp. 23–37, 1999.
5. K. Sakamoto, H. Gouzu, K. Komiya, D. Kiga, S. Yokoyama, T. Yokomori, and M. Hagiya: "Molecular Computation by DNA Hairpin Formation," Science, vol. 288, pp. 1223–1226, 2000.
6. T. Head, M. Yamamura, and S. Gal: "Aqueous Computing: writing on molecules," Proc. Congress on Evolutionary Computation 1999, IEEE Service Center, Piscataway NJ, pp. 1006–1010, 1999.

7. J. Oliver: "Computation With DNA: Matrix Multiplication," DNA Based Computers 2, DIMACS Series in Discrete Mathematics and Theoretical Computer Science Volume 44, pp. 113–122, 1999.
8. M. Yamamoto, J. Yamashita, T. Shiba, T. Hirayama, S. Takiya, K. Suzuki, M. Munekata, and A. Ohuchi: "A Study on the Hybridization Process in DNA Computing," DNA Based Computers 5, DIMACS Series in Discrete Mathematics and Theoretical Computer Science Volume 54, pp. 101–110, 2000.

Another Realization of Aqueous Computing with Peptide Nucleic Acid

Masayuki Yamamura, Yusuke Hiroto, and Taku Matoba

Interdisciplinary Graduate School of Science and Engineering,
Tokyo Institute of Technology, 4259 Nagatsuta, Yokohama 226-8502, Japan,
my@dis.titech.ac.jp,
http://www.es.dis.titech.ac.jp

Abstract. Head proposed a framework of aqueous computing as a code design free molecular computing. Aqueous computing handles an aqueous solution of general-purpose memory molecules with a small set of elementary laboratory operations. It fits to solve a certain pattern of NP complete problems. We focus upon scaling the address space up and propose another biomolecular realization. Peptide nucleic acid (PNA) is an artificial analogue of DNA. Since PNA-DNA hybrid has higher melting temperature than DNA-DNA case, PNA will take over hydrogen bonds and displace itself into a double strand DNA. This phenomenon can be regarded as an irreversible write-once operation on a memory molecule. PNA brings a much larger address space than natural enzymes can provide. In this paper, we propose elementary operations for aqueous computing with PNA and realize one bit memory for a feasibility study to confirm strand displacement by PNA. We also propose an idea to copy a memory state upon a DNA sequence by using whiplash PCR.

1 Introduction

Molecular computing originated by Adleman's admirable work[1] has several issues to overcome. Code design is one of the hardest tasks. We must design the optimal DNA sequences for each given problems to be solved when we use hybridization of nucleic acids as computing device. Code design can often be harder than the given problem to be solved.

Head proposed a framework of aqueous computing as a code design free molecular computing[2]. Aqueous computing handles an aqueous solution of general-purpose memory molecules with a small set of elementary laboratory operations. A memory molecule has a set of distinguishable positions or addresses and each address associates a bit. Initially, all memory molecules represent the same bit pattern of all 1's, say "1111111···". For an aqueous solution of such memory molecules, a small set of elementary laboratory operations should be prepared as follows.

Pour(n): divide the solution into n tubes. All tubes are expected to have the same distribution of various bit patterns of memory molecules.

Unite: mix n tubes into one. Resulting solution should be randomized well.

N. Jonoska and N.C. Seeman (Eds.): DNA7, LNCS 2340, pp. 213–222, 2002.
© Springer-Verlag Berlin Heidelberg 2002

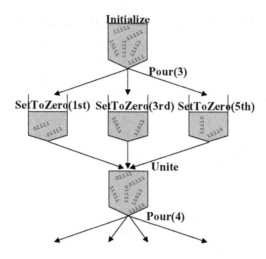

Fig. 1. aqueous algorithm

SetToZero(k): set the kth bit of all memory molecules in that tube to be 0.
MaxCountOfOnes: find the maximum number of 1's in one memory molecule from that tube. Standard laboratory operation like a gel electrophoresis would help this step.

Figure 1 shows a portion of an aqueous algorithm with these elementary operations. It generates bit patterns, which is combinatorial candidates of the solution for given problem, by repeating **Pour**(n), **SetToZero**(k) and **Unite** starting from the same memory molecules. Aqueous computing fits to solve a certain pattern of NP complete problems[2][3][4][5].

Remark that aqueous computing does not require any operation to restore a bit to solve NP complete problems. There are many candidate molecules available to realize such an irreversible write-once operation in the nature. This fact is another merit of aqueous computing to scale the address space up towards practical applications.

Figure 2 shows existing biomolecular realizations of a memory molecule. Both use a plasmid, a circular double strand DNA, as a memory molecule and use standard DNA modification enzymes to realize elementary operations.

CDL writing is developed by people in Leiden university[4]. A memory bit pattern is represented by a set of short fragments inserted in the same restriction sites. A set of restriction enzymes provides the address space. Initially, all restriction sites have short fragments and interpreted as all 1's. **SetToZero**(k) is realized by; (1) cut plasmids by kth restriction enzyme, (2) delete the fragment, and (3) ligate itself at sticky ends. After **SetToZero**(k), kth restriction site of a memory molecule does not have a short fragment and its molecular weight is reduced according to the missing fragment. This is an irreversible write-once operation. **MaxCountOfOnes** is realized by a polyachrylamide gel electrophoresis. Experiments with five bits memories have been reported.

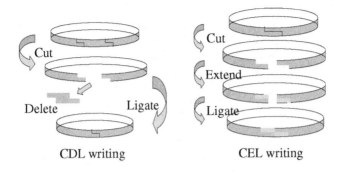

Fig. 2. existing realizations of aqueous memory

CEL writing is developed by people in Binghamton university[3]. A memory bit pattern is represented by a set of carefully chosen unique restriction sites themselves, which remain the same length of 5' overhang. A set of restriction enzymes provides the address space. Initially, all restriction sites exist and interpreted as all 1's. **SetToZero**(k) is realized by; (1) cut plasmids by kth restriction enzyme, (2) extend overhangs by DNA polymerase, and (3) ligate itself at blunt ends. After **SetToZero**(k), kth site of a memory molecule does not code original restriction site and its molecular weight is increased according to extended bases. This is an irreversible write-once operation. **MaxCountOfOnes** is also realized by a polyachrylamide gel electrophoresis. Currently, eight bits memories are developed.

There are thirty to forty restriction enzymes available to provide an address space for these realizations. However, forty is not large enough for practical applications. We focus upon scaling the address space up and propose another biomolecular realization with peptide nucleic acid (PNA). Section 2 proposes a set of elementary laboratory operations. Section 3 realizes one bit memory for a feasibility study to confirm strand displacement by PNA. Section 4 proposes an idea to copy a memory state upon a DNA sequence by using whiplash PCR.

2   PNA Writing

This section summarizes the characteristics of Peptide nucleic acid (PNA) and proposes a set of elementary laboratory operations for aqueous computing.

PNA is an artificial analogue of DNA[6]. Figure 3 compares single stand PNA and DNA. PNA consists of a sequence of four bases like DNA along with a peptide backbone like proteins. PNA can hybridize nucleic acids by hydrogen bonds of compliment bases. PNA has interesting characteristics; (1) its peptide backbone has no electric charge, (2) PNA-DNA hybrid has higher melting temperature than DNA-DNA case, (3) PNA has higher sequence specificity than other nucleic acids, (4) arbitrary short sequences can be synthesized like DNA oligomers, and (5) standard end markers can be attached on both termini.

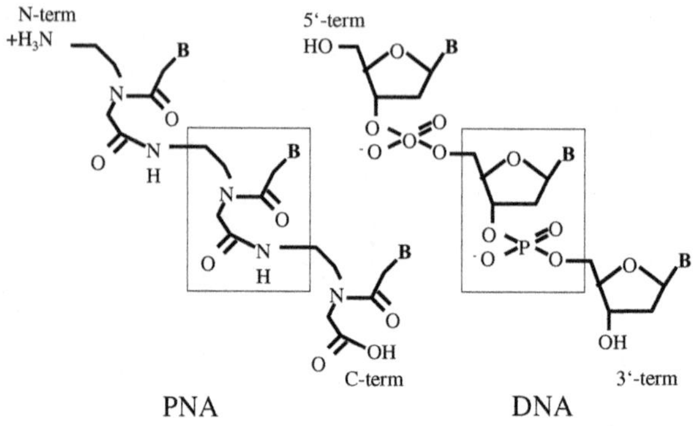

Fig. 3. PNA and DNA

The most interesting phenomenon is strand displacement. Figure 4 shows its conceptual scheme. When a double strand DNA has a complement sequence of a PNA, PNA will take over hydrogen bonds and displace itself into the double strand DNA. It forms DNA-PNA-PNA triple helix if the base sequence satisfies a certain condition. Since PNA-DNA hybrid would denature less than corresponding double strand DNA, it is known that strand displacement is almost irreversible and triple helix formation is totally irreversible. We use this phenomenon as a write-once operation.

We straightforwardly designed a set of elementary laboratory operations for aqueous computing. A memory molecule is a double strand DNA. A memory bit pattern is represented by a set of PNA binding sites on the memory DNA. PNA sequences provide the address space. Initially, all memory DNA has no strand displacement by PNA and interpreted as all 1's. **SetToZero**(k) is realized by merely displacing the memory strand by kth PNA. After **SetToZero**(k), its molecular weight is increased according to displaced PNA. This is an irreversible write-once operation. We call this operation "PNA writing" for short. We show a feasibility study of PNA writing in the next section. **MaxCountOfOnes** also can be realized by a polyachrylamide gel electrophoresis.

There is a cogent alternative to use a single strand DNA instead of a double strand DNA. We only wanted to avoid unexpected secondary structure in a feasibility study. We might change our mind if further investigation proves some merits on using a single strand DNA. In any way, the most important merit of PNA writing is the potentiality of extending the address space more than natural enzymes.

3   Preliminary Experiment

This section shows three preliminary experiments to realize aqueous memory by strand displacement with PNA. First, we confirm hybrid formation with

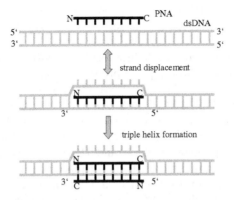

Fig. 4. strand displacement

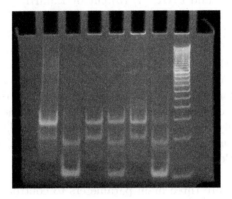

Fig. 5. inhibition of restriction enzyme by hybrid of DNA and PNA

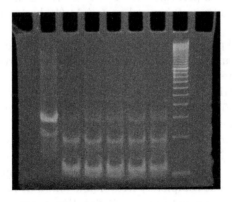

Fig. 6. effect of reaction temperature on strand displacement

double strand DNA and PNA. Next, we look for the optimal condition of strand displacement. Finally, we develop one easy way of **MaxCountOfOnes**.

Experiment 1 The idea to confirm hybrid formation with double strand DNA and PNA is based on Nielsen's work[7]. We prepared 10bps PNA "Lys-TTTTTTTT TT" and 56bps DNA that has the PNA binding site inserted between two BamHI sites "5'-M13M4-GGATCC-AAAAAAAAAA-GGATCC-M13RV-3' ". If the PNA displaces the correct site in the double strand DNA, it will inhibit BamHI activity. We start with single strand DNA, hybridize it to PNA, and then cover it by compliment single strand DNA to make correct hybrid. We want to separate efficiency issue of strand displacement in this experiment. We will discuss on efficiency of displacement in experiment 2.

Figure 5 shows the resulting 10% polyachrylamide gel electrophoresis (PAGE). From the left, lane (1) shows original double strand DNA, (2) (1) cut by BamHI from Takara incubated at 37 °C overnight in 1x reaction buffer K, (3) hybrid of double strand DNA and PNA, (4) (3) cut by BamHI, (5) similar mixture of double strand DNA and non homologous PNA (Lys-TTTTCTTTTT), (6) (5) cut by BamHI, and (7) 20bp ladder (thick band corresponds to 200bps). Each lane contains 5 pmol DNA. We used 50x more PNA to DNA in water.

Lanes (2) and (4) show PNA inhibits BamHI activity. Thin band around 40bps on lanes (1), (3) and (5) are single strand DNAs forming hairpins at two BamHI sites. Double strand DNA looks not perfectly displaced with PNA even start with hybrid of single strand DNA and PNA. Lanes (4) and (6) shows PNA has high affinity. Displaced PNA do not make bands slower. We cannot use PAGE straightforward to realize **MaxCountOfOnes**. We will discuss on realization of **MaxCountOfOnes** in experiment 3.

Experiment 2 There are three main parameters to control for strand displacement; the reaction temperature, the reaction length and the amount ratio of DNA and PNA. We start with 37 °C, 30 minutes with 50x more PNA. We examined the reaction temperature; 37 °C, 55 °C, 73 °C and 91 °C (denatured), the reaction length; 30 min, 2 hrs and over night (16 hrs), and the amount ratio; 5x, 50x and 500x.

As the result, all parameters are not so much sensitive for efficiency of strand displacement. We omit details here. Figure 6 shows a portion of results about reaction temperature. From the left, lane (1) show original double strand DNA, (2) (1) cut by BamHI, (3) double strand DNA incubated with PNA at 37 °C 30 minutes, (4) 55°C, (5) 73 °C, (6) 91 °C and (7) 20bps ladder. Each lane contains 5 pmol DNA. We used 50x more PNA to DNA in water.

Efficiency of strand displacement remains low estimated less than 50%. We could not reproduce Nielsen's perfect results while their recommended condition is 37 °C, 1 hour with 60x PNA. We guess we cannot perform 100% displacement since the reaction is not completely irreversible and falls in certain equilibrium. We need some auxiliary way to make computational steps reliable.

Rose et al. proposes PNA enhanced whiplash PCR[11]. Their work suggests several ideas on ours. One convincing way to improve efficiency is using bis-

PNA, which consists of the same PNAs connected by some flexible molecule, to form DNA-PNA-PNA triple helix. Rose's thermo dynamical analysis suggests this triple helix becomes very stable and consequently improve efficiency.

Experiment 3 The idea to realize **MaxCountOfOnes** is also simple. The left part of figure 7 shows the idea. When PNA displaces double strand DNA, the other strand, which is not hybridized with PNA, becomes free to hybridize another compliment nucleic acids. We can attach some flag molecule to help gel electrophoresis can distinguish the existence of strand displacement.

We prepared flag DNA which have 20bps double strand and sticky end 5'-AAAAAAAAAA-3'. The right part of figure 7 shows results. From left, lane (1) shows original double strand DNA, (2) mixture of (1) with flag DNA, (3) mixture of displaced double strand DNA, which is purified by QIAquick nucleic acid removal kit from Qiagen, and flag DNA. Each lane contains 5 pmol DNA. We used 50x more PNA and 20x more flag DNA to DNA.

Lanes (2) and (3) show clear distinction between clean double strand DNA and PNA displaced double strand DNA. We believe **MaxCountOfOnes** can be realized by this way.

As shown in this section, we could realize one bit memory for aqueous computing.

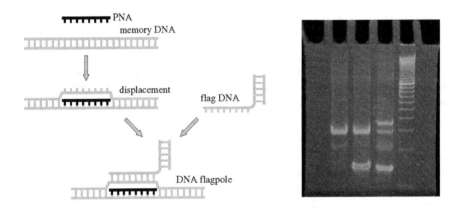

Fig. 7. DNA flagpole for MaxCountOfOnes

4   Memory State Copy

This section proposes an idea to copy the memory state upon a DNA sequence by using whiplash PCR[8][9].

PNA is an artificial material and there are no natural and artificial enzymes to amplify or destroy PNA-DNA hybrid. We cannot amplify memory molecules

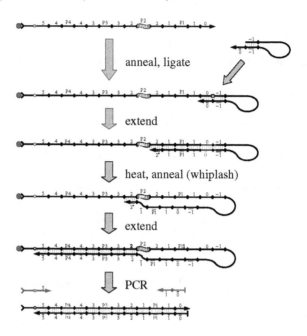

Fig. 8. copy procedure

with PNA displacement. On the other hand, every laboratory processes have a certain errors. Those mean accumulation of errors could be fatal for proposed method when we need much more stages to solve large-scale problems. We should develop extremely precise laboratory operations or another idea.

We propose a procedure to copy the state of a memory DNA with PNA displacement upon a pure DNA strand. This might be too early to consider since we could only developed one bit. However, we believe we should solve such a substantial difficulty as soon as possible.

Figure 8 shows an outline of the copy procedure. We start with a single strand memory DNA with PNA displacement. In previous sections, we proposed a double strand memory DNA. We must remove the covering and from a double strand memory DNA in some way, a magnet and magnetic beads for example. There is an alternative to use a single strand DNA from first to last as we mentioned. Here, we reserve the decision and assume we can successfully get a single strand memory DNA.

The sequence of a single strand memory DNA is carefully designed. It has a set of PNA binding sites. Figure shows such a DNA which have four PNA binding sites P1, P2,$\cdots$,P4. Figure also shows one PNA just hybridizes the memory DNA at the site P2. A PNA site is inserted in the same spacer sequence. Figure shows Pi is inserted in spacer sequence i.

For whiplash PCR, we ligate a single strand memory DNA with a hairpin sequence, add polymerase and start the thermal cycle. In the first cycle, polymerase extend 3' end until the first PNA bound site. PNA blocks polymerase before a few base pairs from the PNA bound site. Figure shows polymerase is blocked in the way of the spacer sequence 2. Remark that extended spacer sequence $\overline{2'}$ is a little shorter than the complete spacer sequence 2.

In the following cycles, the whip happens to skip the PNA bound site since it is inserted in the same spacer sequence. Polymerase can continue to extend remaining sequence if it could successfully skip the PNA bound site. Even if 3' end of whip has an incomplete copy of a spacer sequence like $\overline{2'}$, it will be filled up correctly.

After sufficient cycles of whiplash PCR, we can find a little short sequence is produced so as to skip all PNA bound sites. Since this new sequence consists of pure DNA, we can amplify it by normal PCR. The resulting double strand DNA does not have PNA bound site P2 and one spacer sequence 2, and consequently has smaller molecular weight.

This procedure does not copy a memory molecule itself. Instead, it does copy the state of a memory in the sense that the same PNA can no more displace the strand once that PNA has displaced. For example, PNA P2 cannot displace the amplified memory DNA since P2 has been removed as shown in the figure. In other words, the copy procedure preserves write-once property. We can amplify memory molecules in arbitrary stages of aqueous computing.

We must consider one point that the relationship between the molecular weight and the count of ones by this copy procedure. In order to keep consistency, we must apply the copy procedure just before **MaxCountOfOnes**.

This section shows only an idea. We need a lot of feasibility studies. Whiplash PCR technique can be refined to achieve dozens of steps[10], but it is also known whiplash PCR has difficulty that the back annealing prevent large number of transitions. However, we believe we can develop an efficient procedure since we don't need non-deterministic behavior in the proposed copy procedure. PNA enhanced whiplash PCR also suggests one idea that covering already copied strand by PNA shell would efficiently inhibit back annealing.

There are several complicated constraints to successfully synthesize PNA from the limitation of current manufacturing method. The developer of PNA provides rough heuristic guidelines but we faced some difficulty for almost all the ordered sequence even satisfying the guideline. It could be a major drawback of PNA. However, we estimate current restricted length of codes can provide few hundreds bits that is large enough than existing methods using natural restriction enzymes.

5   Conclusion

Aqueous computing is a framework for code design free molecular computing. We proposed a biomolecular realization of aqueous computing with PNA-DNA hybrid, showed a preliminary experiment with one bit memory, and proposed an idea to copy memory state. We can develop collection of parts, but we believe at least 100 bits class memory is required for practical applications and proposed method has potentiality.

Acknowledgement

Support for this research through Grant-in-Aid for Scientific Research (B) (2) 12480084 is gratefully acknowledged. The first author is very grateful to Professors Tom Head and Susannah Gal of Binghamton University for their collaboration, encouragement and advice.

References

1. L. Adleman, "Molecular computation of solutions to combinatorial problems," Science, 266, 1021-1024 (1994).
2. T. Head, "Circular suggestions for DNA computing," in A. Carbone, M. Gromov, P. Pruzinkiewcz, eds., Pattern Formation in Biology, Vision and Dynamics, World Scientific, Singapore and London, pp.325-335 (1999).
3. T. Head, M. Yamamura, and S. Gal, "Aqueous Computing: writing on molecules," Proc. of CEC99, 1006-1010 (1999).
4. T. Head, G. Rosenberg, R.S. Bradergroen, C.K.D. Breek, P.H.M. Lommerse, H.P. Spaink, "Computing with DNA by operating on plasmids," BioSystems 57, 87-93 (2000).
5. T. Head, X. Chen, M. J. Nichols, M. Yamamura, S. Gal, "Aqueous Solutions of Algorithmic Problems: emphasizing knights on a 3X3," Preliminary Proceedings, 7th International Meeting on DNA Based Computers, University of South Florida, pp.219-230 (2001).
6. B. Hyrup, P. E. Nielsen, "Peptide Nucleic Acids (PNA): Synthesis, Properties and Potential Applications," Bioorganic & Medical Chemistry, 4, 1, 5-23 (1996).
7. P. E. Nielsen, M. Egholm, R. H. Berg, O. Buchardt, "Sequence specific inhibition of DNA restriction enzyme cleavage by PNA," Nucleic Acids Research, 21, 2, 197-200 (1993).
8. Masami Hagiya, Masanori Arita, Daisuke Kiga, Kensaku Sakamoto and Shigeyuki Yokoyama. Towards Parallel Evaluation and Learning of Boolean μ-Formulas with Molecules, Preliminary Proceedings, 3rd DIMACS Workshop on DNA Based Computers, University of Pennsylvania, June 23 - June 25, 1997, pp.105-114.
9. Masami Hagiya, Masanori Arita, Daisuke Kiga, Kensaku Sakamoto and Shigeyuki Yokoyama. Towards Parallel Evaluation and Learning of Boolean μ-Formulas with Molecules. DNA Based Computers III, DIMACS Series in Discrete Mathematics and Theoretical Computer Science, Vol.48, 1999, pp.57-72.
10. K. Komiya, K. Sakamoto, H. Gouzu, S. Yokoyama, M. Arita, A. Nishikawa and M. Hagiya, "Successive State Transitions with I/O Interface by Molecules, DNA6, Sixth International Meeting on DNA Based Computers, Preliminary Proceedings, pp.21-30 (2000).
11. J. A. Rose, A. Suyama, M. Hagiya, R. J. Deaton, "PNA-Mediated Whiplash PCR," Preliminary Proceedings, 7th International Meeting on DNA Based Computers, University of South Florida, pp.311-320 (2001).

Experimental Confirmation of the Basic Principles of Length-only Discrimination

Yevgenia Khodor, Julia Khodor, and Thomas F. Knight, Jr.

Massachusetts Institute of Technology, Artificial Intelligence Laboratory

Abstract. We previously introduced the length-only discrimination (LOD) method for generate and search models of DNA computing. Here we report experimental confirmation of the validity of the basic operations of LOD method. We created a test graph consisting of four nodes and three edges, in which multiple paths are possible. Experimental results indicate that primitive operations required to implement LOD method work efficiently. We conclude that further work is needed to test the efficiency and practicality of applying LOD methodology to larger graphs.

1 Introduction

In his paper "Molecular Computation of Solutions to Combinatorial Problems" [1] Dr. Leonard Adleman presented a DNA-based method for determining the existence of Hamiltonian paths in directed graphs. A directed graph is said to have a Hamiltonian path if and only if there is a path in the graph beginning at vertex v_0 and ending at v_f that enters every other vertex in the graph exactly once. Hamiltonian path problem is a part of the complexity class NP, a class whose membership can be verified in polynomial time. Presently, no polynomial time algorithm for problems in NP is known, and the question of whether P=NP is one of the great open questions of the theory of computation. Hamiltonian path problem is also an example of an NP-complete problem, a subclass of NP problems interesting because their individual complexity is related to the complexity of all the problems in the class in such a way that if a polynomial time algorithm was found for one NP-complete problem, all problems in NP would be solvable in polynomial time.

By utilizing DNA's ability to form double-stranded duplexes with complementary sequenced strands, Adleman developed a "generate and search" approach to molecular computing. Adleman encoded each vertex and edge of a graph with a strand of DNA, then combined the strands and ligated them to form all legal paths in the graph. Finally, he searched through the paths obtained to find one that started and ended in the predetermined vertices and visited every other vertex exactly once. Unfortunately, Adleman's original algorithm was error-prone and time-consuming. The sorting step within his algorithm was proven to be ineffective [3], the flaw further magnified by the PCR amplification.

We previously introduced the length-only discrimination (LOD) method for generate and search models of DNA computing [5]. Using the LOD approach,

N. Jonoska and N.C. Seeman (Eds.): DNA7, LNCS 2340, pp. 223–230, 2002.
© Springer-Verlag Berlin Heidelberg 2002

the scientist is only concerned with the number of base pairs, not their sequence. To implement such an approach, the encodings of vertices must vary in size in such a way that no summation of an equal number of elements would present the same answer. We discuss in Section 2 below how such size difference may be achieved. Instead of utilizing magnetic bead separation in order to discriminate between the various paths of the same length on the base of their sequence, one need only to separate the resulting products by size, for example via agarose gel electrophoresis.

In LOD, as in the original generate-and-search approach introduced by Adleman, we rely on correct paths, and only correct paths forming. LOD approach allows us to easily recognize whether the path we were looking for was actually formed. However, to be sure that the path formed actually corresponds to a Hamiltonian path, we need to be sure that only legal paths in the graph are formed. The experiments presented below are, therefore, aimed at ensuring that the primitive operations required to form paths in an LOD encoding of a graph function properly.

We have conducted a series of experiments to test the primitive operations required for LOD. In the context of our experimental work we have found that when all edges and vertices comprising a path are present, the path is formed; if an edge or a vertex in a path is missing, the path is not formed; and when multiple paths through the graph are possible, and all components of these paths are present in solution, all the paths are formed.

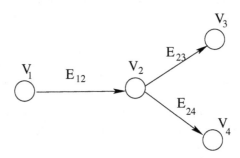

Fig. 1. The test graph.

The structure of the test graph is shown in Figure 1. As one can see from the figure, multiple paths are possible in the graph, namely the paths $V_1 - V_2 - V_3$ and $V_1 - V_2 - V_4$.

In the remainder of this paper we describe an LOD algorithm for the Hamiltonian path problem and the DNA implementation of the test graph we used to test the viability of the basic operations of LOD (Section 2), show that if and only if all edges and vertices comprising a path through the graph are present in solution, then this particular path is formed (Section 3), show that when multiple paths through the graph are possible, and all components of these paths are present in solution, all the paths are formed (Section 4), and conclude that larger graphs, such as Adleman's original graph, need to be explored experimentally before a conclusion on the practicality of the LOD approach can be reached (Section 5).

2 Hamiltonian Path Algorithm and DNA Implementation

The following is an LOD algorithm for solving an instance of the Hamiltonian path problem:

1. Generate all legal paths through the graph.
2. Keep only the paths beginning with v_i and ending with v_f.
3. Keep only those paths that enter each vertex exactly once.
4. If any paths remain, return "yes," otherwise return "no."

The algorithm can be implemented by a set of simple biological operations. Step 1 is implemented by combining together oligonucleotides encoding the vertices and edges of the graph, and using polymerase and ligase to fill in the gaps in the complementary assemblies of these oligonucleotides. Step 2 is implemented by a simple PCR reaction using v_i and the complement of v_f as the primers. Thus only the portions of the paths that start in v_i and end in v_f are amplified. Steps 3 and 4 are implemented simultaneously by size separation of the products of Step 2. In our experiments we used agarose gel electrophoresis, but a number of other approaches are possible.

Because in LOD we encode the vertices in such a way that no summation of lengths of equal number of elements is the same, and because we encode edges to be longer than the longest vertex, the Hamiltonian path, if it exists, would produce a band of characteristic size-the sum of length of all the vertices plus $(n-1)$ edges, where n is the number of vertices. If such a band is present, we return "yes," otherwise we return "no." The method for deriving the length of the vertex encodings is surprisingly simple. It is easy to see that if we need to find n different lengths, than starting with an arbitrary number for the length of the first vertex, we can produce the sequence of length with desired properties by making a gap between the lengths of the ith and the $(i+1)$st vertices be $(n+i)$. This scheme works because any nonhamiltonian path of n vertices necessarily repeats at least one vertex, and our encoding guarantees that no summation of lengths of i non-unique vertex encodings equals the sum of the lengths of any i unique vertex encodings. We also note that the encodings of the edges are longer than any vertex encoding. This ensures that the length of the encoding of any path with $(i+1)$ vertices would be greater than the length of the encoding of any path with i vertices. Table 1 lists the lengths of the particular encodings we used in the test graph shown in Figure 1.

Figure 2 illustrates the encodings of vertices and edges that allow for complementary assembly in Step 1 of the algorithm. For any two vertices v_i and v_j and edge e_{i-j} from vertex v_i to vertex v_j, we produce oligonucleotides v_{n-i-j},

Vertex/Edge	Length (in bps)
V_1	29
V_2	12
V_3	20
V_4	40
E_{12}, E_{23}, E_{24}	63

Table 1. The lengths (in base pairs) of vertices and edges in the encoding of the test graph shown in Figure 1

v_{i-j-m}, and e_{i-j}. Vertex oligonucleotides contain the encoding of the vertex itself surrounded on either side by the complements of portions of the incoming and outgoing edges. Thus, for a vertex with indegree k and outdegree l we produce kl oligonucleotides. For example, 15-mer proximal to the 3' end of v_{n-i-j} is complementary to the 15-mer at the 3' end of e_{i-j}, the edge joining it to the next vertex, v_{i-j-m}. The 5' 15-mer of v_{i-j-m} is complementary to the 5' 15-mer of e_{i-j}. The notation we use assumes that there is an edge e_{n-i} joining v_n to v_i, and there is an edge e_{j-m} joining v_j to v_m. If a vertex has no incoming edges, or no outgoing edges, its encodings only contain complementary portions of the edges in the direction where an edge exists. To label a vertex encoding for a vertex with no outgoing edges, we repeat this vertex's own number in the subscript. For example, an encoding of a vertex v_i with no incoming edges and an outgoing edge to a vertex v_j would be labeled v_{i-i-j}. Labeling of the vertices with no outgoing edges is analogous.

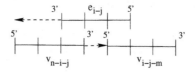

Fig. 2. Complementary assembly of the paths through a graph.

In order to demonstrate viability and validity of LOD method, we need to show that the basic operations required for implementing an LOD algorithm work. In particular, we need to demonstrate that

1. if and only if all edges and vertices comprising a path through the graph are present in the solution this particular path is formed, i.e.
   (i) when all edges and vertices comprising a path are present, the path is formed, and
   (ii) if an edge or a vertex in a path is missing, the path is not formed;
2. when multiple paths through the graph are possible, and all components of these paths are present in solution, all the paths are formed.

3   A Path through the Graph Is Formed if All the Edges and Vertices Comprising This Path Are Present in Solution

In all our experiments we use the thermostable polymerase-ligase system developed by Khodor and Gifford [4]. Both the Taq DNA ligase and Vent DNA polymerase enzymes have been successfully used for large applications separately, and in the programmed mutagenesis experiments conducted by Khodor and Gifford. We do not anticipate any scalability problems with the enzyme system. Figure 3 experimentally demonstrates that a path through the graph can be formed, and that it is only formed when all the components are present.

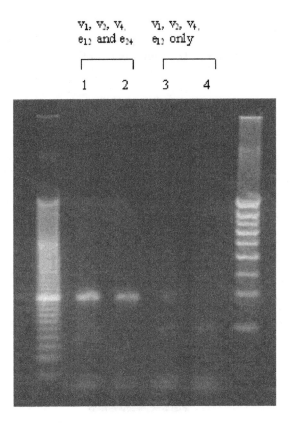

Fig. 3. A path through the graph is formed if all the components of the path are present in solution

We combined vertex oligonucleotides v_{1-1-2}, v_{1-2-4}, and v_{2-4-4} and edge oligonucleotides e_{1-2} and e_{2-4} to form a complete path $v_1 - v_2 - v_4$. As a control, in a separate reaction, we combined the same oligonucleotides with the exception of e_{2-4}. We then followed the algorithm above for solving an instance of Hamiltonian path problem with $v_i = v_1$ and $v_f = v_4$. Lanes 1 and 2 in Figure 3 contain bands of target size, 207 bps, corresponding to the sum of lengths of vertices v_1, v_2, and v_4 and two edges, indicating that when all the encodings of all the vertices and edges comprising a path are present in solution, the path is formed. The band of target size is missing in lanes 3 and 4, which contain those reactions lacking e_{2-4}, indicating that when a component of a path is missing in a reaction, the path is not formed. Thus, the path through the graph is formed iff the encodings of all the components of the path are present in solution.

4   When Multiple Paths Are Possible, All Paths Are Formed

Figure 4 shows that when elements comprising two possible paths through a graph are present in solution at the same time, both paths are formed.

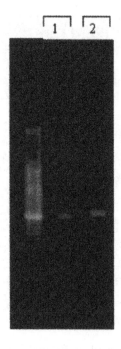

Fig. 4. Multiple paths through the graph can be formed at the same time.

We combined vertex oligonucleotides v_{1-1-2}, v_{1-2-4}, v_{2-3-3}, and v_{2-4-4} and edge oligonucleotides e_{1-2}, e_{2-3}, and e_{2-4} to form both possible paths through the graph ($v_1 - v_2 - v_3$ and $v_1 - v_2 - v_4$). As a positive control, in a separate reaction, we combined oligonucleotides that make up the path $v_1 - v_2 - v_3$. We then followed the algorithm above for solving an instance of Hamiltonian path problem with $v_i = v_1$ and the appropriate $v_f(s)$. Lane 1 in Figure 4 contains a band of target size 187 bps, corresponding to the sum of lengths of vertices v_1, v_2, and v_3 and two edges, representing the path $v_1 - v_2 - v_3$. The bands representing both paths are present in lane 2, although the band representing the longer path is significantly brighter. The result indicates that when multiple paths through the graph are possible, and all components of these paths are present in solution, all the paths are formed.

5   Conclusions

Our experimental results have demonstrated that a number of key aspects of LOD function properly. We have seen that it is possible to create a system where multiple paths can be formed, and the paths are formed only when all the components are present.

Bancroft and colleagues proposed using horizontal chain reactions for the generate and search approach [2]. However, unlike LOD, this approach necessitates processing solution molecules beyond size separation in order to discover whether there is Hamiltonian path in the graph. Both methods of processing proposed have drawbacks. Using a unique restriction enzyme for each edge in the graph imposes the limit on the size of the graph. Readout by degenerative PCR with the primers for each possible edge requires post-processing of the results to deduce the order of the edges in a Hamiltonian path, if such a path exists. It could be argued that such post-processing is akin to solving the original problem itself. In addition, if more than one Hamiltonian path exists, deciphering the output produced by either of the methods described above becomes more difficult. In contrast, using LOD method, whether a Hamiltonian path exists in a graph becomes clear immediately once the size separation step is performed.

Because LOD does not require performing an operation for every vertex in the graph, the amount of time required to perform a computation is constant, regardless of the size of the graph. Therefore, LOD method appears to be more scalable than the original method proposed by Adleman [1]. As with any other self-assembly technique, the larger the number of different elements in the soup at the same time, the greater the probability of assembly error. The initial time necessary for the self-assembly to take place would also increase with the number of components. However, these are concerns with any self-assembly technique and the probability of error in LOD systems may actually be somewhat decreased due to the use of the two-enzyme system. Both the Vent polymerase and the Taq ligase are fairly sensitive to the mismatches occurring near their respective active sites, so the incorrectly pre-assembled paths may never be completed.

One may also be concerned about the sensitivity of the output detection technique for the use with larger graphs. While the increasing sizes of the graphs would mean substantially longer paths, it would also imply that the gaps in sizes between differing paths would also increase. However, since the target size of a Hamiltonian path is known before an experiment is run, we can discriminate between paths with even a very small length difference by running the result of the experiment on an acrylamide gel until the fragments of the target size enter the portion of the gel where single-base discrimination is possible. A variety of other highly sensitive length-based discrimination techniques are now being developed that may be of use for LOD applications.

It is also worth noting that manufacturing oligonucleotides that make up an encoding of a large fully connected graph once would allow us to use subgraphs of it for solving smaller problems without additional set-up time and cost.

While the basic operations of LOD appear to work properly in the test graph we explored in this work, in order to asses the practicality of LOD method for general computation, the method needs to be tested on larger graphs, such as Adleman's original graph.

References

1. Adleman, L. M.: Molecular computation of solutions to combinatorial problems. Science 266 (1994), 1021–1024.
2. Guarnieri, F., Orlian, M., Bancroft, C.: Parallel operations in DNA-based computation. In: Rubin, H., Wood, D.H. (eds.): DNA based computers III :DIMACS Workshop, June 23–25, 1997, Providence, R.I. : American Mathematical Society (1999), 85–100.
3. Khodor, J., Gifford, D.K.: The efficiency of sequence-specific separation of DNA mixtures for biological computing. In: Rubin, H., Wood , D.H. (eds.): DNA based computers III :DIMACS Workshop, June 23-25, 1997, Providence, R.I. : American Mathematical Society (1999), 39–46.
4. Khodor, J., Gifford, D.K.: Design and implementation of computational systems based on programmed mutagenesis. BioSystems 52 (1999), 93–97.
5. Khodor, J., Khodor, Y.: Length-Only Discrimination Approach to Biological Computing. Sixth International Meeting on DNA Based Computers, Leiden, NL, June 2000.

Experimental Construction of Very Large Scale DNA Databases with Associative Search Capability

John H. Reif[1], Thomas H. LaBean[1], Michael Pirrung[2], Vipul S. Rana[2], Bo Guo[1], Carl Kingsford[1], and Gene S. Wickham[1]

[1] Department of Computer Science, Duke University
Durham, NC 27708 USA
{reif, thl, bog}@cs.duke.edu, carlk@cs.princeton.edu,
gwickham@acpub.duke.edu
[2] Chemistry Department, Duke University
Durham, NC 27708 USA
{pirrung, vipul}@chem.duke.edu

Abstract. We describe on-going experiments for executing associative search queries within synthesized DNA databases. Queries are executed by hybridization of a target database strand with a complementary query strand probe. In our initial annealing experiments for processing associative search queries, we employed fluorescently labeled query strands and performed separation of fluorescent versus non-fluorescent beads using Fluorescence Activated Cell Sorting (FACS or flow cytometry). We also tested polymerase chain reaction (PCR) as an output method, and developed a PCR technique for search in the pair-wise constructed library that exploits the particular properties of words in that library. We have also implemented computer software that provides a simulation (viewable on the internet) of the experimental search procedures, as well as a simulation of input/output from conventional 2D images.

1 Introduction

1.1 Overview

All known biological organisms make use of the sequential ordering of monomeric bases in long-chain nucleic acid molecules for storage, processing, and transmission of biological information. Researchers in the field of DNA-based computing are now investigating the possibility of encoding, storing, manipulating, and retrieving non-biological information in DNA sequences. The present study aims to test one such application, specifically, the creation of large databases of DNA sequences and methods for associative search queries within the databases.

The extreme compactness of DNA as a data storage is nothing short of incredible. Since a mole contains 6.02×10^{23} DNA base monomers, and the mean molecular weight of a monomer is approximately 350 grams/mole, then 1 gram of DNA contains 2.1×10^{21} DNA bases. Since there are 4 DNA bases,

N. Jonoska and N.C. Seeman (Eds.): DNA7, LNCS 2340, pp. 231–247, 2002.
© Springer-Verlag Berlin Heidelberg 2002

each DNA base can encode 2 bits, and it follows that 1 gram of DNA can store approximately 4.2×1021 bits. In contrast, conventional storage technologies can store at most roughly 109 bits per gram, so DNA has the potential of storing data on the order of 1012 more compactly than conventional storage technologies.

1.2   Prior Work

Eric Baum [1] first proposed the idea of using DNA annealing to do parallel associative search in large databases encoded as DNA strands. The idea is very appealing since it represents a natural way to execute a computational task in massively parallel fashion. Moreover, the required volume scales only linearly with the data base size. However, there were further technical issues to be resolved. For example, the query may not be an exact match or even partial match with any data in the database, but DNA annealing affinity methods work best for exact matches. Reif and LaBean [2] proposed improved biotechnology methods to do associative search in DNA databases. These methods adapted some information processing techniques (Error-Correction and VQ Coding) to optimize input and output (I/O) to and from conventional media, and to refine the associative search from partial matches to exact matches. Prior to our project, the use of DNA annealing to achieve parallel associative search had not been experimentally implemented.

The current study follows from significant prior work by Lynx Therapeutics on construction of bead-bound DNA libraries [3,4]. The Lynx methods were developed for the purpose of differential expression analysis which is the comparison of the ensemble of mRNA transcribed in different cell types or at different times. The Lynx method begins with the synthesis of a large combinatorial library of oligonucleotides by mix-and-split technique on plastic beads. They are appended with cDNA such that each cDNA is linked with high probability to only a single unique synthetic tag. Differential analysis is accomplished by hybridization of fluorescent labeled probes and sorting by FACS. The prior work made significant strides toward construction of specific purpose DNA-encoded databases. The current study utilizes a similar synthetic technique for construction of immobile libraries, however we have designed modified methods for construction of libraries of much larger size, we are interested in implementation of associative search, and we have allowed for output via not only FACS but by PCR.

There is also considerable prior work on DNA codeword design [5,6,7,8,9,10] and on word design used for surfaced-based DNA computing [11,12]. [13] shows that surface morphology may be an important factor for discrimination of mismatched DNA sequences. A three-base design has been described [14]. Evolutionary search methods for word designs are described in [15]. Laboratory experiments of word designs have been performed [16] and ligation experiments are described by Jonoska and Karl [17]. Wood [18] considers the use of error correcting codes for word design and to decrease mismatch errors. Hartemink et al. [19] described an automated constraint-based procedure for nucleotide sequence selection in word design. We will utilize and improve on these methods for DNA word design, including evolutionary search methods, and error correcting codes.

1.3   Current Work

This paper details a study involving the design, construction, and testing of large databases for the storage and retrieval of information within the nucleotide base sequences of artificial DNA molecules. The database consists of a large collection of single-stranded DNA molecules, either free in solution or immobilized on polymer beads, glass slides, or chips. A database strand carries a particular DNA sequence consisting of a number of sequence words drawn from a predetermined set, or lexicon, of possible words. We use at least 10 times more DNA than the theoretical minimum of one DNA strand per data item, to provide at least 10-fold redundancy, so each database element is represented by approximately 10 identical strands of DNA. This level of redundancy is probably on the lowest end of the range of detectability. These libraries can be used to emulate smaller libraries with much greater redundancy by essentially ignoring the values recorded at some internal positions. We would experience a 12^k fold increase in redundancy if values of k blocks are ignored. This strategy will be explained further below.

The experiments described here involve the hybridization of query strands to database strands such that database strands of interest will become marked and separated from the bulk database following query strand binding. A query strand contains the complement of a portion of a database strand, such that the query strand will specifically hybridize (anneal) and co-localize with its complementary database strand. Although the experimental libraries we constructed essentially only hold only a singe bit of information per database element (the data bit is 1 if and only if the element is contained in the library), we can in principle append to each library element a further DNA sequence providing data values.

One goal of the study is to measure rates of various search errors including: false positives from near-neighbor mismatches, partial matches, and non-specific binding as well as false negatives from limit-of-detection problems. False positives should be minimized by careful design of the word, lexicon, and database elements, as well as experimental tuning of annealing conditions such as temperature-ramp rate, pH, and buffer and salt concentrations. It is desirable to directly measure the limits of detection. It is also useful to construct a database containing words of known, low probability and some of known, high probability; then word strings of known probability can be queried to gauge the ability to retrieve rare sequences within databases of high strand diversity.

1.4   Small Test DNA Library Experiments

For test purposes, we first synthesized by combinatorial, mix-and-split methods a small test library of size 4^6 on plastic microbeads. We used a biased synthesis technique that made certain sequences have very low probability. For testing of annealing stringency on this library, we used fluorescently labeled query strands and then performed separation of fluorescent versus non-fluorescent beads by fluorescence activating cell sorting (FACS). We have completed the construction of this test library and implemented readout by FACS.

1.5   Larger DNA Library Experiments

Then we increased the scale of our experiments and synthesized a larger initial DNA library, again by combinatorial, mix-and-split methods on plastic microbeads. This resulted in an initial library of size $12^7 = 35,831,808$. Each element in the initial database encodes a sequence of 7 base 12 numbers, using a sequence of 7 consecutive 5 base DNA words.

1.6   Extremely Large DNA Library Experiments

We are now constructing a large, diverse library of 12^{15} DNA sequences. Each DNA strand of the library is single-stranded and encodes a number which provides the index to the database element. To encode the base 12 digit in the ith position, we use a distinct 12 element set S_i of DNA subsequences, each of 5 DNA bases in length. The encoding of each k place base 12 digit integer is thus done using a sequence of k consecutive 5 base DNA sequences. In addition each DNA strand of the library has certain flanking subsequences that are used in the synthesis of the library.

Our strategy entails a two phase synthesis of this DNA library:

(1) First we utilize our initial DNA library of size $12^7 = 35,831,808$ which we have already constructed by combinatorial, mix-and-split methods on plastic microbeads.

(2) Then we square the size of the library by combining pairs of the initially synthesized library strands using DNA hybridization and ligation. The resulting DNA data base elements consist of a concatenation of two of the previously constructed strands. The second phase will result in libraries of size $12^{14} = 1.28 \times 10^{15}$.

Although the constructed library essentially holds only a single bit of information per database element (the data bit is 1 if and only if the element is present in the library), we can easily append to each library element a further sequence providing additional data values).

1.7   Associative Search in This Extremely Large DNA Library

For testing of readout of a specific data element in the very large resulting library constructed in the second phase, we will use a two stage process. First, we use FACS to separate out beads whose DNA strands contain a selected suffix. Then we use PCR amplification, exploiting the particular properties of the concatenated words to ensure high fidelity selection of the desired data element. Suppose the database element searched for is indexed by a number written base 12 whose first 7 "digits" base 12 are U and last 7 "digits" base 12 are V. Then that database element is encoded by a DNA word where U is encoded by a 40 base subsequence in the prefix portion of the DNA word, and where V is encoded by a 35 base subsequence in the suffix portion of the DNA word. The PCR amplification can be done by repeated stages, where each stage involves an annealing of a primer U, or V or their complement. The success of the PCR amplification depends critically on annealing stringency of the primers, which is enhanced by the fact that (a) the primers are only 35 bases and (b) that each distinct pair of word sequences within a block differ by at least 3 bases. This PCR amplification can in principle be followed by readout via sequencing and the use of a DNA annealing array (we did not experimentally execute this final readout phase but performed a computer simulation instead.

2   Design and Synthesis of DNA Databases

2.1   Design of DNA Databases

In designing a DNA-encoded database one must consider several important factors including the following. The overall length of the oligonucleotide sequences used for matching is critical because sequence length directly effects the fidelity and melting temperature of DNA annealing. Hamming distance (or number of changes required to morph one sequence into another) is another critical consideration. One would like to maximize the Hamming distance between all possible pairs of encodings in the database in order to minimize near neighbor false-positive matching. One strategy for maintaining sequence distance is to assign block structures to the sequences with sets of allowed words (subsequences) defined for each block (see Figure 2). This strategy also lends itself well to chemical synthesis of the database and will be further described below. Another important consideration is the choice of the words themselves and the grouping of words into sets for use in the blocks. Sentence length, desired library diversity, and word-pair distance constraints all effect the choices of words in the lexicon.

In designing the initial small test DNA-encoded database, first it was decided that 24 residues was a good length for the data-encoding, match region of the sequences (the tag sequence to which the probe sequence on the query strand would bind). Next, a word/block synthesis scheme was decided upon. Figures 1 and 2 show high level views of several important aspects of the database design.

Fig. 1. Schematic Plan for Database Strands. DNA is shown extending toward the left from the spherical resin bead. The 5' and 3' constant regions (identical on each and every bead in the library) contain the proximal and distal sites for ssDNA primer binding for second-strand synthesis, PCR amplification, and sequencing. The database region (rectangular box) contains the variable sequences which are used to encode information. Details of the specific variable regions synthesized are given below.

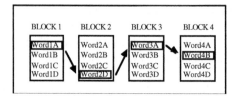

Fig. 2. The Range of Possible Sentences entailed by a Word-Block Construction Scheme. For each block position in the sequence, one word is chosen from the word set and synthesized on the growing DNA strand. Separate reaction vessels are used for each word in the block so that all the word choices are utilized but only one is present on any particular strand. For example, the arrows indicate the trace which results in the sentence: word1A-word2D-word3A-word4B. A particular bead is drawn through a particular path in the set of possible word choices, but all possible paths will be populated with beads therefore all possible DNA sentences will be synthesized. Each bead contains multiple copies of a single DNA sequence (see below for details of the mix-and-split synthesis scheme).

If the set of possible sentences greatly exceeds the number of beads used in the synthesis, then probabilistically, each bead will display a unique sequence (one-bead-one-sequence). In the present case, the number of beads in the synthesis exceeds the number of possible sentences, thus multiple beads containing any particular sentence will be present. On all beads the 5′ and 3′ constant regions contain invariant sequences regardless of the central variable sequence. These constant regions are used for PCR amplification of all database strands present in a test tube. The size of a particular database (its potential information content) is given by the diversity of possible sequences within the specific library design.

The terms 'library' and 'database' are used somewhat synonymously but 'library' refers more to the physical collection of DNA strands, while 'database' refers to the interpretation of that collection of molecules as a means of storing and manipulating information. Diversity refers to the total size of a library or database, that is, the number of unique sequences consistent with the design and synthesis criteria. Library diversity scales with the number of blocks and with the size of the set of words allowed in each block as shown in Figure 3.

By varying the probabilities for different words within a block, the database provided the capability for testing query searches for rare sequences down to one in approximately half a million. From each set of four words in any given block, word probabilities (concentrations) were set at $2x$, x, x, and $1/2x$ by dividing the resin into four, uneven batches and synthesizing a specific word onto each. Thus the most common database entry or "sentence" was present at $(4/9)^6 = 1/130$ randomly selected database strands and the average sentence was be found with probability $(2/9)^6 = 1/8,304$. However, the most rare strand had probability $(1/9)^6 = 1/531,441$. This provided us the opportunity to simulate a library of size $531,441$ where individual strands have probability $1/531,441$. This range of word probabilities provided a very wide range of sentence probabilities while only requiring a 4-fold differential during the division of resin in the oligo synthesis steps.

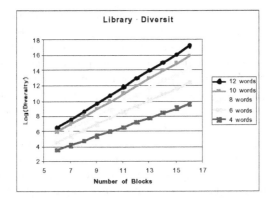

Fig. 3. Library Diversity versus Sentence Length. The figure shows the scaling of library diversity with increasing sentence length (block count) and increasing number of available words within each block. The relationship between words and blocks was described above in Figure 2. Diversity is calculated by raising the word count to the exponential power given by block count (i.e. diversity = [word count]$^{\text{block count}}$). As a simple example, to achieve a total diversity greater than one million with sentences containing 6 blocks, one would require a set of 10 word choices per block. To achieve our goal of a total diversity of 12^{14} with sentences containing 14 blocks (for example) requires a set of 12 word choices per block.

2.2   Word and Block Set Design

The Hamming distance between two DNA words is the total number of base mismatches, for the best possible alignment of the two sequences. A set of programs was written in C^{++} to assist in balancing the following constraints of good DNA code words: 1) minimize the melting temperature difference (T_m) between words so that hybridization of multiple words proceeds simultaneously; 2) maximize Hamming distances between word pairs and between words and complements of word; 3) avoid frame shift binding errors by minimizing overlap between desired words and spurious words straddling boundaries between adjacent blocks.

It was previously noted that a library strand with A, C, G, and T has much greater chance of significant secondary structure than a library strand composed of just A, C, and T [3]. We use only A, C, and T in our DNA code word design. In this document, we represent a DNA code word as a string over the alphabet $\{A, C, T\}$ and assume that the leftmost end of the string corresponds to the 5' end of the associated DNA code word. The number of C residues present per word profoundly affects the T_m value. From the alphabet $\{A, C, T\}$ our first program generates all the words containing the desired number of C's. For example, if we specify the minimum number of C's to be 1, maximum number to be 1, the word length to be four. The following words are generated: *AAAC, AACA, AACT, AATC, ACAA, ACAT, ACTA, ACTT, ATAC, ATCA, ATCT, ATTC, CAAA, CAAT, CATA, CATT, CTAA,*

$CTAT, CTTA, CTTT, TAAC, TACA, TACT, TATC, TCAA, TCAT, TCTA,$ $TCTT, TTAC, TTCA, TTCT, TTTC$. Beginning with this set of words, the second program generates all subsets which satisfy minimum Hamming distance and minimum set size requirements. The third program generates the satisfied library. It addresses the issue of spurious word repeats due to "frame shifts" at boundaries between neighboring blocks. This program takes the following parameters: If we have block A and its adjacent block B. Take one word from A and another from B, we have a total of 16 word pairs that form an 8 base sequence. The shifting distance is the minimum Hamming distance between every word (in A and B) and every pair (the 8 base sequence) by shifting. We also defined a score function that reflects the probability of generating word repeats. This score was minimized in order to minimize duplicated words. Due to the high number of sets generated from the second program, it was not possible to test all possible libraries. This program uses instead a greedy algorithm. It gives us a result that satisfies the specified parameters. We ran the whole program multiple times and each time received distinct results. To choose between these viable results, we used two constrains: i) minimize duplicated words; ii) minimize shifting score (as mentioned above).

To decide which one to use from among the generated sets, we simulated hybridization tests for every sequence in the library as follows: (i). We used a program to generate all the possible sentences; (ii). We concatenated the resulting sentences together to form a huge DNA sequence. (iii). The first sentence, the last sentence and the sentence with the most duplicated words (from i) were used as probes. The software BIND [20] was used to calculate free energies of association and melting temperatures for each potential probe binding sites; results were used to compare exact matches with spurious or partial match sites. The T_m value between the best binding site and the next best binding site is compared. The chosen set had larger overall T_m difference and fewer duplicated words.

2.3   Synthesis of Database

Desired manipulations of the DNA databases require the strands to remain bound to the resin beads upon which they are synthesized. The resin chosen as solid support was TentaGel M NH_2 (Rapp, Inc; see http://rapp-polymere.com/). It is a monodisperse resin consisting of polystyrene microspheres functionalized with amide groups (NH_2). The uniformity of bead size enables application of automatic sorting techniques. Release of the DNA can be made by treatment with acid. For the 20 mm bead size, the NH_2 capacity is 1.0 pmoles/bead (pmole = 10^{-12} mole), which provides attachment of 6.02×10^{11} strands of DNA per bead. One gram of the 20 μm bead resin contains approximately 2.4×10^8 beads. For the 10 μm bead size, the NH_2 capacity is 0.13 pmoles/bead, which provides attachment of 7.83×10^{10} strands of DNA per bead. A gram of the 10 μm bead resin contains approximately 1.95×10^9 beads.

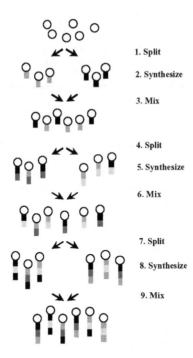

1. Split

2. Synthesize

3. Mix

4. Split

5. Synthesize

6. Mix

7. Split

8. Synthesize

9. Mix

Fig. 4. Flow chart of mix-and-split synthesis scheme. At the beginning of the process (top) bare resin beads are prepared for library construction; this can include synthesis of the 3′ constant region in a single reaction vessel. Step 1: resin is split into separate reactions (the figure shows division into two parts). Step 2: a single, specific sequence word is synthesized on all beads in each of the vessels (one vessel — one word). Step 3: All resin is recombined and mixed so that in the next splitting (Step 4) each vessel will receive beads carrying each of the preceding words (both black and white words from Step 2). The next word is added in Step 5; the entire ensemble is remixed in Step 6 and the process can continue with one splitting step for each block in the design. The final result of the process (bottom of figure) is a library of beads where each bead contains multiple copies of a single sequence (for clarity only a single copy is shown per bead). Note that in this summary figure the word syntheses are shown as single steps while the actual chemistry requires the addition of nucleotide monomers one at a time, so each synthesis step in the figure corresponds to a series of chemical cycles — one cycle for each base in a word.

Figure 4 provides a general outline for the mix-and-split procedure used for the synthesis of combinatorial database libraries. The process can produce vast libraries in which multiple, identical copies of a DNA sequence will be created on any given bead. This one-bead/one-sequence design produces libraries suitable for associative search query experiments because fluorescent probe will localize to and label specific beads carrying copies of target sequence and fail to label beads carrying unrelated sequence.

	Block 1	Block 2	Block 3	Block 4	Block 5	Block 6	
5' CCACATTAATCCTCCACC - Ase I	ATAC CAAT ACTA TTCA	AACT TCTT TTAC CATA	CATT ATCA TCTA AAAC	TATC CAAA CTTT ACAT	AATC ACAA CTAT TTCA	CTAA TACT TTTC AAAC	3' - CACCACCTTTAAACCTCC - bead Dra I

Fig. 5. DNA Sequences for Small Database. The word set design criteria and procedures described above were used to arrive at the word sets listed. The distal and proximal constant regions were designed to contain Ase I and Dra I restriction endonuclease cleavage site (respectively). Restriction sites were included because they may be useful for removing strands from the beads and for possible concatenation or cloning if the necessity should arise. The constant regions each contain 50% C to increase the melting temperature of primer binding.

The database strands were synthesized using an ABI automatic synthesizer and conventional phosphoramidite chemistry. Fresh, dry reagents and solvents were used with coupling times and deprotection conditions designed to optimize synthesis yield. Yields in the high ninety percents were obtained per synthesis cycles. Figure 5 presents detailed sequence information for synthesis of the initial, small library. Figure 6 shows four database sequences which were chosen as targets for experimental queries. The four traces shown in Figure 6 define a set of queries for testing searches over the entire range of difficulty within the small database. Shown below are the target sequences (written 5' to 3') aligned with their complementary probe sequences (written 3' to 5').

Most common sentence/high probability words (1 copy in 130 sentences)
Target 1:   ATAC AACT AAAC TATC AATC CTAA
Probe 1:   TATG TTGA TTTG ATAG TTAG GATT

Moderate sentence probability/constant word probability (1 copy in 8,304)
Target 2:   CAAT TTAC ATCA CTTT ACAA TTTC
Probe 2:   GTTA AATG TAGT GAAA TGTT AAAG

Moderate sentence probability/variable word probabilities (1 copy in 8,304)
Target 3:   ATAC CATA TCTA TATC TTCA TACT
Probe 3:   TATG GTAT AGAT ATAG AAGT ATGA

Least common sentence/low probability words (1 copy in 531,441 sentences)
Target 4:   TTCA CATA CATT ACAT TTCA AAAC
Probe 4:   AAGT GTAT GTAA TGTA AAGT TTTG

We first constructed this small initial library for the first phase. This first phase resulted in initial libraries of size 46 where the most rare strand had probability $(1/9)^6 = 1/531,441$ effectively modeling libraries of size up to $531,441$.

2.4 Experiments in DNA Search and Readout in the Small Library

In initial experiments, fluorescently labeled query strands were used as probe to anneal to and mark microbeads carrying target database strands. Then, fluorescent and non-fluorescent beads were separated by Fluorescence Activated

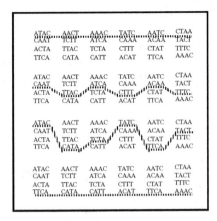

Fig. 6. Target Sequences shown as Traces through Word Sets. The six blocks of four words are shown four times with lines tracing the patterns of synthesis for selected database sequences. Words are listed with the most common subsequence (2X concentration) at the top, followed by the two moderate (1X) concentration words, and finally the least common word (1/2X) at the bottom of each list. The top trace gives the most common sentence, composed entirely of most common words. The second trace shows a sentence of moderate probability in the database, comprised of moderate concentration words. The third trace gives a sentence of moderate probability, composed of common, rare, and moderate words. The final trace shows the rarest sentence, made up of the least common words.

Cell Sorting (FACS). FACS or flow cytometry was developed for the study and purification of specific cell types from complex mixtures of cells. FACS has also been used for sorting diverse libraries of biomolecules immobilized on micro-scale plastic beads [3]. In FACS, a thin stream of individual particles are passed though one or more laser beams, causing light to scatter and fluorescent dyes to emit light. The light signals are converted to electrical impulses, and data about the observed particles are used to direct particle-containing droplets into appropriate down stream receptacles. The electrostatically charged droplets are deflected into a desired vessel by charged plates. Scattered light indicates a cell or bead is present, and fluorescence signal indicates that a bead has been labeled with a probe strand and thus matches the database query. Besides separation of beads FACS also provides a count of the number of beads sorted into each group.

Search experiments with FACS output (Figure 7) provided data in good agreement with expected results. In control runs, bead counts were within a few percentage points of the designed values. As shown in Figure 7, the location of the F peak shifts down field in the experimental versus the control samples. This shift indicates decreased fluorescence intensity for fluorescein attached to annealed probe compared to the same dye directly attached to bead-bound strand. The expected result for a clean separation is an F "island" fairly well removed from the N peak, as seen in the controls. Further experiments will clarify whether the observed results are due to a difference in fluorescence yield of the dye or due to varying levels of non-specific probe binding. The former explanation currently appears most plausible because increases in annealing and washing stringency decrease the entire F peak and do not preferentially affect the down field portion of the peak. Overall, the current results indicate the viability of the query and output methods.

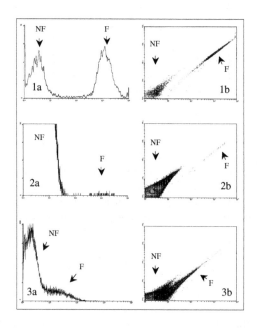

Fig. 7. FACS Data from Query of the Small Library. Data from three experiments is shown in two formats; panels labeled 'a' show histograms of bead count (y-axis) versus fluorescent intensity (x-axis); panels labeled 'b' show fluorescent intensity measures from two separate photomultiplier channels with each dot representing a single bead observation. Groupings of non-fluorescent (NF) and fluorescent (F) beads are indicated. Panels 1a and b show results from a control experiment containing a 50:50 mixture of fluorescently labeled and unlabeled beads. This sample contained $20,000$ beads. Control mixtures were prepared using beads carrying a small test oligonucleotide (NF) and the same small oligo with a molecule of fluorescein covalently attached to the $5'$ end (F). Panels 2a and b show the results of a similar control in which the labeled beads were present at only 1 in $10,000$ beads. A sample containing $150,00$ beads was sorted. Panels 3 a and b show data from a search experiment in which a fluorescently label DNA probe was hybridized with the bead bound library. The sample contained $47,000$ beads. The probe sequence was expected to bind specifically to approximately 1 in 130 database beads (Probe 1, described in the previous section). Depending upon where the boundary between the groupings was drawn, the F (labeled) peak contained between 0.6% and 1.0% of the total beads, in reasonable agreement with the expected value.

2.5   Increasing the Number of Database Strands

The next stage of the project is to increase database size. Note that this database size generated by combinatorial, mix-and-split methods on plastic microbeads is limited by a number of factors, including the total number of beads, which is up to $20,000,000,000$ for $5\,\mu m$ beads, but is approximately $1,000,000,000$ for the $25\,\mu m$ beads used here. The maximum library size generated by combinatorial, mix-and-split methods on $25\,\mu m$ plastic microbeads can thus be at most approximately $100,000,000$. We increased database size by two methods: (i) The number of possible words chosen for each block was increased from 4 to 12 and the length of each word was increased from 4 bases to 5 bases. There is Hamming distance of at least 3 between each pair of distinct words within a block. Seven stages of the combinatorial, mix-and-split synthesis on plastic microbeads then resulted in a library L of size $12^7 = 35,831,808$, where each strand has 35 bases in addition to the fixed flanking sequences at each end. (ii) The size of the library is then squared by combining pairs of the synthesized library strands. A library was designed with 7 blocks each of 12 possible words using procedures similar to those described above for the smaller library. Each DNA strand of the library is single stranded, and encodes a number which provides the index to the database element. To encode the base 12 digit in the ith position, we used a distinct 12 element set S_i of DNA subsequences, each of 5 bases in length. The encoding of each k place base 10 digit integer is thus done using a sequence of k consecutive 5 base DNA sequences. In addition each DNA strand of the library has certain flanking subsequences that are used in the synthesis of the library (a $5'$ *Bam HI* site — CATCGGATCC and a $3'$ *Bgl II* site — AGATCTCACACCCTCCAC). Designed word sets are:

Block 1	Block 2	Block 3	Block 4	Block 5	Block 6	Block 7
AAACC	AATCC	AACCA	AACCT	AATCC	ACACA	AACCA
ACCAA	ACACT	ACATC	ACCTA	ACAAC	ACCAT	ACACT
ACTCT	ATCAC	ACCAT	ACTAC	ACCTT	ATCTC	ACTTC
ATCTC	CAAAC	ATTCC	ATACC	ATCCA	CACAA	ATCAC
CATAC	CCATA	CACTT	CAAAC	CAACT	CATTC	CATAC
CCTTA	CCTAT	CATAC	CCATT	CCATA	CCATT	CCAAA
CTACA	CTCTT	CCAAA	CTCAA	CTCAT	CTAAC	CTCTA
CTCAT	CTTCA	CTACT	CTTCT	CTTTC	CTTCA	CTTCT
TACCA	TACCT	TAACC	TATCC	TACTC	TAACC	TACTC
TCAAC	TCCAA	TCCTA	TCACA	TCCAA	TCCTA	TCCAT
TCCTT	TCTTC	TCTCT	TCCAT	TCTCT	TCTAC	TCTCA
TTTCC	TTACC	TTCAC	TTCTC	TTACC	TTCCT	TTACC

The second phase of construction of the very large library is shown in Figure 8. At the time of this writing the first phase had been successfully completed, while the second phase had yet to be attempted.

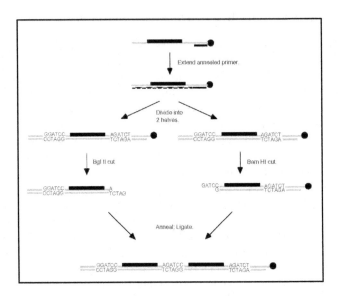

Fig. 8. Procedure for Squaring the Diversity of the Library. The database encoding region of the DNA is shown as a black rectangle, and the resin bead as a black circle. First, a complementary primer is annealed to the bead bound database strand and extended with DNA polymerase to generate the double-stranded library. Next, the library is divided and one half is digested with *Bam HI* while the other half is digested with *Bgl II*. Note that any pair of restriction enzymes which generates compatible sticky-ends could be used. Note also that dephosphorylation of the *Bgl II* cut fraction will prevent tail-to-tail ligation in the subsequent step. Following annealing and ligation of the *Bam HI* and *Bgl II* digested fractions, the double length library is obtained. The resulting sentences consist of 14 words, each of 5 bases, plus the remaining invariant sequences. The new library will have half as many total elements as the starting library, but its diversity will be the square of the starting library's diversity. Even with 10-fold coverage in the starting library there is a high probability that all potential sequences will be represented in the final library.

2.6   Experiments in DNA Search and Readout in the Very Large Libraries

Our libraries grew so large that the exclusive use of FACS readout was no longer sufficient to isolate a desired database entry. However, FACS readout can be employed as a first step, isolating sets of DNA strands with a common suffix. Given a search query, we will first separate out (either using magnetic bead separation or a cell sorter) all sequences matching the last 7 blocks of a sentence. This will greatly decrease the amount of DNA to be searched. A simple design for our hybridization experiments will be to use PCR with primers consisting of two sequences, each consisting of 7 blocks. These can be sequenced to provide output.

3   Computer Simulations

We also made computer simulations of our method for DNA-based associative search. We began with the selection of (digital) multispectral images to form a trial image database, and then preprocessed the image data to construct an "attribute database" which was then converted into a DNA database. Using known parameters for the kinetics of DNA hybridization, we simulated the use of PCR to perform an associative search. The result was used to reconstruct the original image using extensions of techniques described in [2]. These simulations can be viewed on the internet at the following sites.

DNA Database Search Simulation
> `http://cgi.cs.duke.edu/~clk/dnasrch/dna_search.cgi`

DNA Database Selection
> `http://cgi.cs.duke.edu/~clk/dnasrch/open_dna_db.cgi`

DNA Image Database Construction Page:
> `http://cgi.cs.duke.edu/~clk/dnasrch/MakeDNADB.html`

More information about searching:
`http://cgi.cs.duke.edu/~clk/dnasrch/help/search_info.html#select`

4   Conclusion and Future Applications

This study explored the use of DNA databases for storage of information and retrieval of that information by associative search techniques. Important procedures in subsequence and sequence design were demonstrated as well as successful application of those designs to combinatorial library synthesis by mix-and-split methods. Query output by bead separation using a fluorescence activated cell sorter was achieved. Test databases of increasing diversity were synthesized, and construction of an extremely large database is now partially complete. Future work includes finishing construction of the large library, further optimization of hybridization conditions for associative searches, sequencing a number of database elements as quality control of the synthesis, evaluating the benefits (increased fidelity?) of increasing the word length from 4 to 5 bases, and examining the limits of detection by searching for very rare database elements.

Bead-bound DNA databases similar to those described here would be useful for applications in which spatial clustering of identical sequences imparts some benefit, for example searches within libraries so huge that a presorting or enrichment procedure would be required as an initial filtering step. It should also be noted that the DNA databases described here can easily be cleaved from the solid support and utilized in soluble form for applications in which solubility would be beneficial. We are currently adapting our database system for construction of large cDNA libraries with synthetic DNA tags on both ends of each element for more complex associative search procedures.

References

1. Baum, E. B., How to build an associative memory vastly larger than the brain, Science **268**, 583–585 (1995).
2. Reif, J.H. and LaBean, T.H. (2000) Computationally Inspired Biotechnologies: Improved DNA Synthesis and Associative Search Using Error-Correcting Codes and Vector-Quantization, Sixth International Meeting on DNA Based Computers (DNA6), DIMACS Series in Discrete Mathematics and Theoretical Computer Science, Leiden, The Netherlands, (June 2000) ed. A. Condon. To be published by Springer-Verlag as a volume in Lecture Notes in Computer Science, (2001). [PostScript: `http://www.cs.duke.edu/~reif/paper/Error-Restore/Error-Restore.ps`]   [PDF: `http://www.cs.duke.edu/~reif/paper/Error-Restore/Error-Restore.pdf`]
3. Brenner, S., Williams, S.R., Vermaas, E.H., Storck, T., Moon, K., McCollum, C., Mao, J., Luo, S., Kirchner, J.J., Eletr, S., DuBridge, R.B., Burcham, T. and Albrecht G. In Vitro cloning of complex mixtures of DNA on microbeads: Physical separation of differentially expressed cDNAs. PNAS, **97**(4), 1665–1670 (2000).
4. Brenner S., Johnson M., Bridgham J., Golda G., Lloyd D.H., Johnson D., Luo S., McCurdy S., Foy M., Ewan M., Roth R., George D., Eletr S., Albrecht G., Vermaas E., Williams S.R., Moon K., Burcham T., Pallas M., DuBridge R.B., Kirchner J., Fearon K., Mao J., and Corcoran K. Gene expression analysis by massively parallel signature sequencing (MPSS) on microbead arrays, Nature Biotech. **18**(6), 630–634, (2000).
5. Marathe, A., Condon, A.E. and Corn, R.M. (2000) On Combinatorial DNA Word Design, DNA Based Computers V, DIMACS Workshop on DNA Based Computers (5th, MIT, June 14–15, 1999) editors E. Winfree and D.K. Gifford, DIMACS Series in Discrete Mathematics and Theoretical Computer Science, Volume 54, pp. 75–90, (2000).
6. Amenyo, J.-T., Mesoscopic computer engineering: Automating DNA-based molecular computing via traditional practices of parallel computer architecture design, Proceedings of the 2nd Annual DIMACS Meeting on DNA Based Computers, (June 1996).
7. Baum, E. B. DNA Sequences Useful for Computation, 2nd Annual DIMACS Meeting on DNA Based Computers, Princeton University, June 1996.
8. Deaton, R., R.C. Murphy, M. Garzon, D.R. Franceschetti, and S.E. Stevens, Jr., Good encodings for DNA-based solutions to combinatorial problems, Proceedings of the 2nd Annual DIMACS Meeting on DNA Based Computers, June 1996.
9. Mir, K.U. A Restricted Genetic Alphabet for DNA Computing, 2nd Annual DIMACS Meeting on DNA Based Computers, Princeton University, June 1996.
10. Garzon, M., R. Deaton, P. Neathery, R.C. Murphy, D.R. Franceschetti, S.E. Stevens Jr., On the Encoding Problem for DNA Computing, 3rd DIMACS Meeting on DNA Based Computers, U. Penn., (June 1997).
11. Gray, J. M. , T. G. Frutos, A.M. Berman, A.E. Condon, M.G. Lagally, L.M. Smith, R.M. Corn, Reducing Errors in DNA Computing by Appropriate Word Design, University of Wisconsin, Department of Chemistry, October 9, 1996.
12. Frutos, A.G., A.J. Thiel, A.E. Condon, L.M. Smith, R.M. Corn, DNA Computing at Surfaces: 4 Base Mismatch Word Design, 3rd DIMACS Meeting on DNA Based Computers, Univ. of Penn., (June 1997).
13. Cai, W., E. Rudkevich, Z. Fei, A. Condon, R. Corn, L.M. Smith, M.G. Lagally, Influence of Surface Morphology in Surface-Based DNA Computing, Submitted to the 43rd AVS Symposium, (1996).

14. Cukras A.R., Faulhammer D., Lipton R.J., and Landweber L.F. Chess games: a model for RNA based computation, Biosystems **52**(1–3), 35–45, (1999).
15. Deaton, R., R.C. Murphy, J.A. Rose, M. Garzon, D.R. Franceschetti, and S.E. Stevens, Jr., A DNA Based Implementation of an Evolutionary Search for Good Encodings for DNA Computation, ICEC'97 Special Session on DNA Based Computation, Indiana, April 1997.
16. Kaplan, P., G. Cecchi, and A. Libchaber, DNA based molecular computation: Template-template interactions in PCR, The 2nd Annual Workshop on DNA Based Computers, Amer. Math. Society, (1996).
17. Jonoska, N., and S.A. Karl, Ligation Experiments in Computing with DNA. ICEC'97 Special Session on DNA Based Computation, Indiana, (April 1997).
18. Wood, D. H., Applying error correcting codes to DNA computing, 4th DIMACS International Meeting on DNA-Based Computing, Baltimore, (June 1998).
19. Hartemink, A., David Gifford, J. Khodor,, Automated constraint-based nucleotide sequence selection for DNA computation, 4th Int. Meeting on DNA-Based Computing, Baltimore, Penns., (June 1998).
20. Hartemink, A. and Gifford, D. (1997) Thermodynamic Simulation of Deoxyoligonucleotide Hybridization for DNA Computation. Proceedings of the 3rd DIMACS Workshop on DNA Based Computers, University of Pennsylvania (June 1997).

Operation of a Purified DNA Nanoactuator

Friedrich C. Simmel and Bernard Yurke

Bell Labs, Lucent Technologies, Murray Hill, NJ 07974
{simmel, yurke}@lucent.com

Abstract. During the self-assembly and operation of DNA-based nano-mechanical devices like the previously reported molecular tweezers or actuators, unwanted dimerization can occur. Here we show that in the case of the DNA nanoactuator dimerization predominantly occurs at the assembly stage. Correctly formed molecular devices can be purified and subsequently operated without interference by dimers.

1 Introduction

Owing to its unique molecular recognition properties, DNA is an attractive self-assembly molecule for the construction of nanoscale objects [1,2,3,4]. Moreover, it has recently been shown that DNA-based nanomechanical devices can be used to induce motion on a nanometer scale. This can either be achieved by buffer-mediated conformational changes (the B-Z transition [5]) or by base-pairing interactions between DNA strands. With the latter approach, the construction and operation of several simple nanomachines has been demonstrated so far: Molecular "tweezers" [6,7], "actuator" [8] and "scissors" [9]. All machines have a tendency to form dimers or multimers during their assembly from single strands or during their operation. This might reduce or even destroy the functionality of these devices, especially when scaling to higher concentrations. In the case of the molecular tweezers, dimers are formed during the operation of the machine. At concentrations of $\approx 1\mu$M, some 80% of the tweezers are operating properly, whereas 20% are linked together in multimers [7]. In contrast, for the DNA actuator dimer formation seems to occur predominantly at the assembly stage. For this device, the amount of dimer formation is only about 10% [8]. Possible further multimerization of nanoactuators during their operation seems to be inhibited. Here we show that these properties can be exploited to purify actuator devices after their assembly and to operate the purified nanomachines without interference by dimers.

2 Operation Principle

The DNA nanoactuator is assembled from two single strands of DNA, a 40 base long strand A and an 84 base long strand B (sequences are given in the appendix). These strands hybridize together to form a ring-like structure consisting of two double-stranded (ds) 18 base-pair (bp) long arms connected by a four base long

N. Jonoska and N.C. Seeman (Eds.): DNA7, LNCS 2340, pp. 248–257, 2002.
© Springer-Verlag Berlin Heidelberg 2002

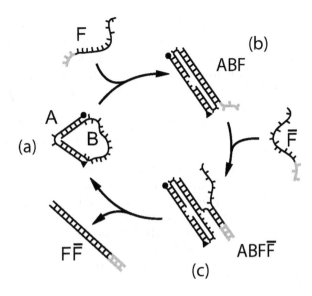

Fig. 1. Operation cycle of the DNA actuator: (a) After self-assembly from strands A and B the actuator is in its relaxed state. (b) Hybridization with the fuel strand F leads to the formation of the straightened complex ABF. The fuel strand has a short overhang section (depicted in gray) which remains un-hybridized at this stage of the operation cycle. (c) The removal strand F̄ — the complement of F — can attach to the unhybridized section of F and remove F from the complex ABF by branch migration. After completion of the strand displacement, the waste product FF̄ detaches and the original relaxed state of the actuator (a) is restored.

single-stranded (ss) hinge and a 44 base long single-stranded section connecting the other ends of the arms (Fig.1 (a)). The double-stranded regions are much shorter than the persistence length of dsDNA ($\approx$ 100 bp) [10], whereas the single-stranded sections are comparable to or longer than the persistence length of ssDNA (a few bases [10,11]). Therefore, the dsDNA sections of the machine can be regarded as mechanically stiff elements connected by rather flexible ssDNA sections. Motion is induced by the introduction of a DNA "fuel strand" F: The first 40 of the 48 bases of strand F are complementary to the central 40 base long section of the single-stranded loop of the nanoactuator. Upon hybridization of these strands, the straightened DNA complex ABF is formed (Fig. 1 (b)). The original state AB of the actuator can be restored by the introduction of a removal strand F̄ which is completely complementary to F. F̄ first attaches to the eight unhybridized bases at the end of F and then competes with AB for binding to F in a three-strand branch migration process (Fig. 1 (c)) [12,13]. Since F̄ has the advantage of being firmly attached to the overhang section of F, it finally wins this competition and completely wrests F from AB. A waste product FF̄ is formed and the actuator returns to its relaxed state. The DNA actuator is a thermodynamic engine in the sense that it can be cyclicly driven through its different mechanical states, thereby consuming chemical energy and producing a waste product. The path from AB to ABF is thermodynamically different from the path from ABF to AB which means that the machine is capable of performing work.

This controllable conformational change can be utilised to push two molecular components attached to the ends of the actuator's arms apart from each other. In our first experiments [6,7,8], these components were fluorescent dyes forming a fluorescence resonance energy transfer (FRET) pair (TET (tetrachlorofluorescein) and TAMRA (N,N,N',N'-tetramethyl-6-carboxyrhodamine), symbolized by a triangle and a circle in the Fig. 1). FRET [14,15] is a nonradiative energy transfer process occuring between fluorescent dyes with overlapping emission and absorption bands. An excited FRET "donor" can transfer its excitation energy to the FRET "acceptor" without emission of a photon. Therefore, the fluorescence of the donor is effectively quenched in the presence of the acceptor. This process is only effective at distances of a few nanometers and has a steep distance dependence. With the help of FRET, the motion of the DNA actuator can be monitored: in the relaxed state the two dyes have a mean distance of $5 - 6$ nm. At this distance the donor fluorescence is partially quenched by the acceptor. When the actuator is straightened by the fuel strand the distance between the dyes increases to 13.6 nm and pushes the dye pair out of the FRET regime: the donor fluorescence increases to its "free" value. Removal of the fuel strand restores the relaxed state, accompanied by a drop of the donor fluorescence to its starting value (cf. [8]).

3   Dimer Formation

Not all of the strands A and B properly assemble into nanoactuators. They can also form the whole range of multimers A_nB_n with $(n > 1)$. Figures 2 (a) and (b) show the simplest possible dimer A_2B_2 and trimer A_3B_3 which can be formed during the assembly of the nanoactuators. When strands A and B are not mixed in stoichiometry, "odd" multimers like AB_2, A_2B etc. are also observed.

To identify dimer bands in gel electrophoresis experiments, in addition to A and B we designed strands α, β_{12} and β_{21}, which can only form dimer and higher multimer complexes $A_n(\beta_{12})_n(\beta_{21})_n\alpha_n$. α is a 40 base long strand like A, but with a different sequence. β_{12} and β_{21} are similar to strand B, but form one dsDNA arm with A and the other arm with α. Therefore under stoichiometric conditions only dimers and multimers are formed which are structurally identical to the multimers $A_{2n}B_{2n}$ with $n \geq 1$.

It is also conceivable that additional multimers are formed during the operation cycle. Two such possibilities are depicted in Figure 2 (c,d). In the case of Fig. 2 (c), two actuators AB would compete for binding to one fuel strand F. This structure, however, is unstable and would decay into a straightened actuator ABF and a relaxed structure AB by branch migration. Similarly, a

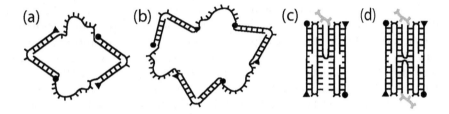

Fig. 2. Possible multimers formed from strands A and B: (a) A_2B_2 and (b) A_3B_3 can form during assembly of the actuator. (c) and (d) contain fuel strands F and can occur during actuator operation. The latter two multimers are unstable and dissociate into monomers by branch migration.

Holliday-like complex formed by two AB and two fuel strands (Fig. 2 (d)) would dissociate into two ABF complexes. Additional formation of dimers of the first type (Fig. 2 (a,b)) during operation can only occur if correctly assembled actuators dissociate and reassociate as dimers. Due to the relative stability of the 18bp duplexes (the "arms" of the actuator) at room temperature this is highly unlikely (see Section 5).

From previous experiments [8] we deduced that at $1\mu M$ concentrations approximately 10% of all A strands are bound in multimers, whereas 90% assemble properly into actuators. No indication of any additional dimer formation during the operation of the molecular machine was found in these experiments. In the next section we exploit this result to purify the properly formed actuators after their assembly and operate the purified devices.

4   Experiments and Results

DNA strands A,α,B,β_{12},β_{21}, F and $\bar{F}$ were purchased from IDT DNA Inc.. Strands A and α were labeled with the FRET pair TET and TAMRA at their 5' and 3' ends, respectively. The lyophilized strands were redissolved in TE buffer (Tris EDTA, pH 8.3) to a nominal concentration of 25 μM. The correct stoichiometric ratios of all strands were obtained from fluorescence titrations of the strands utilising FRET. TET was excited at 514.5 nm with light from an Argon ion laser. Fluorescence intensity was measured with a Si photodiode after filtering with a 20 nm bandpass filter centered at 540 nm. Gel electrophoresis experiments were performed to purify the nanoactuator and to check its operation. All gels were non-denaturing polyacrylamide gradient gels (10-20%) purchased from Zaxis Inc. The gel runs were performed at 20°C using TBE (Tris-borate, EDTA, pH 8.5) as running and reservoir buffer. To purify the actuators, strands A and B were mixed in stoichiometric amounts and incubated at room temperature for 30 minutes to allow for near-completion of the self-assembly process. Subsequently, the sample was loaded into a gel. After completion of the gel run, the bands corresponding to correctly assembled actuators were cut out and eluted from the gel in a large volume (several ml) of TE/salt buffer (TE buffer and 1M sodium chloride) for 48 h. The solution containing the eluted actuators was concentrated twice using Millipore Microconcentrators Microcon YM-3 following the manufacturer's protocol. The final concentration was determined in absorbance measurements performed with a Hitachi GeneSpec III UV-vis spectrophotometer. To avoid problems with hypochromicity due to base stacking (nanoactuators contain both single- and double-stranded sections), the concentration was derived both from the DNA absorbance maximum at 260 nm and from the TET absorbance maximum at 521 nm. Both values yielded the same result.

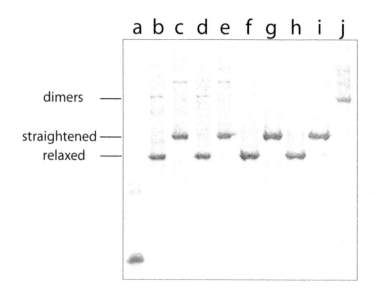

Fig. 3. Gel electrophoresis on a 10-20% polyacrylamide gradient gel. Migration direction is from top to bottom. The lanes contain: (a) strand A, (b) strands A+B, (c) A+B+F (straightened), (d) A+B+F+$\bar{F}$ (relaxed), (e) A+B+F+$\bar{F}$+F (straightened again), (f) purified actuator AB$_{pure}$, (g) AB$_{pure}$+F, (h) AB$_{pure}$+F+$\bar{F}$, (i) AB$_{pure}$+F+$\bar{F}$+F, (j) dimers formed by A+α+β_{12}+β_{21}.

The purified actuators were then compared with a "crude" mixture of actuators and dimers in another gel electrophoresis run(Figure 3). The samples to be loaded into the gel were prepared from microliter amounts of 25μM strand solutions mixed in stoichiometry. All reactions were given at least 20 min to proceed to near-completion. The amount of purified actuator was adjusted to the concentration of A strands in the non-purified samples, i.e. all lanes contain the same amount of dye and corresponding crude and purified lanes (i.e., b and f, c and g, etc.) contain the same total amounts of strands.

To obtain Figure 3, the gel was transilluminated with Argon laser light (514.5 nm) and photographed through a 40 nm bandpass filter centered at 550 nm. The photo was subsequently improved with standard image processing software. Lane (a) contains strand A alone, lane (j) only dimers formed with the dimer strands A, α, β_{12} and β_{21}. Lanes (b)–(e) of Figure 3 contain the crude actuator and lanes (f)–(i) are the corresponding lanes for the purified device. In lane (b) the result of the assembly process can be seen: Most of the A and B strands form the actuator monomer, but there is also a significant amount of dimer. In contrast, the purified sample in lane (f) contains almost no dimers. In lanes (c), (d) and (e) as well as in (g), (h) and (i) the actuator is straightened, relaxed and straightened again as described in Section 2. The formation of the higher molecular weight complex ABF leads to a shift of the band, when the actuator is straightened. Removal of the fuel strand restores the original state AB and the band shifts back. Fuel strands can also attach to dimers and therefore the dimer lanes in (c)–(e) shift in parallel with the actuator lanes.

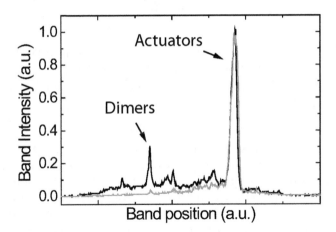

Fig. 4. Intensity profile of lanes b (black) and f (gray) in Fig. 2 from top to bottom of the gel. The profiles are normalized to intensity of the band with the strongest fluorescence.

In Figure 4 the intensity profile of lane (b) is compared to that of lane (f). The amount of dimers in lane (f) is one order of magnitude smaller than in lane (b), also the fluorescence background is significantly lower. From Figure 4 one can deduce that in the crude sample, approximately 20% of strand A are improperly assembled into dimers, whereas in the purified sample this fraction is only 2.5%. These 2.5% are a contamination of unknown origin. In fact, in other purification attempts we found no indication of dimers at all (see Fig. 6). The value for the crude actuator is higher than that reported previously [8]. However, in [8] this value was obtained in a different set of experiments at significantly lower concentrations — higher concentrations favour multimerization.

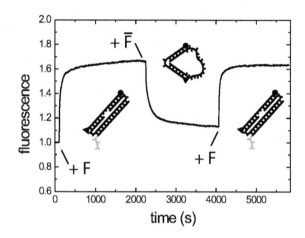

Fig. 5. Fluorescence signal during the operation of the purified DNA actuators. The moments at which fuel or removal strands are added are indicated. The fluorescence intensity is normalized to the fluorescence of the relaxed actuators at the beginning of the measurement.

We also performed kinetics measurements with the purified DNA actuators. In Fig. 5 the fluorescence signal of purified actuators at $1\mu M$ concentration is monitored during one and one-half operation cycles (Fig. 1). Addition of fuel strand F straightens the actuator and reduces FRET efficiency, thereby increasing the fluorescence signal. Addition of the removal strand $\bar{F}$ restores the original configuration and the fluorescence drops due to more efficient FRET. For the straightening reaction the time for half-completion $t_{1/2}$ is ≈ 23 s, for the relaxing reaction $t_{1/2} \approx 50$s. These values are close to the values previously found for the non-purified actuators at the same concentration [8]. This result indicates that there is no significant speedup or otherwise enhanced performance of the purified samples.

5   Discussion

Our experimental results confirm that dimer formation predominantly occurs during the self-assembly of the DNA actuator and not during its operation. The amount of dimer formation deduced from the experiments is consistent with previous findings. Using standard molecular biology protocols, purification of the correctly assembled devices is possible. Driving the purified nanomachines through their operation cycle reveals almost perfect behaviour. This also shows that the actuators retain their structural integrity throughout the purification process and over at least several days. Thermodynamically, the purified sample is in a nonequilibrium state. In equilibrium one would expect a distribution over all possible multimers [16,17]. However, to relax to equilibrium, the actuators would have to dissociate first and reassociate partly as multimers. At room temperature, this process is kinetically hindered: From kinetic data of DNA duplex dissociation [18] one can estimate the half-life of an 18bp duplex (one of the arms of an actuator) to be on the order of 10-100 years at 25°C.

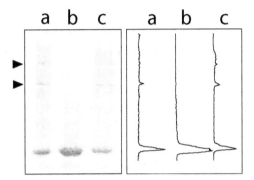

Fig. 6. Annealing of purified actuators. Left: Photograph of the gel. Right: Intensity profiles. Lane (a) contains non-purified actuators taken from the actuator mixture before purification. Lane (b) contains the purified actuators (notice: no dimers at all). Lane (c) contains strands from the purified sample after annealing to 100°C and cooling back to room temperature. DNA amounts and concentrations (as deduced from the absorbance at 260 nm) are the same in all lanes.

To speed up this process, we annealed a sample of purified actuators by heating it to 100°C and slowly cooling it back to room temperature. Indeed, the resulting gel run for this sample shows a dimer distribution like that for the non-purified actuators (Fig. 6).

6   Conclusion

We have presented a simple molecular machine — a nanoactuator — made from and powered by DNA. During the self-assembly of the machine, apart from correctly assembled actuators also dimers are formed. We show that the correctly assembled devices can be purified and subsequently operated in pure form. The ability of purifying molecular machines may become important in future applications in nanotechnology when issues like assembly yield or error rates have to be adressed.

Acknowledgments

We wish to thank Allen P. Mills Jr. and Yuan-Chin Ching for assistance in setting up the experiments. F.C.S. gratefully acknowledges financial support through the Feodor Lynen program of the Alexander von Humboldt foundation.

Appendix

$A = $ 5′ TGCCTTGTAAGAGCGACCATCAACCTGGAATGCTTCGGAT 3′

$\alpha = $ 5′ GCGATGAAGTGTCCACCTATTTCCGTATGGAGCAGAATCC 3′

$B = $ 5′ GGTCGCTCTTACAAGGCACTGGTAACAATCACGGTCTATGCG
GGAGTCCTACTGTCTGAACTAACGATCCGAAGCATTCCAGGT 3′

$\beta_{12} = $ 5′ GGTCGCTCTTACAAGGCACTGGTAACAATCACGGTCTATGCG
GGAGTCCTACTGTCTGAACTAACGGGATTCTGCTCCATACGG 3′

$\beta_{21} = $ 5′ AGGTGGACACTTCATCGCCTGGTAACAATCACGGTCTATGCG
GGAGTCCTACTGTCTGAACTAACGATCCGAAGCATTCCAGGT 3′

$F = $ 5′ AGTTCAGACAGTAGGACTCCCGCA
TAGACCGTGATTGTTATGCTACGA 3′

$\bar{F} = $ 5′ TCGTAGCATAACAATCACGGTCTA
TGCGGGAGTCCTACTGTCTGAACT 3′

References

1. J. Chen and N. C. Seeman. Synthesis from DNA of a molecule with the connectivity of a cube. *Nature* 350: 631–633 (1991).
2. Y. Zhang and N. C. Seeman. The construction of a DNA truncated octahedron. *J. Am. Chem. Soc.* 160: 1661–1669 (1994).
3. C. Mao, W. Sun, and N. C. Seeman. Assembly of Borromean rings from DNA. *Nature* 386: 137–138 (1997).
4. E. Winfree, F. Liu, L. A. Wenzler, and N. C. Seeman, Design and self-assembly of two-dimensional DNA crystals. *Nature* 394: 539–544 (1998).
5. C. Mao, W. Sun, Z. Shen, and N. C. Seeman. A nanomechanical device based on the B-Z transition of DNA. *Nature* 397: 144–146 (2000).
6. A. J. Turberfield, B. Yurke, and A. P. Mills Jr. DNA hybridization catalysts and molecular tweezers. *DNA Based Computers V: DIMACS Workshop DNA Based Computers, June 14–15, 1999* DIMACS Series vol. 54: 171–182 (2000).
7. B. Yurke, A. J. Turberfield, A. P. Mills Jr., F. C. Simmel, and J. L. Neumann. A DNA-fuelled molecular machine made of DNA. *Nature* 406: 605–608 (2000).

8. F. C. Simmel and B. Yurke. Using DNA to construct and power a nanoactuator. *Phys. Rev. E* 63: 041913-(1–5) (2001).

9. J. Mitchell and B. Yurke. Molecular scissors. See article in this volume (2001).

10. S. B. Smith, Y. Cui, and C. Bustamante. Overstretching B-DNA: the elastic response of individual double-stranded and single-stranded DNA molecules. *Science* 271: 795 (1996).

11. B. Tinland, A. Pluen, J. Sturm, and G. Weill. Persistence length of single-stranded DNA. *Macromolecules* 30: 5763–5765 (1997).

12. C. Green and C. Tibbetts. Reassociation rate limited displacement of DNA strands by branch migration. *Nucl. Acids Res.* 9: 1905–1918 (1981).

13. I. G. Panyutin and P. Hsieh. The kinetics of spontaneous DNA branch migration. *Proc. Natl. Acad. Sci. USA* 91: 2021–2025 (1994).

14. Th. Förster, Zwischenmolekulare Energiewanderung und Fluoreszenz. *Annalen der Physik* 6: 55–75 (1948).

15. L. Stryer and R. P. Haugland. Energy transfer: a spectroscopic ruler. *Proc. Nat. Acad. Sci. USA* 58: 719–726 (1967).

16. H. Jacobson and W. H. Stockmayer. Intramolecular Reaction in Polycondensations. I. The Theory of Linear Systems. *J. Chem. Physics* 18: 1600–1606 (1950).

17. A. A. Podtelezhnikov, C. D. Mao, N. C. Seeman, and A. Vologodskii. Multimerization-cyclization of DNA fragments as a method of conformational analysis *Biophys J.* 79: 2692–2704 (2000).

18. L. E. Morrison and L. M. Stols. Sensitive Fluorescence-Based Thermodynamic and Kinetic Measurements of DNA Hybridization in Solution. *Biochemistry* 32: 3095–3104 (1993).

DNA Scissors

James C. Mitchell* and Bernard Yurke**

Bell Laboratories, Lucent Technologies,
600 Mountain Avenue, Murray Hill, NJ 07974
yurke@lucent.com

Abstract. Designed strands of DNA were used to construct a nanomachine with the appearance of a pair of scissors. Further strands of DNA were then employed to close and reopen the handles of the scissors with resultant closing and reopening of the jaws with a change in angle of $10°$. It was further shown that it is possible to open the handles wider than their equilibrium position, with resultant widening of the jaws.

1 Introduction

The famous "DNA double-helix" is formed when two anti-parallel strands twist together. This hybridisation happens when the bases on opposite strands form Watson-Crick base-pairs, i.e. A complements T and G complements C. If the complementarity is perfect, the double-helix is sufficiently stable that sections of up to 100 base pairs may be treated as straight, rigid rods (so long as certain sequences, such as A-tracts, are avoided) with regards to thermal fluctuations [1]. Watson-Crick complementarity [2] has been used to design single-strands such that, when placed in solution together, complementary regions on the different strands twist together to form a linked set of rigid double-helices, with the designed structure. This is known as "self-assembly." Structures such as 2-D arrays [3], cubes [4] and even trefoil knots [5] and Borromean rings [6] have been formed in this way. In addition to the self-assembly of static structures, DNA has also been utilised to create devices that can undergo conformational changes [7]. Recently, a DNA machine in the form of a pair of tweezers was reported [8]. The arms of the tweezers can be closed and reopened by the addition of a closing strand followed by its complement. The drop in free energy when two complementary strands of DNA hybridise is used to drive the closure of the tweezers. Strand displacement by branch migration is then used to reopen the tweezers.

Here we report upon the next step in this technology, a scissors-like structure in which a pair of jaws are closed and reopened via handles that are operated by a closing strand and a removal strand. The handles are closed by the binding of the closing strand, TY, to the handle-tails, and are reopened when the closing strand is removed by its complement, RY. The closure and opening of the handles

---

* Current address: Department of Physics, University of Oxford, Clarendon Laboratory, Parks Road, Oxford, OX1 3PU, UK.
** To whom correspondence should be addressed.

N. Jonoska and N.C. Seeman (Eds.): DNA7, LNCS 2340, pp. 258–268, 2002.
© Springer-Verlag Berlin Heidelberg 2002

is coupled, through a junction, to the movement of a pair of jaws in the manner of a pair of scissors or tongs. The significance of this is that, on a nanometre length scale, it is possible to obtain precision movements of these machines at a small distance (~ 13 nm) away from where the closing force is being applied. This means that any two things that we may wish to attach to the ends of the jaws, and whose separation we wish to control, will not get entangled in the closing mechanism. This is a crucial step closer towards using DNA machines in real applications.

2 DNA Scissors

2.1 Mechanism for Operation

Table 1 shows the sequences used for the X and D versions of the scissors (the only difference between the two versions being the sequences used). Each of the scissors is constructed through the hybridisation of complementary regions of the four strands QY1, QY2, SY1, and SY2 with each other. As in Table 1, an (X) or (D) in front of these strand names will be used when it is necessary to distinguish to which of the two versions a strand belongs. The four double-stranded, rigid arms forming the handles and jaws of one of the scissors are connected together via "C3" spacers consisting of three CH_2 units, which are incorporated internally in the QY1, QY2, SY1, and SY2 strands. This spacer was chosen because it has enough single bonds to allow a $180°$ change in direction (important for the QY strands), while being short enough to be "hinge-like" rather than "spring-like".

Figure 1 shows the mechanism used to open and close the scissors. In the open configuration, there is nothing holding the two handles together. However, when the closing strand, TY, is added, it picks up the loose tails on the handles, hybridising with them, until finally the handles are pulled together. The scissors are now in the closed configuration. To reopen the scissors, the complement, RY, of the closing strand is added, which hybridises to the loose tail of TY and then gradually peels TY off from the scissors by a process known as "branch migration." The mechanism for all this is identical to that for the tweezers [8]. However, what happens to the jaws is less than obvious. If the C3 spacers between the "rods" are modelled as springs, the jaws should spring open when the handles are closed, but if the spacers are modelled as hinge-like, then the jaws should close, albeit to a lesser extent than the handles.

2.2 FRET Observation of Closure and Opening of Scissors

The jaws' behaviour was investigated using a phenomenon called "Fluorescence Resonance Energy Transfer" (FRET) between pairs of 5'-TET and 3'-TAMRA dyes [8]. TAMRA absorbs light at the fluorescence emission wavelength of TET. Therefore, if a TAMRA dye is in close proximity to a TET dye when TET is being made to fluoresce by an incident laser beam, fluorescence quenching occurs. The closer TAMRA is to TET, the less fluorescence is measured.

Table 1. Oligonucleotide sequences, written 5' to 3'. "X" in a sequence denotes a C3 spacer. QY1 and QY2 have 5'-TET and 3'-TAMRA dyes attached for some experiments, in which case they are called QY1F and QY2F.

(X/D)QY1
TGCCTTGTAAGAGCGACCATXCAACCTGGAATGCTTCGGAT
(X/D)QY2
GCGATGAAGTGTCCACCTATXTTCCGTATGGAGCAGAATCC
(X)SY1
GGAGTCCTACTGTCTGAACTAACGXATCCGAAGCATTCCAGGTTGXATAGGTGGACACTTCATCGC
(X)SY2
GGATTCTGCTCCATACGGAAXATGGTCGCTCTTACAAGGCAXCTGGTAACAATCACGGTCTATGCG
(X)TY
CGCATAGACCGTGATTGTTACCAGCGTTAGTTCAGACAGTAGGACTCCTGCTACGA
(X)RY
TCGTAGCAGGAGTCCTACTGTCTGAACTAACGCTGGTAACAATCACGGTCTATGCG
(D)SY1
GCAGGCTTCTACATATCTGACGAGXATCCGAAGCATTCCAGGTTGXATAGGTGGACACTTCATCGC
(D)SY2
GGATTCTGCTCCATACGGAAXATGGTCGCTCTTACAAGGCAXCAGCTAGTTTCACAGTGGCAAGTC
(D)TY
GTACTACGGACTTGCCACTGTGAAACTAGCTGCTCGTCAGATATGTAGAAGCCTGC
(D)RY
GCAGGCTTCTACATATCTGACGAGCAGCTAGTTTCACAGTGGCAAGTCCGTAGTAC
DTXY
TAGAGCCTCGCATAGACCGTGATTGTTACCAGCTCGTCAGATATGTAGAAGCCTGC
DTYX
CTGGACATGACTTGCCACTGTGAAACTAGCTGCGTTAGTTCAGACAGTAGGACTCC

In this experiment, these dyes were placed either on the handles, by using TET- and TAMRA-labelled QY1 strands (i.e. QY1F), as in (X)B and (D)B solutions, or on the jaws, using similarly labelled QY2 strands (i.e. QY2F), as in (X)A and (D)A solutions. The (X)A and (X)B solutions were placed in cuvettes, temperature-controlled at 20°C, with 514.5nm laser light incident upon them. The fluorescence was then measured using a Si photodiode, which detected the light reemitted at 90° to the incident laser beam. The fluorescence from each of the solutions was plotted on the same axes against time. The closing strand, (X)TY, and its complement, the opening strand, (X)RY, were added one after the other at 30-minute intervals. These graphs, after correction for a dilution factor, are shown in Figure 2. The dilution of the scissors results from the fact that $5\mu l$ of fluid is added to the sample volume each time the (X)TY and (X)RY strands are added. Since the dilution factor has been removed, the remaining fluctuation in the fluorescence of each configuration is due to imprecise stoichiometry, which could be improved simply by using an ever-increasing excess of (X)TY and (X)RY strands. Similar results (not shown) were found for the

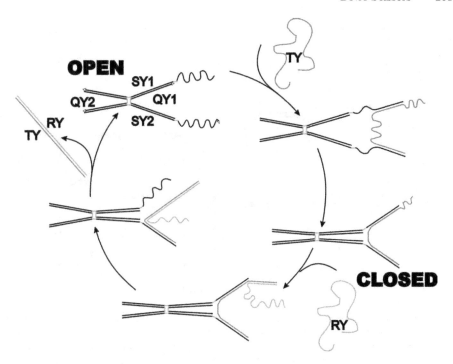

Fig. 1. Scissors operation — The short grey lines represent flexible C3 spacers (three CH_2 units) which have been included to break up the helices into separate "rods" and importantly, to allow QY1 and QY2 to bend at acute angles. Hence, the junctions are NOT Holliday junctions. The 3' end of each strand has been labelled with an arrowhead on the open configuration diagram.

D versions of the scissors (solutions DA and DB) with closing and reopening strands, (D)TY and (D)RY.

Furthermore, it is possible, using the calibration data in reference [10], to measure the distance between the TET and TAMRA dyes, and, hence, to determine the angles subtended by the jaws and by the handles. These were found, respectively, to be 26.0° and 46.6° for open scissors and 15.4° and 15.3° for closed scissors. These values are averaged over the two sequences, and errors are dominated by pipetting errors, which are thought to be within 10%.

2.3   Gel Electrophoresis Data

A polyacrylamide electrophoresis gel was run to show that the closing and reopening of the scissors is reversible, as shown in Figure 3. Lanes 1+2, 3+4 and 5+6 show open and closed scissors that have gone through 0, 1 and 2 closing and opening cycles, respectively. The bottom band, which appears in lanes 3-6 is the TY-RY waste product. This band is darker in lanes 5 and 6, because these scissors have undergone an extra closing and reopening cycle. The closed band

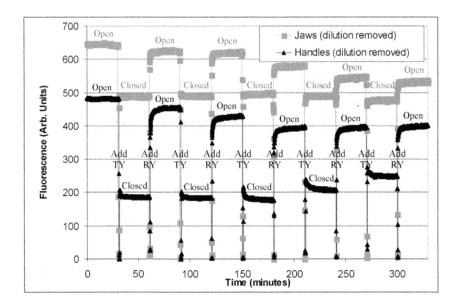

Fig. 2. Cycling scissors (fluorescence) — the high values on each data set correspond to when the jaws (grey squares) or handles (black triangles) are open; the low values when they are closed. The fluorescence drops to zero when TY or RY is added, because the pipette blocks the laser beam.

in lanes 2, 4, 6 and 8 migrates more slowly than the open band (lanes 1, 3, 5, and 7).

Lanes 7 and 8 are identical to lanes 1 and 2, except that the D sequences have been used to make the scissors instead of the X sequences. All gel results were independent of this choice. The existence of a single strong band for the closed scissors (lanes 2 and 8) suggests that TY binds to both handle-tails of the scissors. If it were sometimes to bind to one handle-tail, and sometimes to the other, two non-identical species would form (the tail on the TY breaks the symmetry — see Figure 1), which would migrate at different rates.

3   Control Experiments

3.1   Designed Dimer Cannot Form Closed Scissors

It was necessary to confirm that the band so-labelled in Figure 3 is indeed the closed scissors structure and not the dimer structure that may form when two separate open scissors structures are linked together by closing strands. For this purpose, two open scissors structures with different sequences were designed (X and D versions) with pseudoclosing strands (DTXY and DTYX) that link the X and D versions together as shown in Figure 4, rather than closing the handles of either the X or D scissors in the conventional way. This structure is designed to

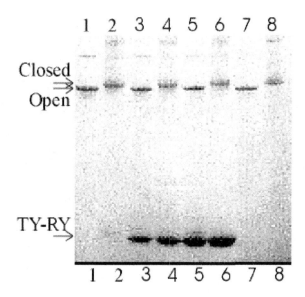

Fig. 3. 10-27% gradient polyacrylamide gel showing closure and reopening of scissors.

be the same as the dimer structure described above. It should be impossible for a subset of the strands making up this dimer structure to instead form a closed scissors structure. Therefore, by comparing the behaviour of the dimer standard with a closed scissors solution, the two would hopefully be shown to be different, hence demonstrating that the closed scissors are not dimers.

The dimer standard was run alongside a closed scissors solution in a gel to show that the dimer bands migrated slower than the band thought to be the closed scissors. It was seen that the most intense band of the dimer standard matched the faint band above the closed scissors band, which was expected to be the dimer structure.

However, the dimer standard solution unexpectedly managed to form a faint band at the same level as the closed scissors, as shown in Figure 5. We were able to establish that this band consists of a pair of open scissors with a pseudoclosing strand attached. The supposedly free end of this strand loosely associated with the other handle of the scissors, to which it was not complementary. This effect was shown to disappear when the temperature was increased to 40°C. Details of how this was established may be found in the appendix. To sum up, the closed scissors solution forms a structure that is not formed by the dimer solution, namely an open scissors structure that then has a closing strand firmly hybridised to both handle-tails. This supports our assertion that the change in fluorescence shown in Figure 2 is not a result of the formation and then break up of a dimer structure and hence must be due to the moving together and apart of TET and TAMRA dyes found on the same pair of scissors.

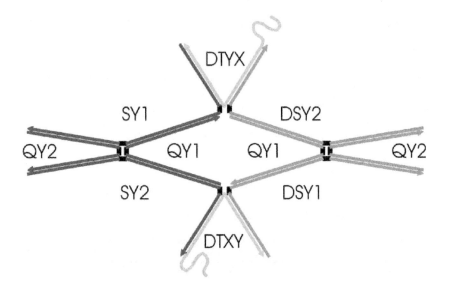

Fig. 4. Dimer Structure.

3.2   There Is a Specific Closing Strand for Each Pair of Scissors

There should be a specific interaction between the closing strand, (X)TY, and the X scissors, and also between (D)TY and the D scissors. To show this, the fluorescence of a handle-labelled scissors solution, (D)B, was monitored when strands of identical length to the correct closing strand, and with sequence complementary to one handle-tail, but not to the other, were added. In general, it was found that the fluorescence simply dropped by the dilution factor. This shows that the strand with sequence complementary to both the handle-tails is needed to close the scissors, and that some other strand, identical in all respects barring sequence, will not pull the handles together, nor will it interact directly with the dyes, to give the same drop in fluorescence. However, one undesigned interaction did manifest itself as an unexpected drop in fluorescence. In particular, when DTYX was added to X scissors, or DTXY was added to D scissors (which is the equivalent interaction, structurewise), there was a drop in fluorescence similar to that due to the correct closing of the scissors. The same effect was found in both handle- and jaw-labelled scissors, to the same extent, in each case, as for the normal closing of the scissors. This latter point is significant, as it rules out the possibility of a direct interaction with the dyes, hence the handles must be pulled closed by the pseudoclosing strand. However, because of base mismatches, this undesigned interaction should be weaker than that of the designed interactions. This was demonstrated by monitoring the fluorescence of a solution of this structure as a function of temperature. As shown in Figure 6 (black triangles) the fluorescence of the DB+DTYX complex returned to the dilution factor value at a temperature of 40°C, signalling the melting of the undesigned interaction of the DTYX strand with the D scissors. It is not until 65°C that a further drop in

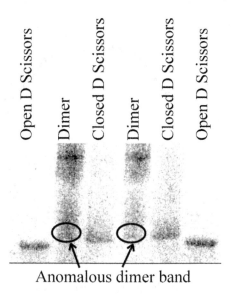

Fig. 5. 4-20% gradient polyacrylamide gel showing an anomalous band in the dimer standard lanes.

fluorescence occurs due to melting of the scissors. That the undesigned interaction is weaker than the designed interactions is also supported by the fact that the effect disappeared when the strand complementary to the other handle-tail was added.

3.3 Number of Strands Necessary for Formation of Scissors

Polyacrylamide gels were run confirming that all four and five strands, respectively, were necessary to form the open and closed scissors structures. All bands in lanes that lacked a strand migrated much faster than the lanes with a complete set of strands.

4 Jaws Can Also Be Prised Farther Apart

An interesting effect was found when unlabelled D scissors were added to jaw-labelled X scissors along with the DTXY and DTYX pseudoclosing strands. The fluorescence actually increased by about 3% (after removal of the dilution factor), suggesting that the unlabelled D scissors were forcing apart the handles of the X scissors, with a resultant widening of the jaws of the X scissors. Similarly, when both X and D scissors have labelled jaws, there was an 8% increase in fluorescence (after removal of the dilution factor), as one would expect from the symmetry of the structure, as X forces open D and vice-versa.

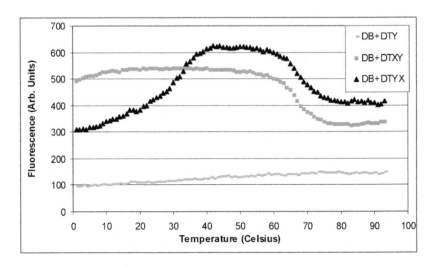

Fig. 6. 6 Graph showing that the pseudo-closing strand DTYX only closes the D scissors below 40°C. Above this temperature, the binding is well-behaved. At 65°C the fluorescence drops due to the dissociation of the pseudo-closing strand. Note that in the present case the reporter dyes are located on the handles of the scissors (labelled DB), whereas, in the case of Figure 7, the dyes are located on the jaws (labelled DA). This accounts for the difference in the shape of the corresponding curves between the two figures. No significance should be attached to the relative fluorescence of different solutions, as the different solutions were excited by different intensity laser beams.

5   Conclusions and Possible Application

We have shown that a scissors-like structure may be devised out of DNA, where the jaws, which close and reopen as a result of the closure and reopening of the handles, are displaced by a short distance (13nm) from those handles. It is this separation of force application from resultant motion that could make the scissors a more practical tool than the tweezers [8]. For example, this structure has the potential to be used for experiments where the separation of two molecules, which would be attached to the jaws, needs to be controlled and varied to high precision. For instance, by attaching an enzyme to one jaw and its substrate to the other jaw, one may be able to measure that enzyme's affinity for its substrate. This could be done by devising a set of alternative closing strands, which due to a certain number of non-complementary nucleotides in the middle, would only partially close the handles. Thus one could move the jaws closer together in discrete steps by adding strands, followed by their respective removal strands, which have fewer and fewer non-complementary nucleotides, until the handles are fully closed.

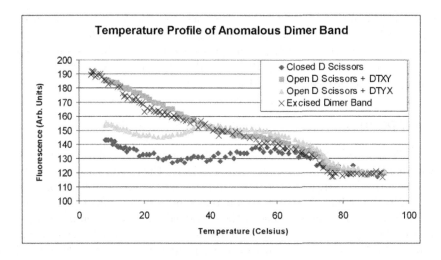

Fig. 7. Temperature Profile of Excised Anomalous Dimer Band.

6 Methods

PAGE-purified oligonucleotides with C3 spacers and dye-labelled HPLC-purified oligonucleotides were supplied by Integrated DNA Technologies. Stock solutions were prepared by resuspension of the lyophilised oligonucleotides to a concentration of 25 μM in TE buffer (10mM Tris pH 8.0, 1mM EDTA). Fluorescence quenching was used as an indicator to titrate most strands in SPSC buffer (50mM monobasic/dibasic sodium phosphate, pH 6.5, 1M NaCl). The exceptions were DTY-DRY, DTY-DSY2, QY1-CQY1, and QY2-CQY2 (where CQY1 and CQY2 are the complements of QY1 and QY2, respectively, used purely to calculate stoichiometry) for which a PAGE titration was carried out. Stoichiometric ratios of strands were used in all cases, unless the strand has been specified to be in excess. The 514.5nm line of an argon ion laser (chopped at 130Hz) was used to excite TET fluorescence, which was then selected by an interference filter with bandpass 10nm centred at 540nm and detected by a Si photodiode and phase-sensitive detector. The error in relative concentrations is thought to be around 10%, mainly due to pipetting errors.

For fluorescence measurements, open scissors solutions were mixed from 3μl QY2F and stoichiometric volumes of the other strands, made up to 50μl with SPSC buffer. Closing and opening strand solutions were made up from stoichiometric volumes (or excesses, depending on the experiment) of stock solution, made up to a convenient volume with SPSC buffer. After all additions, reactants were mixed by rapid pipetting up and down, lasting less than 5s. All measurements, except the temperature runs, were taken at 20°C.

Polyacrylamide Gel Electrophoresis (PAGE) was carried out with SIGMA pre-cast mini-polyacrylamide gels. Neat stock solutions were used in the gels, the volumes varying due to the limits imposed by the depth of the well (but always stoichiometric across lanes on any one gel). Each reaction mixture was allowed at least 20 minutes to proceed to near-completion before being loaded into the gel. The DNA bands in the gels were made visible as shadows cast on a fluorescent substrate when illuminated with 254nm UV light. These were photographed to provide Figures 3 and 5.

7 Appendix

A closed scissors solution was run in a gel alongside a dimer solution. There was a faint, anomalous band in the dimer lane, level with the main closed scissors band, suggesting that contrary to design, the dimer strands may be able to form closed scissors. To investigate this, the anomalous band was cut out of the gel and its DNA extracted. The fluorescence of the extracted DNA was then measured as a function of temperature and was compared to that of the closed scissors solution, as shown in Figure 7. The temperature dependence of the excised anomalous band was found not to be the same as for the closed scissors, but rather to be the same as for an open pair of D scissors with DTXY attached, as shown in Figure 7. This supports the assertion that this anomalous band in the dimer lane is not the same as the closed scissors. Hence, the gel shown in Figure 5 demonstrates that the labelled band is indeed the closed scissors structure and not some form of unwanted dimer.

References

1. Smith, S. B., Finzi, L., and Bustamante, C., 1992 "Direct mechanical measurements of the elasticity of single DNA groups by using magnetic beads," Science 258, pp. 1122–1126.
2. Watson, J. D. and Crick, F. H. C., 1953 "Molecular structure of nucleic acids; a structure for deoxyribose nucleic acid," Nature 171, pp. 737–738.
3. Winfree, E., Lui, F., Wenzler, L. A., and Seeman, N. C., 1998 "Design and Self-assembly of two-dimensional DNA crystals," Nature 394, pp. 539–544.
4. Chen, J. and Seeman, N. C. 1991 "Synthesis from DNA of a molecule with the connectivity of a bcube," Nature 350, pp. 631–633.
5. Ming Du, S., Wang, H., Tse-Dinh, Y.-C., and Seeman, N. C., 1995 "Topological Transformations of Synthetic DNA Knots," Biochemistry 34, pp. 673–682.
6. Mao, C., Sun, W., and Seeman, N. C., 1997 "Assembly of Borromean rings from DNA," Nature 386, pp. 137–138.
7. Mao, C., Sun, W., Shen, Z. and Seeman, N. C., 1999 "A nanomechanical device based on the B-Z transition of DNA," Nature 397, pp. 144–146.
8. Yurke, B., Turberfield, A. J., Mills Jr., A. P., Simmel, F. C., and Neumann, J. L., 2000 "A DNA-fuelled molecular machine made of DNA," Nature 406, pp. 605–608.
9. Stryer, L. and Haugland, R. P., 1967 "Energy Transfer: a spectroscopic ruler," Proc. Natl. Acad. Sci USA 58, pp. 719–726.
10. Simmel, F. C., and Yurke, B., 2001 "DNA molecular motors" Proc. SPIE, "Industrial and Commercial Applications of Smart Structures Technologies 2001," to be published.

A Realization of Information Gate by Using *Enterococcus faecalis* Pheromone System

Kenichi Wakabayashi and Masayuki Yamamura

Tokyo Institute of Technology, 4259 Nagatsuta, Kanagawa, Japan
wakabayashi@es.dis.titech.ac.jp

Abstract. In this paper, we introduce a novel signal element by using bacterial pheromones. In multicellular organism, every cell can communicate and exchange information with other cells. Bacteria also have such mechanisms. *Enterococcus faecalis*, one of the gram-positive bacteria, has a unique pheromone system. Male cells are stimulated by pheromones from female cells, and they give their plasmid to female cells through conjugation phenomenon. The variety of pheromones and their inducible activities of plasmid transfer inspire us that *Enterococcus faecalis* can serve as a pheromone-dependant DNA transporter. We show a design to realize logically controllable Information Gates by using *Enterococcus faecalis* and show an experimental plan. It is still on going project, but we can show the feasibility that bacterial pheromone system would provide alternative methodologies in molecular computing research.

1 Introduction

The recent studies of molecular computing show various methods to make computational elements with bioorganic materials. Head made a computer-memory with plasmid DNA [1]. This plasmid memory is accessed by restriction enzymes and designed to prevent re-cutting off by enzymes. Ehrenfeucht introduced the computing system using ciliates [2]. Ciliates can execute DNA recombination that cannot be easily carried out *in vitro*. Bloom's *LMBMC* is computing system based on liposome [3]. Liposome has potential to provide the space with various traffics of computational information as well as living-cells. The computing method mediated by cell fusion is also attractive. Abelson proposed *Cellular Computing*, which is aim for constructing a digital logic circuit with living-cells by regulating concentrations of DNA-binding proteins [4]. DNA-binding proteins are major controllers of gene transcriptions and therefore have potential to manipulate the cellular signaling.

Cellular computational system is one of the alternative methodologies in molecular computing. The cells, which are enveloped in membranes, can behave as functional devices that can localize some computational information in them. Cells are able to be stocked up on a large scale as well as proteins and DNA. It implies that we can perform the massive parallel processing with these materials. In addition, cells have many control-regulated functions that we cannot

N. Jonoska and N.C. Seeman (Eds.): DNA7, LNCS 2340, pp. 269–278, 2002.
© Springer-Verlag Berlin Heidelberg 2002

easily reproduce *in vitro*. It seems important to introduce organisms to molecular computing.

In this paper, we investigate organisms as molecular devices and show a practical design to construct an Information Gate by using *E. faecalis* pheromone system. In Section 2, we learn pheromone system of *E. faecalis*. In Section 3, we describe a scheme for the *E. faecalis* Information Gate. In Section 4, we show a practical design of the Information Gate and a small experiment to confirm the gate execution. In Section 5, we describe features of the Information Gate and pointed out advantages and some confronting problems about *E. faecalis* conjugation system.

2   Organic Materials in Molecular Computing

2.1   Information Carrier with Organic Materials

As well as many organic cells, even bacteria have communication system [5]. Quorum sensing is the most studied signaling network in bacterial cell-cell communication system [6]. Weiss made a cell-to-cell signal carrier with E.coli by transfection of LuxI/LuxR system derived from *Vibrio fischeri* [7]. This signal carrier consists of two types of cells, pheromone producing and pheromone receiving cells. It can produce luminescence protein only when two types of cells exist together. They predicted that bacteria and bacterial pheromones are useful to be an information carrier.

2.2   Pheromone-induced Conjugation of *Enterococcus faecalis*

We investigate another type of pheromone system: *Enterococcus faecalis* and their pheromones that induce conjugative plasmid transfer [8, 9, 10]. Pheromones are secreted by female *E. faecalis*. When pheromones are received by male *E. faecalis*, they activate transcriptions of some genes on their corresponding plasmid that encode cell surface adhesion molecules. Expressions of the adhesion molecules on the cell surface induce *E. faecalis* cell aggregation. Once cell aggregation is performed, the plasmid is transferred from male to female. This cell-to-cell DNA transfer phenomenon is called as conjugation (Figure 1).

By now, about 20 transferable plasmids are known, and five pheromones are determined. It is confirmed that certain class of plasmid is responsive to only corresponding class of pheromone. The relation between plasmid and pheromone is basically uniqueness. Plasmid-obtained recipient stops producing corresponding class of pheromone and becomes a new donor in turn. It is also known that there are a few pheromone inhibitors secreted by male *E. faecalis* that specifically inhibit pheromone functions in antagonistic manners. All of determined pheromones and inhibitors consist of 7 or 8 residue amino peptides (Table 1).

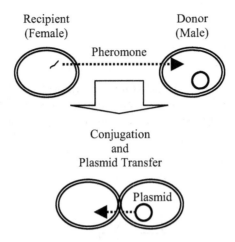

Fig. 1. Conjugative plasmid transfer of *Enterococcus faecalis*. Double ellipses indicate *E. faecalis* cells. Cells that do not have transfer-plasmids are considered as female, as they secret factors called pheromone that get male cells to conjugate and transport their plasmid to female cells.

Table 1. Structures of pheromones and inhibitors. Names start with initial 'c' are pheromones, 'i' are their inhibitors.

Pheromone / Inhibitor	Peptide structure
cPD1	H-Phe-Leu-Val-Met-Phe-Leu-Ser-Gly-OH
iPD1	H-Ala-Leu-Ile-Leu-Thr-Leu-Val-Ser-OH
cAD1	H-Leu-Phe-Ser-Leu-Val-Leu-Ala-Gly-OH
iAD1	H-Leu-Phe-Val-Val-Thr-Leu-Val-Gly-OH
cCF10	H-Leu-Val-Thr-Leu-Val-Phe-Val-OH
iCF10	H-Ala-Ile-Thr-Leu-Ile-Phe-Ile-OH
cAM373	H-Ala-Ile-Phe-Ile-Leu-Ala-Ser-OH
iAM373	H-Ser-Ile-Phe-Thr-Leu-Val-Ala-OH
cOB1	H-Val-Ala-Val-Leu-Val-Leu-Gly-Ala-OH
iOB1	H-Ser-Leu-Thr-Leu-Ile-Leu-Ser-Ala-OH

3   Design of *E. faecalis* Information Gate

Viewing the pheromone system overall, we can regard pheromone and plasmid as a kind of INPUT and OUTPUT. In this system, *E. faecalis* can change of pheromone for plasmid. It inspires us to construct a simple information carrier by using *E. faecalis* pheromones and plasmids. We illustrate a schematic design of Information Gate in Figure 2. The Information Gate is regulated by two INPUT signals (A and B), and discharges an OUTPUT signal (X). Each INPUT signal includes a bit of information (0 or 1). OUTPUT signal is given in the form of plasmid genes.

The Gate system works out by the following steps.

– Entry of INPUT signals A and B into the Information Gate.
– Data processing by *E. faecalis* pheromone system.
– Discharging of OUTPUT signal X from the Information Gate.

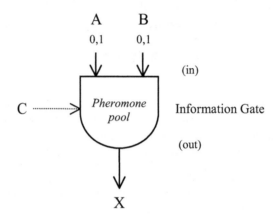

Fig. 2. Schematic representation of *E. faecalis* Information Gate. A, B: INPUT signals (=0 or 1); C: another input (see Table 2); X: OUTPUT in the form of gene expressions.

4   Realization Plan of *E. faecalis* Information Gate

To realize the Information Gate, we describe an experimental plan about *E. faecalis*. Later, we present an instance of the experiment to confirm the practical performance of the system.

4.1   Outlines of Experimental Procedure

1. Two types of cells (A) and (B) that can produce INPUT signals are prepared. INPUT signals are virtually expressed by a combination of two pheromone inhibitors (Table2). *E. faecalis* is thought to be always producing certain class of pheromone until they obtain its corresponding class of plasmid. On the other hand, certain class of inhibitor is not produced unless *E. faecalis* has its corresponding class of plasmid. Because of that, we represent INPUT signals with inhibitors instead of pheromones.

Table 2. INPUT signals and their relating real molecules (a). Both INPUT A and B are symbolized with combination of two inhibitors. Another extra INPUT C includes all four types of pheromones. *I1-I4*: inhibitor, *C1-C4*: pheromone. Intermediate result of operation A+B+C (b).

(a)

Input signal	Relating molecule
A=0	*I3, I4*
A=1	*I1, I2*
B=0	*I2, I4*
B=1	*I1, I3*
C	*C1, C2, C3, C4*

(b)

	B	**0**	**1**
A		*I2, I4*	*I1, I3*
0		*I2, I3, I4*	*I1, I3, I4*
I3, I4		=*C1*	=*C2*
1		*I1, I2, I4*	*I1, I2, I3*
I1, I2		=*C3*	=*C4*

2. Cells (A) and (B) are mixed in any combination ({A, B} ={0, 0}, {0, 1}, {1, 0}, or {1, 1}). In any case, three types of all four pheromone inhibitors are sent to the Information Gate. Other cell type (C) always produces four types of pheromones *C1*, *C2*, *C3*, and *C4*. In fact, mixing of three types of cells (A), (B), and (C) intermediately results in sending three inhibitors and four pheromones into the pheromone pool. That means only one pheromone can avoid inhibitory effect from inhibitors and maintain inducible activity of conjugation. Which pheromone type can avoid inhibition is dependent on mixing pattern of inhibitors, and the type is unique among four combination patterns.
3. These mixtures in the pheromone pool are exposed to donor cells (Figure 3). Four classes of donors are preset in the pheromone pool. Each class of donors has each one of four plasmids, *p1*, *p2*, *p3*, or *p4*. Each plasmid has any functional gene *x1*, *x2*, *x3*, or *x4*. Donor is activated if its specific inhibitor (*I1*, *I2*, *I3*, or *I4*) does not exist in the mixture. Thus, any one of four donors is activated and only plasmids of activated donors are transmitted to recipients through conjugation between donors and recipients.
4. Recipients are separated from donors by addition of Streptmycin (Sm), an antibiotic substance. Strains that we are going to use for recipients have resistance gene against Streptmycin but strains for donors do not.
5. Recipients are grown for hours and total cell numbers are recovered. Recipients produce any proteins according to genes on their plasmids.

4.2 Experimental Confirmation of *E. faecalis* Information Gate

We are now investigating and starting examination of the Information Gate experimentally. We organized a small experiment in order to confirm that the Information Gate will be feasibly executed. We are going to establish four transferplasmids that contain antibiotic resistance genes as markers (pCF10 :: tetracycline resistance, pAD1 :: erythromycin resistance, pPD1 :: chloramphenicol resistance, pOB1 :: kanamycin resistance). We already prepare two plasmids. The other plasmids are now under development in our laboratory.

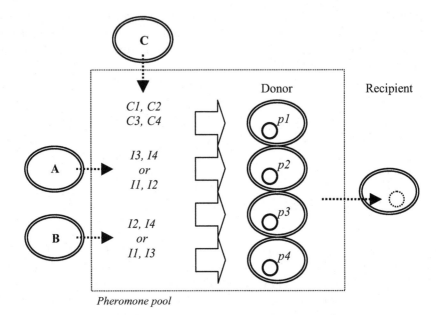

Fig. 3. Model of *E. faecalis* Information Gate. INPUT-cells A, B, and C secret inhibitors and pheromones into the medium. These signaling molecules specifically activate one of four donors. OUTPUT-Plasmids are transferred to recipients through conjugation between active donors and recipients.

We incubated donor cells under presence of pheromones and inhibitors : Pheromones were added to the mixture of donors immediately after addition of 10-fold excess of inhibitors. After incubation at 37°C for 60 min, donors were combined with recipients and incubated for 30min. We spread the cell mixture on THB agar plates that contain antibiotic substance, tetracycline / streptmycin or erythromycin / streptomycin to detect the number of plasmid-positive recipients. Then, we counted the number of bacterial colonies on agar plates (Figure 4).

Under presence of pheromones (C1, C2) and absence of inhibitors, recipients gained both plasmid class p1 (Tc^r) and p2 (Em^r) (control). When we added inhibitors along with pheromones, plasmids of corresponding class were not transferred to recipients. Neither plasmid class was transferred under presence of both inhibitors.

5   Discussions

E. faecalis Information Gate we show in this paper is able to describe each of four possible states that are generated from the pair of INPUT data, and throw back OUTPUT data in the form of gene expression.

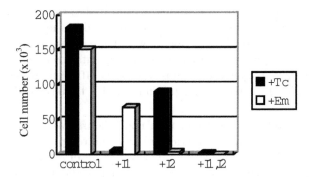

Fig. 4. Inhibition of plasmid transfer by pheromone inhibitors. Pheromones and inhibitors were added to mixture of donor strains OG1RF (p1::Tc^r) and OG1RF (p2::Em^r). 4x10^6 Donors were combined with 100-fold excess of recipients (OG1SSp (Sm^r)). Cell mixture was diluted and spread to THB agar plates. Total colony number was counted after selection with tetracycline / streptmycin (black bar), erythromycin / streptmycin (white bar).

 E. faecalis Information Gate is able to be piled up if we design plasmids so as to produce certain classes of pheromone inhibitors. For example, we can set I3 and I4 inhibitor genes in p1 plasmid. After conjugation, the p1-obtained recipients begin to produce these two pheromone inhibitors. The combination of I3 and I4 indicates an OUTPUT signal 'A=0', which becomes new INPUT signal of next operation. Thus, we can manipulate logical operations repetitively.

 Moreover, we can set signals freely in every four plasmids. For example, if we set OUTPUT signals {0, 0, 0, 1} in the four plasmid {p1, p2, p3, p4} respectively, the Information Gate becomes an 'AND' operator. Thus, we can manage all types of Boolean operation and freely accumulate them at will by designing those four plasmids. Though, it has previously reported about molecular switches or Boolean circuits [11-14], our method would be one of the alternatives.

 It is known that there are many classes of conjugative plasmids isolated from various *E. faecalis* strains. We might could design more extended system with additional a few plasmids.

 Donor strains of *E. faecalis* are highly sensitive to pheromones. The minimum pheromone densities to induce conjugations are 0.03-0.2nM [15]. It is estimated at the best that bindings of two or three pheromone molecules per a cell are enough to induce *E. faecalis* conjugation. This result would suggest that density of pheromones in the medium in natural condition is enough for *E. faecalis* to perform conjugation and this conjugation does not require high cell concentration to ensure high pheromone densities to be induced.

 Pheromone-activated donors do not always transfer plasmids successfully. Rather, most of conjugant is conceivable to fail to transfer plasmids, maybe because transfer phenomenon involves highly probabilistic events. In our experiment, plasmid transfer rate was approximately 1–4%. Conceptually about 10^2 donors ensure the gate execution so that one plasmid at the lowest can be transferred. It is preferable to combine donors with large excess of recipients in order to increase donor's opportunity to mate with recipients and reduce donor's meaningless homogenous conjugation. Thus, about 10^4-10^5 cells would be required. $0.1 \mu l$ culture of bacteria contains approximately 10^5 bacterial cells in laboratory condition.

The entire gate reaction would take many hours. We would have to wait in every step – pheromone production, pheromone reception, donor activation, conjugation, selection, amplification, and expression of acquired phenotypes. It is not determined yet what is the best performance about how many hours we should take to these reactions. However, apparently there is a limitation, perhaps a few hours in total are the best. Therefore, it seems not very significant to make efforts to speed up the gate. We are rather interested in utilizing the role of *E. faecalis* as a nano-scale information carrier.

The early project of us will propose only transfecting the marker genes onto plasmids, and therefore would enable no more than 1 bit Boolean circuit. However, the Information Gate has some expansivity, which can be written more additional information beside the marker genes on plasmids. Each plasmid serves as a set of memories. We can regard *E. faecalis* as a 'capsule' to store them and the gate system would provide cell-to-cell memory transfer. We can derivate information from a capsule to another capsule, can switch information from one to another, and in future, may continue next operations by initiating gene expressions whose directive genes are also written beside of data on plasmids. Thus, *E. faecalis* will become an information carrier and contribute to robotizations of processes in molecular computation.

Technologies of realizing plasmid-memory succeeded in mounting a few binary digits by now [1]. If we could mount memories tentatively up to hundreds of bits on a plasmid, we would able to apply the gate to more practical problems.

Unlike F plasmid of *E. coli*, conjugation system of *E. faecalis* is inducible by pheromone, includes homogenous conjugation, and carries highly aggregates [16]. Most of these plasmids do not transmit to other bacteria except for *E. faecalis* because this conjugation system is specific in *E. faecalis*.

Compared to the standard *E. coli*, *E. faecalis* has experimentally some difficulty. For example, it needs special vectors designed for *E. faecalis*. Because ordinary vectors are one for *E. coli*, they are unable to replicate in *E. faecalis*. When we perform gene transfection, it needs electroporation, which is more complicating method than DNA-Ca phosphate co-precipitation. However that may be, *E. faecalis* pheromone system seems to have many redeeming features than that of other bacteria or even some of the organism in terms of both pheromone and conjugation system.

Pheromone inhibitor is the key parameter that determines the workability of the Information Gate. It is required that pheromone inhibitors work sufficiently so that the Information Gate is feasibly executed without error. However, because pheromones and inhibitors competitively bind same pheromone receptors, it is impossible to inhibit pheromone activity completely unless the amount (or activity) of inhibitors are far dominant compared to those of pheromones. The secretion amount of pheromones and inhibitors are diverse among various *E. faecalis* strains, and not all of them ensure that inhibitors wipe out the effect of pheromones completely. For example, inhibitors secreted by OG1RF (pCF10) are approximately equal amount of corresponding class of pheromones secreted by OG1RF (none) [**17**]. In fact, in order to solve this problem, *E. faecalis* evolutionally developed mechanism, called pheromone shutdown, that *E. faecalis* suppress pheromone production itself concurrently with inhibitor production. After obtaining new conjugative plasmid, *E. faecalis* suppress the pheromone activities in such double regulation systems. Because pheromone shutdown system is concerned to multiple processes, consists of various molecules, and is not clearly established their associations, it is difficult to manipulate them. Instead, we need to devise another strategy to suppress pheromone activities. One of the answers is to increase the amount of inhibitors by up-regulating gene transcriptions in such ways that we would insert a forcible promoter on upstream of inhibitor gene or merely extend the plasmid copy number. It is confirmed in our experiment that the gate will feasibly be executed if the amount of inhibitors are thoroughly high compared to that of pheromones.

As many authors suggested, molecular computing has various prospects and is not necessary to be confined to NP-complete problems [18]. Possible directions are wide-ranging. Molecular computational algorithms are also likely to adapt to biology, medicine, nanotechnology and others as well as mathematical problems. We will continue to investigate the prospect of the computing system with these nano-scale bio organisms.

6   Acknowledgement

All of *E. faecalis* strains and plasmids used in this project are obtained from H. Nagasawa laboratory. We thank J. Nakayama for his contributions and helpful advises.

References

1. Head, T.: Circular suggestions for DNA computing. in Carbone A, Gromov M, Pruzinkiewcz P, eds., *Pattern Formation in Biology, Vision and Dynamics, World Scientific, Singapore and London.* (1999) 325–335.
2. Ehrenfeucht, A., Harju, T., Petre, I., and Rozenberg, G.: Patterns of micronuclear genes in ciliates. *7th International Meetings on DNA Based Computers.* (2001) 33–42.
3. Bloom, B., and Bancroft, C.: Liposome Mediated Biomolecular Computation. *DNA Based Computers V.* (1999) 75–84.
4. Abelson, H., Knight, T.F., and Sussman, G.J.: Amorphous Computing. *White paper.* (1995) October.

5. Head, T.: Communication by documents in communities of organisms, Millennium III, Winter issue 199/2000, Published by Current Media Group in collaboration with the Romanian Academy, the Romanian Association for the Club of Rome and the Black Sea University Foundation. ISSN 1454-7759.

6. Bassler, B.L.: How Bacteria talk to each other: regulation of gene expression by quorum sensing. *Current Opinion in Microbiology.* **2** (1999) 582–587.

7. Weiss, R., and Knight, Jr., T.F.: Engineered communications for microbial robotics. *6th DIMACS workshop on DNA based computers.* (2000) 5–19.

8. Clewell, D.B., and Weaver, K.E.: Sex pheromones and Plasmid transfer in *Enterococcus faecalis. Plasmid* **21** (1989) 175–184.

9. Clewell, D.B.: Bacterial Sex Pheromone-Induced Plasmid Transfer. *Cell.* **73** (1993) 9–12.

10. Wirth, R.: The sex pheromone system of *Enterococcus faecalis. European Journal of Biochemistry.* **222** (1994) 235–246.

11. Ogihara, M., and Ray, A.: DNA-based parallel computation by "counting." *3rd DIMACS workshop on DNA based computers.* (1997) 265–274.

12. Liu, Q., Wang, L., Frutos, A.G., Condon, A.E., Corn, R.M., and Smith, L.M.: DNA computing on surfaces. *Nature* **403** (2000), 175–179.

13. LaBean, T.H., Winfree, E., Reif, J.H.: Experimental progress in computation by self-assembly of DNA tiling. *5th DIMACS workshop on DNA based computers* (1999), 121–138.

14. Winfree, E., Eng, T., and Rozenberg, G.: String tile models for DNA computing by self-assembly. *6th DIMACS workshop on DNA based computers.* (2000) 65–84.

15. Nakayama, J.: Chemistry and Genetics on Sex Pheromone Signaling in *Enterococcus faecalis. Nippon Nogeikagaku Kaishi* **73** (11) (1999), 1155–1166.

16. Andrup, L., and Anderson, K.: A comparison of the kinetics of plasmid transfer in the conjugation systems encoded by the F plasmid from Escherichia coli and plasmid pCF10 from *Enterococcus faecalis. Microbiology* **145** (1999), 2001–2009.

17. Nakayama, J., Dunny, G.M., Clewell, D.B., and Suzuki, A.: Quantitative Analysis for pheromone inhibitor and pheromone shutdown in *Enterococcus faecalis. Developments in Biological Standarization* **85** (1995), 35–38.

18. Hagiya, M.: Towards Autonomous Molecular Computers. Genetic Programming (1998) *Proceedings of the Third Annual Conference*, 691–699.

Patterns of Micronuclear Genes in Ciliates

Andrzej Ehrenfeucht[1], Tero Harju[2], Ion Petre[2], and Grzegorz Rozenberg[1,3]

[1] Department of Computer Science, University of Colorado,
Boulder CO 80309-0347 USA
`andrzej@cs.colorado.edu`
[2] Department of Mathematics, University of Turku and
Turku Centre for Computer Science
Turku FIN 20014 Finland
`{harju, ipetre}@cs.utu.fi`
[3] Leiden Institute of Advanced Computer Science, Leiden University,
Niels Bohrweg 1, 2333 CA Leiden, The Netherlands
`rozenber@liacs.nl`

Abstract. The process of *gene assembly* in ciliates is one of the most complex examples of DNA processing known in any organism, and it is fascinating from the computational point-of-view — it is a prime example of DNA computing *in vivo*. In this paper we continue to investigate the three molecular operations (*ld, hi,* and *dlad*) that were postulated to carry out the gene assembly process in the intramolecular fashion. In particular, we focus on the understanding of the IES/MDS patterns of micronuclear genes, which is one of the important goals of research on gene assembly in ciliates. We succeed in characterizing for each subset $\mathcal{S}$ of the three molecular operations those patterns that can be assembled using operations in $\mathcal{S}$. These results enhance our understanding of the structure of micronuclear genes (and of the nature of molecular operations). They allow one to establish both *similarity* and *complexity* measures for micronuclear genes.

1 Introduction

Ciliates are an ancient group of organisms, comprising at least $10,000$ genetically different organisms. A unique feature of ciliates is their nuclear dualism: they have a germline nucleus (*micronucleus*) used in cell mating, and a somatic nucleus (*macronucleus*), providing RNA transcripts to operate the cell. The DNA in the micronucleus is hundreds of thousands base pairs long, with genes occurring individually or in groups, dispersed along the DNA molecule, and separated by stretches of spacer DNA. The DNA molecules in the macronucleus are gene-size, on average about 2000 base pairs long. The micronuclear gene consists of gene segments, called MDSs, separated by multiple, noncoding segments called IESs. During sexual reproduction a macronucleus develops from a micronucleus, and during this process IESs are excised, and MDSs are ligated to form transcriptionally competent genes. This gene assembly process is very interesting from a computational point-of-view. One of the amazing features of this process

N. Jonoska and N.C. Seeman (Eds.): DNA7, LNCS 2340, pp. 279–289, 2002.
© Springer-Verlag Berlin Heidelberg 2002

is that ciliates apparently know the *linked list data structure* and use it (for the assembly of macronuclear genes) by a very elegant pattern matching process.

The computational beauty of gene assembly in ciliates has been brought to the attention of the DNA computing community by L. Landweber and L. Kari (see, e.g., [7] and [8]). The focus of their research is the computational power of molecular operations used in gene assembly. This power is measured using the yardsticks of computation theory — thus, e.g., it is shown in [8] that using two operations postulated in [7] one can simulate Turing machines.

A different line of research on the computational nature of gene assembly in ciliates has been initiated in [6] and [12]. Here one investigates the *assembly process itself*. A typical question is: "What are the assembly strategies for a given micronuclear gene?" This research has led to a formal system for gene assembly in ciliates (see [12]). The usefulness of this system has been demonstrated by applying it to *all* known cases of gene assembly in ciliates, obtaining in this way *one uniform* explanation of all these cases. Moreover, it has been proved in [3] that every micronuclear gene (pattern) can be assembled using the molecular operations postulated by this formal system — in this sense, this system can be applied to all the micronuclear genes that will be found in the future. The major difference between the set of two molecular operations considered by Kari and Landweber, and the set of three molecular operations considered here, is that the model based on the former is *intermolecular*, while the model based on the latter is *intramolecular*. Also, from the point of view of rewriting systems, our rewriting rules are always length decreasing, while the rules used by Landweber and Kari are both length increasing and length decreasing.

These two lines of research complement each other, and together they form a broad framework for the understanding of the "computational abilities" of ciliates.

Formal systems for gene assembly, based on string rewriting and on graph rewriting were further investigated in [1,3,4,5], providing useful technical tools for the investigation of the gene assembly process. We use some of these tools in this paper to learn more about the patterns of micronuclear genes (and about the nature of molecular operations). We believe that understanding the structure of micronuclear genes is one of the important goals of research on gene assembly. For each subset S of the set of the three molecular operations, we investigate the micronuclear gene patterns that can be assembled using S — as a matter of fact we succeed in obtaining full characterizations for each S.

Due to lack of space, we omit here the proofs of our results, referring for details to the full paper [2].

2   Gene Assembly in Ciliates

The structure of a gene in a micronuclear chromosome consists of MDSs (macronuclear destined sequences) separated by IESs (internally eliminated sequences). During sexual reproduction, the micronuclear genome is converted into the macronuclear genome. In this process, the IESs become excised, and the

MDSs are spliced in the orthodox order $M_1, M_2, \dots, M_\kappa$ suitable for transcription (see, e.g., [9], [10], and [11]). Each MDS M_i has the structure $M_i = (p_i/\overline{p}_i, \mu_i, p_{i+1}/\overline{p}_{i+1})$, except for M_1 and M_κ, which are of the form $M_1 = (b/\overline{b}, \mu_1, p_2/\overline{p}_2)$, and $M_\kappa = (p_\kappa/\overline{p}_\kappa, \mu_\kappa, e/\overline{e})$. We refer to each double strand $p_i/\overline{p}_i$ as a *pointer*, and to μ_i as the *body* of M_i - we call $p_i/\overline{p}_i$ the *incoming pointer* of M_i, and $p_{i+1}/\overline{p}_{i+1}$ the *outgoing pointer* of M_i. The markers b and e (and their "complements" $\overline{b}$ and $\overline{e}$) are not pointers — they are just symbolic markers designating the locations where an incipient macronuclear DNA molecule will be excised from the macronuclear genome.

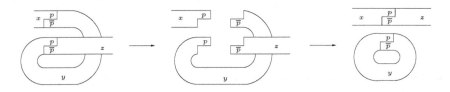

Fig. 1. The *ld*-excision operation applied on the direct repeat pattern (p, p).

Throughout the paper, we will represent single DNA strands by strings, always written in the 5'-3' direction (from left to right). Then, combining single strings into "double strings" written one above the other, we obtain the notation for double stranded DNA molecules — the upper string denotes one of the two strands written in the $5' - 3'$ direction (from left to right). Also, for a DNA molecule p (single or double stranded), $\overline{p}$ denotes its inversion. Note however that when we write $\overline{p}$ as a single string, then it is also written in the $5' - 3'$ direction. Hence, e.g., for $ACTGAT$ we write its inversion as $ATCAGT$; then the double stranded molecule formed by these two strands is written as $\frac{ACTGAT}{TGACTA}$.

In the macronucleus, the MDSs $M_1, M_2, \dots, M_\kappa$ are spliced together by "gluing" each M_j with M_{j+1} on the pointer $p_{j+1}/\overline{p}_{j+1}$.

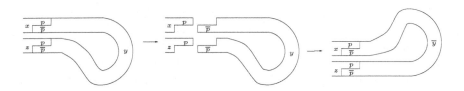

Fig. 2. The *hi*-excision/reinsertion operation applied on the inverted repeat pattern $(p, \overline{p})$.

It is argued in [6] and [12] that the gene assembly process in ciliates is accomplished through the following three molecular operations: *ld-excision* (or *ld*, for short), *hi-excision/reinsertion* (or *hi*, for short), and *dlad-excision/reinsertion* (or *dlad*, for short). We refer to [6] and [12] for the description of these operations, while giving here Figures 1, 2, and 3 illustrating the domain and the effect of each of them.

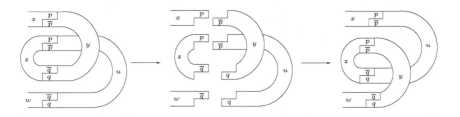

Fig. 3. The *dlad*-excision/reinsertion operation applied on the alternating direct repeat (p, q, p, q).

3   Formalizing the Gene Assembly in Ciliates

Three levels of abstraction were considered in [4], [6] and [12] in order to formalize the MDS/IES structure of genes: MDS descriptors, legal strings and legal graphs (in this order of increasing abstraction). Then three types of formal systems were investigated: MDS descriptor reduction systems (operating on MDS descriptors), string pointer reduction systems (operating on legal strings), and graph pointer reduction systems (operating on legal graphs). These formal systems were used to investigate the gene assembly process on various levels of abstraction. Surprisingly enough, it has turned out that, as far as assembly strategies are concerned, all three systems are equivalent and translatable into each other ([1]). For the purpose of this paper, it suffices to recall only the modeling of the gene assembly process through string pointer reduction system.

3.1   Preliminaries on Strings

For strings u, v, we say that v is a *substring* of u if $u = w_1vw_2$, for some strings w_1, w_2. Also, v is a *cyclic substring* of u if v is a substring of u, or $v = v_1v_2$ and $u = v_2wv_1$ for some strings v_1, v_2, w. We say that v is a *scattered substring* of u if there are strings $v_1, \ldots, v_n, u_1, \ldots, u_{n+1}, n \geq 1$, such that $v = v_1 \ldots v_n$ and $u = u_1v_1u_2v_2 \ldots u_nv_nu_{n+1}$. A *proper* substring (cyclic, scattered substring) of u is a substring (cyclic, scattered substring, resp.) v of u such that $v \neq \lambda$ and $v \neq u$ (where λ denotes the empty string).

A *legal string* is a nonempty string $u \in \Pi^*$ such that for each $p \in \Pi$, if p occurs in u, then u contains exactly two occurrences from the *pointer set* $\{p, \overline{p}\}$.

If u contains one occurrence of p and one of $\overline{p}$, then p (and $\overline{p}$) is said to be *positive* in u, otherwise p (and $\overline{p}$) is *negative* in u. We say that u contains a pointer set $\{p, \overline{p}\}$ if u has an occurrence of either p, or $\overline{p}$. A legal string with no proper legal substrings is called *elementary*.

For a string $u = x_1 x_2 \ldots x_n$, with $x_i \in \Pi$ for all $1 \le i \le n$, the *inverse* of u is the string $\overline{u} = \overline{x}_n \ldots \overline{x}_2 \overline{x}_1$. For a pointer $p \in \Pi$ such that $\{x_i, x_j\} \subseteq \{p, \overline{p}\}$, for some positive integers $i < j \le n$, the *p-interval* of u is the substring $x_i x_{i+1} \ldots x_j$. Two distinct pointers $p, q \in \Pi$ *overlap* in u if the p-interval of u overlaps with the q-interval of u (recall that for $i_2 \ge i_1$, the substrings $x_{i_1} \ldots x_{j_1}$ and $x_{i_2} \ldots x_{j_2}$ of u overlap if $i_1 < i_2 < j_1 < j_2$, or $i_2 < i_1 < j_2 < j_1$). We denote by $O(p)$ ($O^+(p)$, $O^-(p)$, resp.) the set of (positive, negative, resp.) pointers overlapping with the pointer p in u. For technical reasons, we always include p in $O(p)$. If p is a positive pointer, then $p \in O^+(p)$, and if p is negative, then $p \in O^-(p)$.

Let u be a legal string and p, q two pointers in u. Then p is *connected* to q in u, if there are pointers $r_1, \ldots r_n$, $n \ge 2$, such that $r_1 = p$, $r_n = q$, and $r_{i+1} \in O(r_i)$, for all i.

3.2   Micronuclear Patterns

The process of gene assembly begins with the micronuclear precursor and it leads to a composite MDS containing all MDSs spliced in the orthodox order $M_1 M_2 \ldots M_\kappa$. For the purpose of computational analysis of the gene assembly process, one can represent a macronuclear gene (as well as its micronuclear and the intermediate predecessors) by the sequences of its MDSs only.

For a specific gene, the number κ of its micronuclear MDSs is fixed (the gene is of *size* κ). We assume that κ is fixed for the sequel of this paper.

We use the alphabet $\Theta = \{M_{i,j}, \overline{M}_{i,j} \mid 1 \le i \le j \le \kappa\}$ to denote the MDSs and their inverses. The letters $M_{i,i}$ may also be written as M_i and they denote the micronuclear MDSs, also called *elementary MDSs*. The letters $M_{i,j}$ denote the *composite MDSs* formed during the assembly process by splicing the MDSs $M_i, M_{i+1}, \ldots, M_j$. We say that an MDS sequence $M_{i_1,j_1} \ldots M_{i_n,j_n}$, $n \le \kappa$, is *orthodox* if $i_1 = 1$, $i_l = 1 + j_{l-1}$, for all $2 \le l \le n$, and $j_n = \kappa$. As an example, $M_{1,3} M_{4,5} M_{6,9}$ is an orthodox sequence of size 9.

We simplify our notations by using positive integers $2, 3, \ldots$ and $\overline{2}, \overline{3}, \ldots$, to denote pointers. Hence, the MDSs are represented as $M_i = (i, \mu_i, i+1)$, $\overline{M}_i = (\overline{i+1}, \overline{\mu}_i, \overline{i})$. We get in this way the alphabets $\Delta = \{2, 3, \ldots\}$ and $\overline{\Delta} = \{\overline{2}, \overline{3}, \ldots\}$ and then $\Pi = \Delta \cup \overline{\Delta}$ denotes the set of pointers. We use the "bar operator" to move from Δ to $\overline{\Delta}$ and the other way around. Hence, for $z \in \Pi$, $\overline{z}$ is the *partner* of z, where $\overline{\overline{z}} = z$. For any $z \in \Pi$, we call $\{z, \overline{z}\}$ *the pointer set* of z (and of $\overline{z}$).

A sequence over Θ is a *micronuclear pattern* if it is a permutation of an orthodox sequence, possibly with some of its elements inverted. In its macronuclear form, an assembled gene has no pointers. On the other hand, its micronuclear precursor has pointers that flank IESs. Thus, the gene assembly process can be seen as a succession of pointer set removals by molecular operations, and so, it can be analyzed by tracing the sequences of pointers present in the intermediate molecules. Consequently, we can simplify the formal framework by denoting each MDS by the ordered pair of its pointers or markers, i.e., if $M = (p, \mu, q)$, then M is represented as (p, q), and its inversion $\overline{M} = (\overline{q}, \overline{\mu}, \overline{p})$, as $(\overline{q}, \overline{p})$. Moreover, the model can be further simplified by removing the parenthesis from the descriptors and deleting the markers. We obtain in this way legal strings, as the formalism for describing the sequences of pointers present in the micronuclear and the intermediate molecules.

Hence, for a micronuclear pattern δ, by writing the sequence of pointers and deleting the markers, we obtain the *legal string u corresponding to δ*. We denote this mapping by π: $\pi(\delta) = u$. Formally, $\pi(M_i) = i(i + 1)$, for $2 < i < \kappa$, $\pi(M_1) = 2$, $\pi(M_\kappa) = \kappa$, $\pi(\overline{M}_i) = \overline{\pi(M_i)}$, for $1 \leq i \leq \kappa$. We say that the legal string u is *realistic* if there is a micronuclear pattern δ such that $u = \pi(\delta)$.

3.3   The String Pointer Reduction System

The *string pointer reduction system* over Π (**SPRS** for short) consists of three sets of reduction rules operating on legal strings, corresponding to the micronuclear operations of *ld*, *hi*, and *dlad* (in [1] and [4] these string reduction rules were denoted by **snr**, **spr**, and **sdr**, resp.).

For $p, q \in \Pi$, with $p \neq q$,

1. the **ld**-rule for p is defined by $\mathbf{ld_p(u_1ppu_2)} = \mathbf{u_1u_2}$,
2. the **hi**-rule for p is defined by $\mathbf{hi_p(u_1pu_2\overline{p}u_3)} = \mathbf{u_1\overline{u}_2u_3}$, and
3. the **dlad** rule for p, q is defined by $\mathbf{dlad_{p,q}(u_1pu_2qu_3pu_4qu_5)} = \mathbf{u_1u_4u_3u_2u_5}$.

Let $\mathcal{LD}$ be the set of **ld**-operations, $\mathcal{LD} = \{\mathbf{ld_p} \mid \mathbf{p} \in \Pi\}$, $\mathcal{HI}$ the set of **hi**-operations, $\mathcal{HI} = \{\mathbf{hi_p} \mid \mathbf{p} \in \Pi\}$, $\mathcal{DLAD}$ the set of **dlad**-operations, $\mathcal{DLAD} = \{\mathbf{dlad_{p,q}} \mid \mathbf{p, q} \in \Pi, \mathbf{p} \neq \mathbf{q}\}$.

Let $\mathcal{S} \subseteq \{\mathcal{LD}, \mathcal{HI}, \mathcal{DLAD}\}$ be a subset of string reduction rules. For a legal string u and a sequence of reduction rules $\rho_1, \dots, \rho_l$ from $\mathcal{S}$, $D = (u; \rho_1, \dots, \rho_l)$ is a *reduction of u by $\rho_1, \dots, \rho_l$ in $\mathcal{S}$*, if ρ_1 is applicable to u, and ρ_i is applicable to $\rho_{i-1} \dots \rho_1(u)$, for all $1 < i \leq l$. We say that D is *successful* in $\mathcal{S}$ if $\rho_l \dots \rho_1(u) = \lambda$ — we also say then that u is *successful* in $\mathcal{S}$. Note that in the above $\rho_i \neq \rho_j$, whenever $i \neq j$. The notion of a successful assembled micronuclear pattern is defined similarly for the set of molecular operations *ld*, *hi*, and *dlad*.

The following is an informal version of a result from [1] and [4].

Theorem 1.

(1) *Let τ be a micronuclear gene. Each sequence ζ of molecular operations that leads to a successful assembly of τ into its macronuclear form is uniquely translatable into the sequence of operations from* **SPRS** *which together with the legal string representing τ forms a successful reduction in* **SPRS**.

(2) *Let π be a legal string representing a micronuclear gene τ. Each sequence ρ of operations from* **SPRS** *such that (π, ρ) is a successful reduction in* **SPRS** *is uniquely translatable into the sequence ζ of molecular operations that leads to a successful assembly of τ into its macronuclear form.*

4   Successful Patterns for Realistic Strings

A central question for understanding the role of the set {**ld, hi, dlad**} of molecular operations in the process of assembly is: what kind of micronuclear patterns can be successfully assembled by various subsets of this set? In this section, we provide a complete answer to this question. The main "translating vehicle" for the investigation of this question is Theorem 1: for each subset of operations, we characterize the realistic strings successful in **SPRS** using the corresponding subset of reduction rules — then the characterization results are translated into micronuclear patterns using Theorem 1.

4.1   Preliminaries

We first consider the set of legal strings having legal substrings, reducing the problem in this case to two smaller instances of the same problem.

Lemma 1. *Let $u = \alpha v \beta$ be a legal string such that v is also a legal string. Then u is successful if and only if v and $\alpha\beta$ are successful strings.*

A set $P = \{\alpha_1, \dots, \alpha_m\}$ $(m \geq 1)$ of strings over Π is *legal* if any string $\alpha_{i_1} \dots \alpha_{i_m}$ is legal, where $i_1, \dots, i_m$ is a permutation of $1, \dots, m$. Let α_P be such a string. A pointer p is said to *occur* in P if it occurs in α_P. Moreover, p is *negative (positive, resp.) in* P, if it is negative (positive, resp.) in α_P.

If u is a string of length 2, $u = pq$, then p is *left* in u and q is *right* in u.

A legal set $P = \{\alpha_1, \dots, \alpha_m\}$ is a *disjoint cycle* if the following conditions are satisfied for some pointers $p_1, \dots, p_m$:

(i) $|\alpha_i| = 2$, for all $1 \leq i \leq m$;
(ii) $\alpha_1 = p_1 p_2$, and for each $2 \leq i \leq m$, α_i contains one occurrence from $\{p_i, \bar{p}_i\}$ and one occurrence of p_{i+1}; moreover $p_{m+1} \in \{p_1, \bar{p}_1\}$;
(iii) for each $2 \leq i \leq m$, if p_i is negative in P, then p_i occurs in both α_{i-1} and α_i; moreover, p_i is left in α_i if and only if p_i is right in α_{i-1};
(iv) for each $2 \leq i \leq m$, if p_i is positive in P, then $\bar{p}_i$ occurs in α_i and p_i occurs in α_{i-1}; moreover, $\bar{p}_i$ is left in α_i if and only if p_i is left in α_{i-1}.

Clearly, if all pointers of P are negative (in P), then P is a disjoint cycle if and only if $P = \{p_1 p_2, p_2 p_3, \dots, p_m p_1\}$, for some $p_1, p_2, \dots, p_m \in \Pi$, $m \geq 1$.

Let $P = \{\alpha_1, \alpha_2, \dots, \alpha_m\}$ be a disjoint cycle and u a legal string. We say that u *contains* the disjoint cycle P if all strings of P are *nonoverlapping* cyclic substrings of u. If u does not contain P, we say that u *avoids* P.

4.2   ld-successful Patterns

Theorem 2. *A realistic string u corresponding to a micronuclear pattern of size κ has a successful reduction in $\mathcal{LD}$ if and only if $u \in \mathcal{L} \cup \bar{\mathcal{L}}$, where*

$$\mathcal{L} = \{2233 \dots \kappa\kappa\} \cup \{p(p+1)(p+1) \dots \kappa\kappa 22 \dots (p-1)(p-1)p \mid 2 \leq p \leq \kappa\}.$$

Theorem 2 shows that the micronuclear patterns successful in $\mathcal{LD}$ correspond to the cyclic shifts of the orthodox sequences. Thus, the *R1* gene of the *O.nova* ([11]), described by the pattern $M_1M_2M_3M_4M_5M_6$, can be successfully assembled using $\mathcal{LD}$ only, while $M_1M_3M_2$ cannot be assembled in this way.

4.3 {ld, hi}-successful Patterns

Lemma 2. *Let u be an elementary legal string.*

(i) *If $|u| \leq 2$, then u is successful in $\mathcal{LD} \cup \mathcal{HI}$;*
(ii) *If $|u| > 2$, then u is successful in $\mathcal{LD} \cup \mathcal{HI}$ if and only if u contains at least one positive pointer.*

The following result follows then by Lemma 1 and Lemma 2.

Theorem 3. *A realistic word u has a successful reduction in $\mathcal{LD} \cup \mathcal{HI}$ if and only if for all legal substrings v of u, if $v = v_1u_1v_2 \ldots v_ju_jv_{j+1}$, where each u_i is a legal substring, then $v_1v_2 \ldots v_{j+1}$ either contains a positive pointer, or it is reducible in $\mathcal{LD}$.*

Thus, the pattern $M_1\overline{M}_2M_3$ has a successful reduction in $\mathcal{LD} \cup \mathcal{HI}$, but no successful reductions in either $\mathcal{LD}$, or $\mathcal{HI}$. On the other hand, the pattern M_3 $M_1M_7M_4\overline{M}_5M_6M_9M_8M_2$, described by the realistic string $34278456\,\overline{56}798923$, has no successful reduction in $\mathcal{LD} \cup \mathcal{HI}$.

4.4 {ld, dlad}-successful Patterns

Theorem 4. *A realistic string u has a successful reduction in $\mathcal{LD} \cup \mathcal{DLAD}$ if and only if all the pointers in u are negative.*

Thus, the α-*TP* gene of *O.nova* ([11]), described by the pattern $M_1M_3M_5M_7$ $M_9M_{11}M_2M_4M_6M_8M_{10}M_{12}M_{13}M_{14}$ has a successful reduction in $\mathcal{LD} \cup \mathcal{DLAD}$, but it cannot be assembled in either $\mathcal{LD}$, or $\mathcal{DLAD}$. On the other hand, the pattern $M_1\overline{M}_2M_3$ cannot by assembled using $\mathcal{LD} \cup \mathcal{DLAD}$ only.

As a consequence of Theorems 3 and 4, we obtain the universality of the set {ld, hi, dlad} of operations, a result already known from [3].

Corollary 1 ({ld, hi, dlad}-successful patterns). *Any realistic string has a successful reduction in $\mathcal{LD} \cup \mathcal{HI} \cup \mathcal{DLAD}$. Moreover, any elementary realistic string can be successfully assembled using only two operations: either **ld** and **hi**, or **ld** and **dlad**.*

The last claim of Corollary 1 is not true for all micronuclear genes. E.g., the *actin I* gene of *O.trifallax* ([11]), described by the pattern $M_3M_4M_6M_5M_7M_9$ $M_{10}\overline{M}_2M_1M_8$, cannot be assembled either in $\mathcal{LD} \cup \mathcal{HI}$ (because of the legal substring 567567), or in $\mathcal{LD} \cup \mathcal{DLAD}$ (because of the positive pointers). Nevertheless, it can be successfully assembled in $\mathcal{LD} \cup \mathcal{HI} \cup \mathcal{DLAD}$. The same holds for the *actin I* gene of *O.nova*, given by the pattern $M_3M_4M_6M_5M_7M_9\overline{M}_2M_1M_8$.

4.5   dlad-successful Patterns

Theorem 5. *A realistic string u has a successful reduction in $\mathcal{DLAD}$ if and only if u consists of negative pointers only and it avoids disjoint cycles.*

According to Theorem 5, the pattern $M_2 M_1 M_3$ has a successful reduction in $\mathcal{DLAD}$, while the pattern $M_1 M_2 M_3$ has no such reduction in $\mathcal{DLAD}$.

4.6   hi-successful Patterns

The characterization result for this case can be proved in the same manner as Lemma 2 and Theorem 5.

Lemma 3. *An elementary legal string has a successful reduction in $\mathcal{HI}$ if and only if it contains at least one positive pointer and avoids disjoint cycles.*

Theorem 6. *A realistic word u has a successful reduction in $\mathcal{HI}$ if and only if for all legal substrings v of u, if $v = v_1 u_1 v_2 \ldots v_j u_j v_{j+1}$, where each u_i is a legal substring, then $v_1 v_2 \ldots v_{j+1}$ contains at least one positive pointer and avoids disjoint cycles.*

Thus, the pattern $M_1 M_4 \overline{M}_2 M_3$ has a successful reduction in $\mathcal{HI}$, while the patterns $M_1 \overline{M}_2 M_3$ and $\overline{M}_1 M_2 M_4 M_3$ have no such reductions in $\mathcal{HI}$.

4.7   {hi, dlad}-successful Patterns

Using Theorems 5 and 6, we can prove the following result.

Theorem 7. *A realistic string u has a successful reduction in $\mathcal{HI} \cup \mathcal{DLAD}$ if and only if u avoids disjoint cycles.*

Theorem 7 has the following corollary.

Corollary 2. *An elementary realistic string u has a successful reduction in $\mathcal{HI} \cup \mathcal{DLAD}$ if and only if either u has a successful reduction in $\mathcal{HI}$, or u has a successful reduction in $\mathcal{DLAD}$.*

The pattern $\mu = M_1 \overline{M}_2 M_4 M_3$, described by the elementary realistic string $u = 2\,\overline{3}\,\overline{2}434$ is successfully reducible in $\mathcal{HI} \cup \mathcal{DLAD}$: $\mathbf{dlad_{3,4}}(\mathbf{hi_2}(\mu)) = \lambda$. By Corollary 2, it also has a successful reduction, either in $\mathcal{HI}$, or in $\mathcal{DLAD}$. Indeed, $\mathbf{hi_2}(\mathbf{hi_4}(\mathbf{hi_3}(\mu))) = \lambda$.

5   Discussion

In this paper we have investigated the IES/MDS patterns of micronuclear genes in ciliates. For each subset S of the set $\{ld,hi,dlad\}$ of molecular operations, we gave a characterization of all the patterns that can be assembled into a macronuclear gene using S. We believe that these results provide a valuable insight into the micronuclear gene patterns, *and* into the nature of molecular operations that we consider. One can use our results to determine a similarity and to classify the complexity of micronuclear genes. For example, two seemingly different micronuclear genes may be considered similar, if, e.g., both of them can be assembled using only *hi* operation. Also, the genes that need all three operations in any assembly strategy may be considered more complex than the genes needing only one or two of them. In this sense, (see Section 4.4) the micronuclear *actin I* gene of *O.nova* is more complex than the micronuclear α-*TP* gene of *O.nova*. Also, one can order our three operations according to their complexity (e.g., $ld < hi < dlad$) and use this ordering to get a comparative classification of the complexity of *all* micronuclear genes.

It has been demonstrated in [12] that all known cases of gene assembly in ciliates can be accomplished using only the so-called *simple* versions of *ld, hi*, and *dlad*. Therefore the logical next step is to characterize the micronuclear patterns that can be assembled by the simple versions of our operations.

Acknowledgments. Tero Harju, Ion Petre, and Grzegorz Rozenberg gratefully acknowledge support by TMR Network GETGRATS. The authors are indebted to David M. Prescott for his invaluable guidance in biology of ciliates.

References

1. Ehrenfeucht, A., Harju, T., Petre, I., Prescott, D.M., and Rozenberg, G., Formal systems for gene assembly in ciliates, to appear in Theoret. Comput. Sci. (2001).
2. Ehrenfeucht, A., Harju, T., Petre, I., Prescott, D.M., and Rozenberg, G., Patterns of micronuclear genes in ciliates, manuscript.
3. Ehrenfeucht, A., Petre, I., Prescott, D. M., and Rozenberg, G., Universal and simple operations for gene assembly in ciliates, in C. Martin-Vide and V. Mitrana (eds.) *Where Mathematics, Computer Science, Linguistics and Biology Meet,* Kluwer Academic Publishers, Dordrecht/Boston, 329–342 (2001).
4. Ehrenfeucht, A., Petre, I., Prescott, D.M., and Rozenberg, G., String and graph reduction systems for gene assembly in ciliates, to appear in Math. Struct. in Comput. Sci. (2001).
5. Ehrenfeucht, A., Petre, I., Prescott, D.M., and Rozenberg, G., Circular gene assembly in ciliates, manuscript (2001).
6. Ehrenfeucht, A., Prescott, D.M., and Rozenberg, G., Computational aspects of gene (un)scrambling in ciliates. In *Evolution as Computation*, L. Landweber and E. Winfree (eds.), 45–86, Springer Verlag, Berlin, Heidelberg (2001).
7. Landweber, L.F. and Kari, L., The evolution of cellular computing: nature's solution to a computational problem. Proceedings of the 4th DIMACS meeting on DNA based computers, Philadelphia, 3–15 (1998).

8. Landweber, L.F. and Kari, L., Universal molecular computation in ciliates. In *Evolution as Computation*, Landweber, L. and Winfree, E. (eds.), Springer Verlag, Berlin, Heidelberg, to appear (1999).

9. Prescott, D.M., Cutting, splicing, reordering, and elimination of DNA sequences in Hypotrichous ciliates. BioEssays, 14 (5): 317–324 (1992).

10. Prescott, D.M., The unusual organization and processing of genomic DNA in Hypotrichous ciliates. Trends in Genet. 8:439–445 (1992).

11. Prescott, D.M. and DuBois, M., Internal eliminated segments (IESs) of Oxytrichidae, J. Eukariot. Microbiol. 43, 432–441 (1996).

12. Prescott, D.M., Ehrenfeucht, A., and Rozenberg, G., Molecular operations for DNA processing in Hypotrichous ciliates, to appear in European Journal of Protistology (2001).

Peptide Computing - Universality and Complexity*

M. Sakthi Balan, Kamala Krithivasan, and Y. Sivasubramanyam

Department of Computer Science and Engineering,
Indian Institute of Technology, Madras
Chennai - 600036, India
kamala@iitm.ernet.in,
{sakthi,siva}@cs.iitm.ernet.in

Abstract. This paper considers a computational model using the peptide-antibody interactions. These interactions which are carried out in parallel can be used to solve NP-complete problems. In this paper we show how to use peptide experiments to solve the Hamiltonian Path Problem (HPP) and a particular version of Set Cover problem called Exact Cover by 3-Sets problem. We also prove that this of model of computation is computationally complete.

1 Introduction

Recently the computing world has turned its attention on natural computing of which one model is the biological computing. Biological way of computing was experimently done in a biological laboratory by *L. Adleman* [1]. [1] gives the practical way of finding a Hamiltonian path for a given instance of graph. Some of the other papers which followed [1] are [8,7]. A complete survey can be found in [14,11,13].

In a similar way to DNA hybridization, antibodies which specifically recognize peptide sequences can be used for calculation [2]. [2] introduces the concept of peptide computing via peptide-antibody interaction and solves the well known NP Complete problem namely the satisfiability problem.

Peptide is a sequence of amino acids attached by covalent bonds called peptide bonds. A peptide consists of recognition sites called *epitopes* for the antibodies. A peptide can contain more than one epitope for the same or different antibodies. For each antibody which attach to a specific epitope there is a binding power associated with it called as *affinity*. If more than one antibody participate in recognition of its sites which overlaps in the given peptide, then the antibody with more affinity gets the higher priority.

In the model proposed in [2] the peptides represent the sample space of a given problem and antibodies are used to select certain subsets of this sample space, which will eventually give the solution set for the given problem. Similar

---

* Financial support from Infosys Technologies Limited, India is acknowledged

N. Jonoska and N.C. Seeman (Eds.): DNA7, LNCS 2340, pp. 290–299, 2002.
© Springer-Verlag Berlin Heidelberg 2002

to DNA-computing, parallel interactions between the peptide sequences and the antibodies should make it possible to solve NP-complete problems.

The advantages of computing using the peptides in comparison with DNA-computing are that at each position 20 different building blocks instead of 4 are possible. In contrast to a DNA strand which has only one possible reverse complement of the same length, in peptide-antibody interaction

1. different antibodies can recognize different regions or structures of the same epitopes, and
2. the binding affinity of different antibodies recognizing the same epitope can be different.

These two factors add additional flexibility for doing computation with peptides [5,6].

The main disadvantage in DNA computing is that under slight differences in the melting temperature of oligonucletides of similar sequence and a relative high error rate, only the perfect complementary oligonucleotide can be used to bind to its target. These above facts restrict the use of DNA-computing significantly [3,4].

Without additional enzymes the hybridization of DNA molecules gives only a yes or no solution. This can be used to solve all problems in NP [10,7]. In contrast to this, the epitopes of peptides are recognizable by more than one antibody and with a different affinity. This facilitates encoding of logical gates and to do calculation efficiently.

It should also be noted that the set of peptides and antibodies satisfy the following properties:

1. Under easily achievable conditions (Temperature, pH value, salt concentration) each antibody binds reliably to its peptide (epitope).
2. Under easily achievable conditions (Temperature, pH value, salt concentration) each antibody reliably dissociates from its peptide. This can be achieved by using a second antibody with a higher affinity for the same epitope that competitively removes the first antibody or by using an excess of the free epitope itself. Antibodies can be generally dissociated by changing the pH value. If necessary, all antibodies bound to the epitopes become covalently attached to their epitopes.
3. Under neither of the conditions above does any antibody bind to another peptide (epitope).

The specificity of epitope recognition by antibody need not be absolute. Sometimes the antibody may recognize a different site which is very similar to its own binding site. This issue is called as *cross-reactivity*. This paper does not address the issue of cross-reactivity between antibodies and peptides.

In this paper we give algorithms using peptide-antibody interactions to solve two of the well known NP-complete problems namely the Hamiltonian path problem and a variant of the set cover problem called the exact cover by 3-sets problem. In both the problems we form the peptides in such a way that the set of peptides represent the sample space of the problems. The antibodies are formed according to the problem input and used in such a way that it selects the correct peptide i.e. the solution for the problem. We also show that this model of computation is computationally complete by simulating a Turing Machine with the peptide and antibody interaction.

2   Notations and Definitions

For a finite sequence $M = m_1, m_2, \ldots, m_n$, we define a doubly duplicated sequence of M, MM as

$$MM = m_1, m_1, m_2, m_2, \ldots, m_n, m_n.$$

We define a doubly duplicated permutation of a set S, $\sigma(SS)$ as,

$$\sigma(SS) = \{mm \mid m \text{ is a permutation of the set } S\}.$$

A Turing Machine is denoted by TM. For details about TM, the reader is referred to [12]. We also denote the peptide-antibody interactions simply as peptide systems.

3   Solving Hamilton Path Problem

Let $G = (V, E)$ be a directed graph. Let $V = \{v_1, v_2, \cdots, v_n\}$ be the vertex set of the graph and $E = \{e_{ij} \mid v_j \text{ is adjacent to } v_i\}$ be the edge set. Without loss of generality we take v_1 as the source vertex and v_n as the end vertex. Our problem is to test whether there exists a Hamiltonian path between v_1 and v_n.

For each $v_i \in V$, choose an epitope. We denote the set of all chosen epitopes as $EP = \{ep_j \mid 1 \leq j \leq n\}$. The antibodies and peptides with epitopes from the set EP are formed as follows:

1. Each peptide has an epitope corresponding to the source vertex (ep_1) at the top and an epitope corresponding to the end vertex (ep_n) at the bottom.
2. In between the above two epitopes all doubly duplicated permutations of other epitopes corresponding to the vertices excluding the source and end vertices are formed.
3. Form antibodies A_{ij}, $1 \leq i, j \leq n$ which binds to the site $ep_i ep_j$, provided v_j is adjacent to v_i in G.
4. Form antibodies B_{ij}, $1 \leq i, j \leq n$ which binds to the site $ep_i ep_j$, provided v_j is not adjacent to v_i in G.
5. Form labeled antibody C, which binds to the sites $ep_1 \sigma(\{ep_j \mid 2 \leq j \leq n - 1\})ep_n$, where $\sigma(M)$ is the set of all doubly duplicated permutations over the set M.
6. The affinity of the antibodies B_{ij} are greater than C.
7. The affinity of the antibody C is greater than A_{ij}.

The algorithm is as follows:

1. Take all the peptides formed as said in the above way in an aqueous solution.
2. Add antibodies A_{ij}, $1 \leq i, j \leq n$.
3. Add antibodies B_{ij}, $1 \leq i, j \leq n$.
4. Add labeled antibody C.
5. If fluorescence is detected then there exists a Hamilton path in the graph G, otherwise there exists no such path. If the peptides are bound to an addressed chip the solution can be immediately read.

We illustrate the above algorithm for solving the Hamilton path problem by means of the following example where the graph G is as in Fig. 1

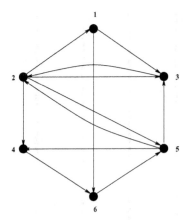

Fig. 1. The graph G

The peptide reactions are shown in Figure 2. The smoothly curved antibodies denote the antibodies B_{ij} and the sharply curved antibodies the A'_{ij}s.

We can easily see that this model requires

1. number of peptides $= (n - 2)!$,
2. length of peptides $= \mathcal{O}(n)$,
3. number of antibodies $= \mathcal{O}(n^2)$.

The number of biological steps needed to solve this problem is constant. First step detects all the possible edges in the graph in parallel. Second step detects all the invalid edges and the third step finds the correct solution, if it exists. Fourth step is for detecting whether there exists a solution and if it exists to read the solution.

4   Solving Exact Cover by 3-Sets Problem

The exact cover by 3-sets problem is a particular version of the set cover problem [9].

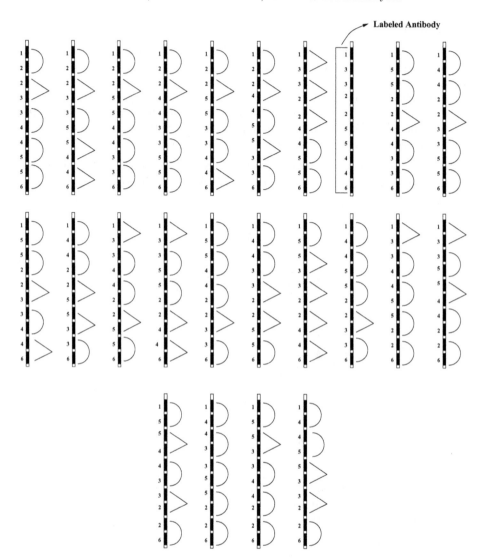

Fig. 2. The state of the peptides after the last step in the algorithm

Instance: A finite set X with $|X| = 3q$ and a collection C of 3-element subsets of X.

Question: Does C contain an exact cover for X, that is, is there a sub collection $C' \subseteq C$ such that every element of X occurs in exactly one member of C'.

We solve this NP-complete problem using the peptide systems.

Let $X = \{x_i\}_{i=1}^n$. The peptide formation is as follows:

1. For each x_i we choose a specific epitope ep_i in the peptide.
2. For every possible permutation of the set $\{ep_i\}$ there is a peptide in which every three sequence of ep_i's say, $ep_i ep_j ep_k$ are followed by a specific epitope ep_{ijk}.
   For example if $X = \{x_1, x_2, \ldots, x_9\}$ one particular peptide with the epitopes is shown in Fig. 3.

ep_1 ep_7 ep_9 ep_{179} ep_2 ep_6 ep_4 ep_{264} ep_3 ep_5 ep_8 ep_{358}

Fig. 3. Sample Peptide

3. Form antibodies A_{ijk} which binds to the site $ep_i ep_j ep_k$ provided $\{x_i, x_j, x_k\} \in C$.
4. Form antibodies B_{ijk} which binds to the site $ep_i ep_j ep_k$ provided $\{x_i, x_j, x_k\} \notin C$.
5. Form colored antibody C whose binding sites are the whole of peptides constructed in the above way.
6. The affinity of the antibodies B_{ijk}'s is greater than that of C
7. The affinity of the antibody C is greater than that of A_{ijk}'s

The algorithm is as follows:

1. Take all the formed peptides in an aqueous solution.
2. Add all the A antibodies.
3. Add all the B antibodies.
4. Add the C antibody.
5. If fluorescence is detected then there exists an exact cover with all the three elements sets in the cover being disjoint. If the peptides are bound to an addressed chip the solution can be immediately read.
   This model requires

   (a) number of peptides $= n!$,
   (b) length of peptides $= \mathcal{O}(n)$,
   (c) number of antibodies $= \mathcal{O}(n^3)$.

The number of biological steps needed to solve this problem is constant. First step recognizes all the valid sets belonging to the given collection. Second step recognizes all the invalid sets and the third step finds the correct solution, if it exists. Fourth step is for detecting whether there exists a solution and if it exists to read the solution.

5   Universality Result

Theorem 1. *A Turing Machine can be simulated by a peptide system.*

Proof. Let $M = (Q, \Sigma, \delta, s_0, F)$ be a Turing Machine. Let $Q = \{q_1, q_2, \ldots, q_m\}$ and $\Sigma = \{a_1, a_2, \ldots, a_l\}$, where Q denotes the finite set of states and Σ the tape alphabet and $\not{b}$ the blank symbol. Without loss of generality we assume that the Turing Machine halts when it reaches a final state. Let $s(n)$ denote the space complexity of the Turing Machine M and assume that it is apriori known.

In order to simulate the moves of the Turing Machine M we form the peptides and the antibodies as follows:

1. Set of epitopes $E = E_Q \cup E_\Sigma$ where

$$E_Q = \{ep_i^Q \mid 1 \leq i \leq s(n)\}$$

   and

$$E_\Sigma = \{ep_i^\Sigma \mid 1 \leq i \leq s(n)\}$$

2. Set of antibodies $A = A_Q \cup A_\Sigma$ where

$$A_Q = \{A_i^q \mid 1 \leq i \leq s(n), q \in Q\}$$

   and

$$A_\Sigma = \{A_i^a \mid 1 \leq i \leq s(n), a \in \Sigma \cup \{\not{b}\}\}$$

We assume that the antibodies $A_i^{q_f}$ where $q_f \in F$ are labeled.

The configuration of the TM M is encoded in the peptide through antibodies which binds to their specific epitopes.

Suppose the initial configuration of TM is

$$q_0 a_{i_1} a_{i_2} \ldots a_{i_k} \not{b} \not{b} \ldots \not{b}$$

then the initial configuration of the peptide will be (see Fig. 4 and Fig. 5. Fig. 4 shows the formation of epitopes in the peptide and Fig. 5 shows the initial state of the peptide)

$$A_1^{q_0} A_1^{a_{i_1}} [\ ] A_2^{a_{i_2}} [\ ] \ldots [\ ] A_k^{a_{i_k}} [\ ] A_{k+1}^{\not{b}} [\ ] \ldots [\ ] A_{s(n)}^{\not{b}}$$

where [ ] denotes the free epitope that is, it has no antibody attached to it.

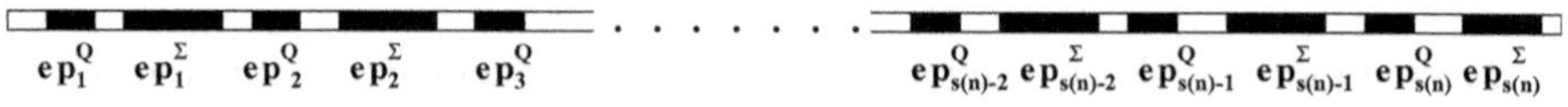

Fig. 4. Peptide with state and symbol epitopes

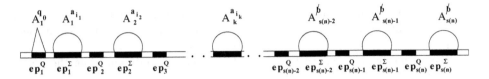

Fig. 5. Initial configuration of the peptide

The state transition is as follows:
Right move Suppose M moves from

$$\ldots \ldots a_{i_{k-1}} q a_{i_k} a_{i_{k+1}} \cdots \cdots$$

to

$$\ldots \ldots a_{i_{k-1}} a'_{i_k} q' a_{i_{k+1}} \cdots \cdots$$

using the transition $\delta(q, a_{i_k}) = (q', a'_{i_k}, R)$. Then correspondingly the following steps (see Fig. 6 and 7) are carried out

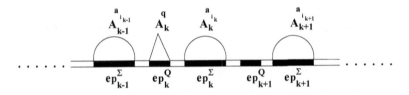

Fig. 6. Peptide before applying the rule

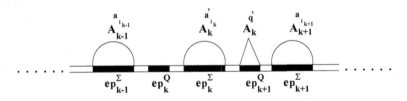

Fig. 7. Peptide after applying the rule

1. add excess of free epitopes ep_k^Σ and ep_k^Q.
2. add the antibodies $A_k^{a'_{i_k}}$ and $A_{k+1}^{q'}$ to the peptide.

Where the peptide is with the configuration corresponding to the left side of the above transition (before applying $\delta(q, a_{i_k}) = (q', a'_{i_k}, R)$).

Applying the rule $\delta(q, a_{i_k}) = (q', a'_{i_k}, R)$ the Turing Machine moves right with the state q' rewriting the symbol a_{i_k} by a'_{i_k}. To encode this move in peptide,

1. the antibody corresponding to the state q has to be removed and the appropriate state corresponding to q' has to be bound in the correct place in the peptide.
2. the antibody $A_k^{a_{i_k}}$ has to be removed and the antibody encoding the new symbol a'_{i_k} i.e., $A_k^{a'_{i_k}}$ has to be bound in the appropriate place in the peptide.

These two steps are carried out by,

1. adding of excess free epitopes ep_k^{Σ} and ep_k^{Q} which removes the antibodies $A_k^{a_{i_k}}$ and A_k^q.
2. The second step adds the antibodies $A_k^{a'_{i_k}}$ and $A_{k+1}^{q'}$ which encodes the change of the state of the Turing Machine correctly.

Left move Left move is very similar to the right move. Suppose M moves from

$$\ldots\ldots a_{i_{k-1}} q a_{i_k} a_{i_{k+1}} \ldots\ldots$$

to

$$\ldots\ldots q' a_{i_{k-1}} a'_{i_k} a_{i_{k+1}} \ldots\ldots$$

using the transition $\delta(q, a_{i_k}) = (q', a'_{i_k}, L)$. Then correspondingly the following steps are carried out

1. add excess of free epitopes ep_k^{Σ} and ep_k^{Q}.
2. add the antibodies $A_k^{a'_{i_k}}$ and $A_{k-1}^{q'}$ to the peptide.

Where the peptide is with the configuration corresponding to the left side of the above transition (before applying $\delta(q, a_{i_k}) = (q', a'_{i_k}, L)$).

The end of computation can be recognized by checking for the antibodies $A_i^{q_f}$ where $q_f \in F$ in the peptide sequence. If fluorescence is detected then it shows the existence of a labeled antibody i.e., the final state in the peptide. □

For each move in the Turing Machine M, two moves namely removal of antibodies and adding of antibodies have to be done. So, if we assume that M takes $\mathcal{O}(t(n))$ then the peptide system also takes the same amount of time. We can also take note that the length of the peptide needed for simulating a TM is $\mathcal{O}(s(n))$, number of peptides is exactly one and amount of antibodies is $\mathcal{O}(m.s(n) + l.s(n))$.

6   Conclusion

In this paper we use the natural interactions between the peptides and antibodies for computing. We give two examples of solving NP-complete problems: Hamiltonian path problem and a version of set cover problem using the peptide-antibody interactions. In the penultimate section we show that this model of computation is computationally complete. We also showed in the examples that the number of biosteps needed is a constant.

There are three things to look at the peptide systems from the complexity point-of-view. One is the length of the peptide, second the number of different peptides needed for the problem, and last the amount of antibodies needed. These three measures probably will constitute the core issue from the complexity point of view. Other measures could be the number of different epitopes, the maximum number of antibodies attaching to a particular epitope, etc.

References

1. L.M. Adleman, *Molecular computation of solutions to combinatorial problems*, Science **266** (1994), 1021–1024.
2. Hubert Hug and Rainer Schuler, *Strategies for the developement of a peptide computer*, Bioinformatics, to appear.
3. Y. Aoi, T. Yoshinobu, K. Kinoshita, and H. Iwasaki, *Ligation errors in DNA computing*, Biosystems **52** (1999), 181–187.
4. A.G. Frutos, Q. Liu, A.J. Thiel, A.M. Sanner, A.E. Condon, L.M. Smith, and R.M. Corn, *Demonstration of a word design strategy for DNA computing on surfaces*, Nucl. Acids Res. **25** (1997), 4748–4757.
5. S. Hashida, K. Hashinaka, and E. Ishikawa, *Ultrasensitive enzyme immunoassay*, Biotechnol. Annu. Rev. **1** (1995), 403–451.
6. K. Shreder, *Synthetic haptens as probes of antibody response and immunorecognition*, Methods **20** (2000), 372–379.
7. Richard Lipton, *DNA Solution of Hard Computational problems*, Science **268** (1995), 542–545.
8. Donald Beaver, *Universality and Complexity of Molecular Computation*, "ACM Symposium," 1996.
9. Michael R. Garey and David S. Johnson, "Computers and Intractability A Guide to the Theory of NP-Completeness," W.H. Freeman and Company, 1979.
10. D. Rooß and K.W. Wagner, *On the power of DNA-computing*, Information and Computation 131 (1996), 95–109.
11. Nadia Pisanti, *DNA computing: a survey*, in "Bulletin of the European Association for Theoretical Computer Science," no. 64, 1998, pp. 188–216.
12. John E. Hopcroft and Jeffrey D. Ullman, "Introduction to Automata Theory, Languages, and Computation," Addison-Wesley, 1979.
13. Gh. Păun, G. Rozenberg, and A. Salomaa, "DNA Computing," Springer-Verlag, 1998.
14. L. Kari, *DNA computing: tommorrow's reality*, in "Bulletin of the European Association for Theoretical Computer Science," no. 59, 1996, pp. 256–266.

Programmed Mutagenesis Is a Universal Model of Computation

Julia Khodor and David K. Gifford

Laboratory for Computer Science
Massachusetts Institute of Technology
{jkhodor, gifford}@mit.edu

Abstract. Programmed mutagenesis is a DNA computing system that uses cycles of DNA annealing, ligation, and polymerization to implement programatic rewriting of DNA sequences. We report that programmed mutagenesis is theoretically universal by showing how Minsky's 4-symbol 7-state Universal Turing Machine [11] can be implemented using a programmed mutagenesis system. Each step of the Universal Turing Machine is implemented by four cycles of programmed mutagenesis, and progress is guaranteed by the use of alternate sense strands for each rewriting cycle. The full version of the proof will appear in the special issue of TOCS.

1   Introduction

Programmed mutagenesis is a DNA computation method that implements the sequence specific rewriting of DNA molecules. Programmed mutagenesis is an in-vitro mutagenesis technique based on oligonucleotide-directed mutagenesis [3]. Like oligonucleotide-directed mutagenesis, programmed mutagenesis does not mutate existing DNA strands, but instead uses DNA polymerase and DNA ligase to create copies of template molecules with engineered mutations. Every time a programmed mutagenesis reaction is thermal cycled a rewriting event occurs. Because the technique relies on sequence-specific rewriting, multiple rules can be present in a reaction at once, with only certain rules being active in a given rewriting cycle. Furthermore, the system's ability to accommodate inactive rules allows it to proceed without human intervention between cycles. We have previously demonstrated the experimental practicality of the key primitive operations required for implementation of programmed mutagenesis systems [7].

Certain DNA computing systems are known to be restricted in the types of computations they can perform. For example, the "generate-and-test" methods proposed by Adleman [1] and Lipton [9] are useful for certain types of combinatorial problems, but these DNA computing methods can not be used to implement general computation.

We show that programmed mutagenesis is capable of general computation by showing how it can be used to implement a Universal Turing Machine. A Turing machine is an abstract model of a programmed computer. A Universal

N. Jonoska and N.C. Seeman (Eds.): DNA7, LNCS 2340, pp. 300–307, 2002.
© Springer-Verlag Berlin Heidelberg 2002

Turing Machine is a machine that is capable of simulating any Turing machine, given the description of that machine and the input to the machine[14]. We provide a constructive reduction of a Universal Turing Machine to programmed mutagenesis, and show how to encode the tape of a Turing machine into a DNA molecule.

An example of a programmed mutagenesis system that implements a unary counter is shown in Figure 1. The first mutation is introduced when oligonucleotide M-1 binds to the initial template and creates the product of cycle 1. This first mutation enables oligonucleotide M-2 to bind the product of cycle 1 in the next cycle. The sequence change caused by M-2 enables oligonucleotide M-1 to bind to the product of cycle 2, and to create a new molecule, the product of cycle 3. As shown in Figure 1, "outer" oligonucleotides are used to create full-length products and the strand that polymerizes from the outer primer is ligated to a mutagenic primer by ligase. All of the enzymes used in the system are thermostable which allows the system to be thermal cycled.

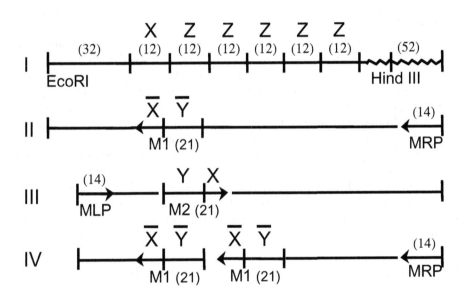

Fig. 1. Schematic representation of the unary counter. M-1 and M-2 are mutagenic rule oligonucleotides; MRP and MLP are perfectly matched outer oligonucleotides. Note that a rule incorporated on a previous cycle becomes part of the template for the following cycle.

In the remainder of this paper we scetch out the proof that programmed mutagenesis is a universal model of computation (Section 2), and compare programmed mutagenesis with other DNA computing systems (Section 3).

2   Programmed Mutagenesis Systems Are Universal

We begin this section with a formal definition of a Turing machine. We discuss challenges in encoding a Turing machine in DNA in Section 2.2, give a proof outline in Section 2.3.

2.1   Turing Machines and an Example of a Universal Turing Machine

A Turing machine is a general model of computation first introduced by Alan Turing in 1936 [14]. The Turing machine model uses an infinite tape as its unlimited memory and finite number of states and symbols to encode and compute a problem. A Turing machine has a head which can move both left and right on the tape, reading and rewriting only the symbol it is currently pointing to. The behavior of a Turing machine is governed by its transition function, which, given the current state and symbol being read, determines the new state the machine will enter, the new symbol to be written on the tape, and the direction of motion of the tape head (left or right).

A Turing machine begins computation with its head on the leftmost cell of the input string, and proceeds by following the transition function. If it ever enters a halting state, it halts. If the head ever reaches the last tape cell on either side, and the transition function indicates moving off the tape, machine appends a single tape cell with a blank symbol on it. An instantaneous description of a Turing machine is its configuration– the current state of the finite control, the current tape contents, and the current head location.

A Universal Turing machine is a single Turing machine U, with the property that for each Turing machine T which computes a Turing-computable function f, there is a string of symbols d_T such that if the output of T on input x is $f(x)$, then the output of U on input $x\ d_T$ is also $f(x)$. The smallest known Universal Turing Machine was described by Marvin Minsky [11], and it has 4 symbols and 7 states.

2.2   A Formal Model of Programmed Mutagenesis

An important observation about programmed mutagenesis is that the rules become part of the template for the next cycle. Thus, the rules do not take the familiar form of antecedent $\rightarrow$ consequent, but rather the rules in solution are consequents, searching for their antecedents. In fact, this property is inherent in the underlying biological technique of oligonucleotide-directed (or site-directed) mutagenesis. It is, therefore, intuitive that in modeling Minsky's Universal Machine, we need to explicitly place state onto the tape (and into the rules). The decision of exactly how to do that is an essential part of this proof and is explained below.

As described above, programmed mutagenesis relies on a distance constraint based on mismatches in rewrite rules to sequence the steps of a program. The number of mismatches is not the only factor determining whether a given rule may bind to and rewrite a given sequence of DNA, but it is the most influential. The distance constraint is reinforced by three biochemical factors. These three factors are the destabilizing effect of mismatches on duplex stability, polymerase effects, and ligase effects. Polymerase and ligase can not function if mismatches are too close to their action sites. Other factors influencing oligonucleotide's ability to bind a given sequence of DNA, extend a given sequence, or serve as a suitable template for a ligation reaction include mismatch geometry, type of mismatch, enzymes and buffers used, and other biochemical parameters. It is impractical to try to model all the parameters influencing the efficiency of binding, extension, and ligation of DNA rewrite rules, in part because insufficient information exists to construct a reliable model. Therefore, our formal model below uses the number of mismatches as the sole determining factor for a primer (rewrite rule) to bind a given DNA sequence, extend, and to be ligated into a longer strand.

We will formally model programmed mutagenesis as a DNA-based string rewrite system in which a single strand of DNA is rewritten in a sequence-specific manner with the use of a set of DNA rewrite rules to produce a single strand of DNA as output. Our formal model of programmed mutagenesis describes rewrite rules that are either 41 or 28 bases long. Also part of the model are the following four assumptions:

1. For a rule of length 41, the allowed mismatch distance is 4 mismatches.
2. For a rule of length 28, the allowed mismatch distance is 3 mismatches.
3. Any DNA sequence for which the distance in mismatches to its target on the template is above the specified number of mismatches will not bind and extend.
4. If only one rule can bind to a spot, that rule executes. If more than one rule can bind to a spot, equal percentages of each rule execute (thus creating parallel branches of computation). Thus, a copy operation occures only if there are no rules that can affect a rewrite.

It is possible to formulate alternative formal models of programmed mutagenesis that use rules of different lengths and different distance constraints. We have experimentaly found above length and distance constraints to be reasonable.

2.3   Proof Outline

We have designed an encoding in our model of programmed mutagenesis to directly simulate Minsky's Universal Turing Machine. There are three key ideas in this construction.

1. We explicitly represent the location of the head on the tape, by extending the alphabet to include a state-symbol pair for every possible state of the machine and every symbol used. We further expand the alphabet, but this is the essential decision.
2. We introduce scratch space (#) between each two tape symbols. This allows us to have read and write parts of each rule. We use the scratch space to

transmit information by first writing to it the information on the new state and the direction of movement. We then use the freshly written information as the anchor for the next step of the rule.

3. We execute each rule of the machine by four rewrite rules.
   - read the current state and symbol and save the information about the new state and the direction of the rule in the scratch space; begin the transition to the new tape symbol.
   - finish the transition to the new tape symbol.
   - read the new state info (in the scratch space), begin transition to the the new state/symol pair.
   - finish transition to the the new state/symbol pair; return the scratch space to its original state.

The rules are executed over four biological cycles. Only the first three biological cycles, however, include rewrite events, while the fourth simply copies the template strand using the outside primer. Furthermore, the rules act on alternative strands of DNA (3′-to-5′ for cycles 1 and 3 and 5′-to-3′ for cycle 2), and we use this property to drive the computation forward.

This proces allows us to have one-to-one rules, because each antecedent takes a different path to the consequent. We extend the alphabet to allow for this. To be precise, we have one-to-one rules for three out of four steps. Step 2 can be and is executed as many-to-one.

We use 101 oligonucleotides to implement the 27 rewrite rules that constitute the transition function of Minsky's machine. The instanteneous description of the DNA encoding of Minsky's machine at the beginning of each computational step is the single instance of a state/symbol character on the tape, indicating the precise head position and the state of the finite control, and all the other symbols on the tape. We represent states and symbols by nucleotide sequences and enact state transitions by primer extension reactions.

A simple way to detect when the computation is complete is to sense the appearance of the sequence which in the next step leads to the halting state on the tape with a molecular beacon. Molecular beacons are hairpin structures with a flourophore and a quencher on the 5′ and 3′ ends respectively in which a perfect match for the loop portion of the molecule forces the hairpin to open, thus separating the flourophore from the quencher and creating a fluorescent signal [15].

3   Programmed Mutagenesis and Other DNA Computing Systems

We call a computing system *composable* when a first computation results in a single molecule that can be used directly by a second computation as input without modification. Programmed mutagenesis systems are composable because both the input and output from a computation can be represented as a single strand of DNA. Programmed mutagenesis is the first composable system to be proven universal.

Our approach builds on other models proposed for DNA-based computing. These can be subdivided into five categories.

- Generate-and-search approach was pioneered by Adleman [1] in 1994. In this approach all possible candidate solution molecules are generated, and then filters are applied to select only those molecules that contain solutions. This was the first experimental demonstration of any computational system based on DNA, but this approach is not universal.
- Splicing systems were pioneered by Tom Head [5] in 1987. When these systems have finite sets of axioms and rules which define splicing, they are limited to generating regular languages [12]. If both of these sets remain finite, the only way to increase the computational power of the system is to introduce certain control mechanisms. Splicing systems utilizing a number of such control mechanisms have been shown to be universal. However, the control systems proposed require researcher's intervention and are vulnerable to experimental error. Furthermore, these mechanisms might prove difficult to implement in a laboratory.
- Self-assembly of two-dimensional DNA structures was shown to be universal by Winfree [16] in 1995. Winfree proposed to use ligation after self-assembly to produce a single reporter strand of DNA. This reporter strand would be used for reading the output of the self-assembly computation. A challenge in the experimental implementation of this approach might be that larger computations, which include more different varieties of tiles, will require longer time intervals for each step of the assembly because self-assembly relies on the correct tile assembly followed by the correct computational structure assembly. Further, minimizing the error rate also requires longer time intervals per assembly step. Recently, the first experimental demonstration of computation by self-assembly [10] has been produced. This is a significant advancement in DNA computing because it demonstrates the first DNA-only universal computational system which requires only ligase to create an output molecule. Certain proposed implementations of self-assembly systems can be composable, but the one experimentally implemented to date is not.
- Autonomous string systems based on hairpin formation were introduced by Hagiya et al. [4]. In these systems each molecule works not only as data carrier but also as a computing unit. Some interesting experimental systems based on this principle have been constructed [8], [13]. However, these systems are not, by themselves, universal. They are also not composable.
- Programmed mutagenesis is an example of a string-rewrite system for DNA computing. Both Beaver [2] and Kari et al [6] have also proposed to use systems based on string-rewriting. While both of these models were proven to be theoretically universal, both have drawbacks:

1. Beaver's model proposes to use the substitution operation, that is rules of the form xyz → xuz. This model, however, requires separations after each substitution step to guard against the possibility of a rule performing a mixture of substitution and deletion, such as rewriting the sequence xyyz

on the template into the sequence xuz, or insertion, such as rewriting the sequence xz on the template into the sequence xuz. Since separation operations are required, this model is also vulnerable to experimental error.

2. Kari et al [6] proposed to use insertion/deletion ("insdel") systems to implement universal computation. While the insdel systems represent a theoretically interesting computational model, new techniques will be required to implement them in practice. With current techniques there are two problems with the proposed rules for the insdel systems. First, there is no way to control the length of the deleted sequences in the insdel systems. The deletion rules in the insdel systems are of the form xzy → xy, which with the currently available techniques have to be implemented by an oligonucleotide encoding sequence xy. This oligonucleotide would bind any occurrence of the contexts x and y, and would, therefore delete any sequence embedded between these contexts. Thus, if the template read xzyy, the rule xy would, with about equal probability, delete sequences z and zy. Second, there is no way to prevent insertion rules in the insdel systems from performing substitutions. For example, an insertion rule of the form xz → xyz, represented by an oligonucleotide encoding the sequence xyz, and given a template xzzzz would be approximately equally likely to perform an insertion (rewriting the template into xyzzzz), or any of the three possible substitutions (rewriting the template into xyzzz, xyzz, or xyz).

An additional interesting property of programmed mutagenesis is that all the machinery necessary to implement it is present in a cell. We have also demonstrated that the key aspects of programmed mutagenesis function properly [7]. Furthermore, once the reaction is put together, and thermal cycling begins, programmed mutagenesis requires no intervention on the part of a researcher.

References

1. Adleman, L. M., 1994, *Molecular Computation of Solutions to Combinatorial Problems*, Science, 266, 5187, 1021–1024
2. Beaver, D., 1995, *Molecular Computing*, Department of Computer Science and Engineering Technical Report, Penn State University.
3. Ausubel, I., Frederick, M. (eds.), 1997 *Current Protocols in Molecular Biology, Section 8.5* (John Wiley & Sons, Inc.)
4. Hagiya, M., Arita,, M., Kiga, D., Sakomoto, K., and Yokoyama, S., 1999, *Towards parallel evaluation and learning of Boolean mu-formulas with molecules*, DIMACS Series in Discrete Mathematics and Theoretical Computer Science, 48, 57–72.
5. Head, T., 1987, *Formal language theory and DNA: an analysis of the generative capacity of specific recombinant behaviors*, Bulletin of Mathematical Biology, 49, 737–759.
6. Kari, L., Thierin, G., 1996, *Contextual Insertions/Deletions and Computability*, Information and Computation 131, 47–61.

7. Khodor, J., Gifford, D. K., 1997 *Design and Implementation of Computational Systems Based on Programmed Mutagenesis*, Proceedings of 3rd DIMACS workshop on DNA-based computers.
8. Komiya, K., Sakamoto, K., Gouzu, H., Yokoyama, S., Arita, M., Nishilkawa, A., and Hagiya, M., 2000, *Successive state transitions with IO interface by molecules*, Proceedings of 6th DIMACS workshop on DNA-based computers.
9. Lipton, R., 1995, *DNA solution to computational problems*, Science, 268, 5210, 542–545.
10. Mao, C., LaBean, T.H., Reif, J.H., Seeman, N.C., 2000 *Logical computation using algorithmic self-assembly of DNA triple-crossover molecules*, Nature 407, 493–496.
11. Minsky, M.L., 1967, *Computation: Finite and Infinite Machines* (Prentice-Hall, Inc.).
12. Paun, Gh., Rozenberg, G., Salomaa, A., 1998, *DNA Computing* (Springer-Verlag).
13. Sakomoto, K., Gouzu, H., Komiya, K., Kiga, D., Yokoyama, S., Yokomori, T., and Hagiya, M., 2000, *Molecular computation by DNA hairpin formation*, Science, 288, 1223–1226.
14. Turing, A.M., 1936, *On Computable Numbers, with an Application to the Entscheidungsproblem*, Proceedings of the London Mathematical Society, Series 2, 42, 230–265.
15. Tyagi, S., Kramer, F.R., 1996, *Molecular Beacons: Probes that Flouresce upon Hybridization*, Nature Biotechnology 14, 303–308.
16. Winfree, E., 1995, *On the computational power of DNA annealing and ligation*, DNA Based Computers: DIMACS Workshop, 27, 199–221.

Horn Clause Computation by Self-assembly of DNA Molecules

Hiroki Uejima[1], Masami Hagiya[1], and Satoshi Kobayashi[2]

[1] Department of Computer Science,
Graduate School of Information Science and Technology, University of Tokyo,
7-3-1 Hongo, Bunkyo-ku, Tokyo 113-0033, Japan.
{uejima, hagiya}@is.s.u-tokyo.ac.jp
[2] Department of Computer Science and Information Mathematics,
The University of Electro-Communications,
1-5-1 Chofugaoka, Chofu, Tokyo 182-8585, Japan.
satoshi@cs.uec.ac.jp

Abstract. Kobayashi proposed Horn clause computation by DNA molecules, which is more suitable for expressing complex algorithms than other models for DNA computing. This paper describes a new implementation of Horn clause computation by DNA. It employs branching DNA molecules for representing Horn clauses. As derivations are realized by self-assembly of such molecules, the implementation requires only a constant number of laboratory operations. Furthermore, it deals with first-order Horn clauses with some restrictions. In order to realize first-order logic, we implement variable substitutions by string tiling proposed by Winfree, et al. As we show the computational power of a Horn clause program in our model, we give another proof that a polynomial number of operations using self-assembly of DNA molecules can compute any problem in NP.

1 Introduction

Various computational models have been proposed to achieve generality and expressiveness in computation by DNA molecules. One of them is Horn clause computation proposed by Kobayashi [1,2]. Horn clause computation is more suitable for expressing complex algorithms than other models for DNA computing, because it has close relationship to the high-level programming language PROLOG.

In the implementation by Kobayashi, each molecule represents a fact or a rule of a Horn clause program, and one sequence of operations corresponds to one application of the resolution principle. Although first-order Horn clauses in Kobayashi's implementation are restricted, he proved that they are equivalent to nondeterministic Turing machines in their computational power. However, the number of operations required by Kobayashi's implementation of Horn clause computation is proportional to the number of derivation steps.

N. Jonoska and N.C. Seeman (Eds.): DNA7, LNCS 2340, pp. 308–320, 2002.
© Springer-Verlag Berlin Heidelberg 2002

Tiling [4] is one of the autonomous molecular computations. DNA molecules having a complex structure, such as double-crossover and triple-crossover molecules, are used as tiles. The sides of a tile correspond to the sticky ends of a DNA molecule, and are labeled by the sequences of the sticky ends. Tiles are aligned and assembled with each other at the sides whose labels are complementary. Tiling proceeds autonomously, i.e., we only need to prepare the DNA tiles, and set the appropriate condition for the reactions. The number of required laboratory operations is therefore constant.

Recently, Winfree, et al. [9] proposed another kind of DNA tile, called *string tile*. Several tile layers are compressed into a single string tile. Mathematically, a string tile comprises a directed graph. Edges of the graph correspond to DNA strands in the tile. Vertices are either on the sides of the tile or inside the tile. Self-assembly of the tiles joins their graphs. When the topology and routing of the edges is designed carefully, string tiles having simple graphs can perform surprisingly sophisticated calculations as shown in [9].

This paper describes a new implementation of Horn clause computation by DNA. It employs branching DNA molecules for representing Horn clauses. As derivations are realized by self-assembly of such molecules, the implementation requires only a constant number of laboratory operations. Furthermore, it deals with first-order Horn clauses with some restrictions. In order to realize first-order logic, we implement variable substitutions by string tiling.

In this paper, we first show how to compute propositional Horn clauses by DNA molecules. We then extend the model to first-order logic. For the extension, we show how to implement variable substitutions by string tiling. Finally, we show the computational power of a Horn clause program in our model. As a result, we give another proof that a polynomial number of operations using self-assembly of DNA molecules can compute any problem in NP.

2 Deductive Inference by Self-assembly of DNA

2.1 Simple Boolean Horn Clause Model

Let $V = \{A, B, C, \ldots\}$ be a countable set of Boolean variables. A Boolean formula of the form A is called a *fact* and a formula of the form $A \leftarrow B_1 \wedge B_2 \wedge \cdots \wedge B_n$ is called a *rule*, where $A, B_1, B_2, \ldots, B_n \in V$. We use a comma in place of $\wedge$ as follows: $A \leftarrow B_1, B_2, \ldots, B_n$. A *Horn clause* is either a fact or a rule. [1] A Horn clause of the form $C \leftarrow A, B$ or of the form $C \leftarrow A$ is called a *simple Boolean Horn clause* (an *SBHC*, for short), where $A, B, C \in V$. A finite set P of Horn clauses is called a *Horn program*. If P consists of SBHCs, P is called a *simple Horn program*.

Let P_V be the set of all Horn clauses on V. A mapping $T : 2^{P_V} \to 2^{P_V}$ is defined as follows. Here, we regard a fact A as a rule of the form $A \leftarrow$. For $I \subseteq P_V$, we define:

---

[1] More precisely, facts and rules are *definite clauses*.

$$T(I) = \{A \leftarrow B_1, \ldots, B_{i-1}, C_1, \ldots, C_m, B_{i+1}, \ldots, B_n \mid$$
$$\text{``}A \leftarrow B_1, \ldots, B_n\text{,''} \text{``}B_i \leftarrow C_1, \ldots, C_m\text{''} \in I, 1 \leq i \leq n\}$$

The mapping T means one application of the resolution principle.

Our molecular computation model consists of the following three basic operations:

1. $Detect(P, A)$: Check whether there exists a specified fact A in a given Horn program P, and output YES or NO.
2. $P_1 \cup P_2$: Compute the set union of a given pair of finite sets P_1 and P_2 of Boolean formulae.
3. $T(P)$: Compute $T(P)$ for a given set P of Boolean formulae.

These basic operations make it possible to compute logical consequences of a given simple Horn program P as follows. The following procedure answers whether a given fact A is a logical consequence of P, or not.

$N_0 := P$;
for $i = 1$ **to** n **do**
$N' := T(N_{i-1})$;
$N_i := N' \cup N_{i-1}$;
end
$Detect(N_n, A)$;

For any Horn program P, $T(P)$ denotes the set of clauses derived from two clauses in P by one application of the resolution principle. Therefore, any element of $T(P)$ is a logical consequence of P. It is also obvious that any element of $T(P) \cup P$ is a logical consequence of P. Since P is a finite set, N_n includes every logical consequence of P, when n is larger than the number of the Boolean variables in P. Consequently, N_n includes A, if and only if A is a logical consequence of P.

2.2   DNA Implementation

We can regard each DNA sequence as a string on the alphabet $\Sigma = \{a, c, g, t\}$. In this paper, when we write a single stranded DNA as a string, we always write it from the 5'-end to the 3'-end. For a single stranded DNA x, $\bar{x}$ denotes the reverse complementary sequence of x with respect to Watson-Crick pairing.

Implementation of Clauses Horn clauses are implemented by DNA molecules as illustrated in Figure 1.

Each Boolean variable $A \in V$ is represented by a single stranded DNA $E(A)$, where $E : V \to \Sigma^+$ is a coding function mapping each variable in V to a string on Σ. Each fact or rule is represented by a hybridized DNA molecule made of one or more single stranded DNA molecules, which contain at least one sticky end. A fact A is represented by a self-annealed single strand, which has the sticky end $E(A)$. A rule $A \leftarrow B_1, B_2, \ldots, B_n$. is represented by $n + 1$ hybridized single strands, which have $n + 1$ sticky ends $E(A), \overline{E(B_1)}, \overline{E(B_2)}, \ldots, \overline{E(B_n)}$.

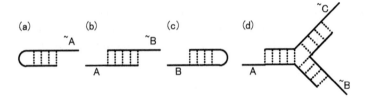

Fig. 1. Examples of a fact molecule, rule molecules and a query molecule. The symbol $\sim B$ denotes the sequence $\overline{E(B)}$ in this figure. (a) The query molecule to detect the fact A. (b) The rule molecule corresponding to the rule $A \leftarrow B$. (c) The fact molecule corresponding to the fact B. (d) The rule molecule corresponding to the rule $A \leftarrow B, C$.

Molecular representations of facts and rules are called *fact molecules* and *rule molecules*, respectively. A fact molecule and a rule molecule, or a pair of rule molecules can be hybridized at the sticky ends to produce one molecule.

A program P is implemented by water solution containing molecules corresponding to its facts and rules. We call a reaction in which two molecules become one molecule by hybridization at their sticky ends a *connection* of the two molecules.

Piotr Wąsiewicz et al. [6,7] proposed a similar DNA implementation for logical inferences via DNA computing. In [6], only linear rule molecules, which are Horn clauses with one premise and one conclusion, are allowed in a logical program. They then extended their implementation so that inference rules can have several premises, by combining linear rules [7]. Their inference system is realized by DNA hybridization like our implementation shown below. A laboratory experiment was also executed for verifying their implementation.

Implementation of Operations The operation $Detect(P, A)$ can be implemented as follows. First, we introduce a *query molecule*, which is a self-annealed single strand with a sticky end as in Figure 1(a). The sticky end of the query molecule that detects the fact A is $\overline{E(A)}$. If the fact A is in P, the query molecule will connect with the corresponding fact molecule, and generates a circular molecule by ligation. Therefore, $Detect(P, A)$ is implemented by the following procedure:

1. By using magnetic beads, keep the molecules which are connected with query molecules at the sticky end $\overline{E(A)}$, and remove the other molecules.
2. Ligate the molecules in the tube.
3. Check whether there exists a circular molecule. (Breaking non-circular molecules by exonuclease)

The computation of the set union of finite sets P_1 and P_2 of Boolean formulae is easily realized by merging the two test tubes.

The computation of the mapping T is implemented by the reaction in which each molecule of a specified Horn program connects with another molecule at one sticky end. If the hybridization proceeds autonomously, the iteration:

for $i = 1$ **to** n **do**
$\quad N' := T(N_{i-1});$
$\quad N_i := N' \cup N_{i-1};$
end

is implemented as time passes. Hence, the above procedure is completed by one laboratory operation.

2.3   Validity of the Implementation

In this section, we examine the validity of the implementation. We assume that two single stranded DNA molecules can hybridize only when they are completely complementary to each other. It is almost obvious that if a molecule produced by the above procedure does not contain a self-loop structure, then there exists a derived Horn clause whose literals correspond to the sticky ends of the molecule.

Now, we examine the case that molecules contain self-loop structures. Let us explain why the operation *Detect* is valid even in this case. First, only a rule molecule can form a self-loop structure because each fact molecule has only one sticky end. Furthermore, the self-loop structure is generated only by hybridization of sticky ends corresponding to the head and an atom in the body of a rule. Second, the sticky end of a query molecule hybridize only with the sticky end corresponding to the head of a rule. As a result,

- Molecules forming self-loop structures are removed in Step 1.
- After Step 1, the remaining molecules do not form self-loop structures.

Therefore, there are no molecules with self-loop structures after Step 1, and we can conclude that the implementation of *Detect* is valid.

3   Substitutions by String Tiling

3.1   Extension for First-Order Logic

We extend the deductive inference by self-assembly of DNA to deal with first-order logic. Let S be a relation symbol, and k be its arity. An expression $S(t_1, \ldots, t_k)$ is called an *atomic formula* (or an *atom*, for short), where t_i ($1 \leq i \leq k$) is a term made of constants, variables and function symbols. If t_i does not contain variables for any i, $S(t_1, \ldots, t_k)$ is called a *ground atom*. The notions of *ground* facts, rules and Horn programs are defined in a similar manner. All variables are considered universally quantified.

We can determine whether a ground fact $S(t_1, \ldots, t_k)$ is a logical consequence of a Horn program P by the following procedure.

$P' := MakeClause(P);$
$N_0 := P';$
for $i = 1$ **to** n **do**
    $N' := T(N_{i-1});$
    $N_i := N' \cup N_{i-1};$
end
$Detect(N_n, S(t_1, \ldots, t_k));$

The operation $MakeClause(P)$ substitutes ground terms for all variables of clauses in P, and returns the instantiated ground clauses.

Any logical consequence of P that contains no variables can be derived from some finite subset P' of ground clauses of P. Hence, on the condition that n is large enough for P and $S(t_1, \ldots, t_k)$, $S(t_1, \ldots, t_k)$ is a logical consequence of P if and only if N_n includes $S(t_1, \ldots, t_k)$.

It is necessary to restrict the form of a clause in a Horn program P in order to implement $MakeClause(P)$ by DNA. The restrictions are as follows.

1. A term in a rule should be of the form $f_1(f_2(\ldots f_n(X) \ldots))$.
2. The arity of a predicate should be at most two.
3. The arity of a function should be one.
4. The variable used in the first argument of an atom should always be X, and the variable in the second argument should always be Y.
5. A fact should contain no variables.

If each clause in P satisfies these conditions, we say that P is *restricted*. In the following implementation by DNA, we deal only with restricted simple Horn programs.

3.2   DNA Implementation

Now, the coding function E is extended so that it maps relations, functions and constants to strings in Σ^+. A ground term $f_1(f_2(\ldots f_n(a) \ldots))$ is then represented by $E(f_1(f_2(\ldots f_n(a) \ldots))) = E(f_1)E(f_2) \ldots E(f_n)E(a)$. A ground atom $S(t_1, \ldots, t_k)$ is represented by $E(S)E(t_1) \ldots E(t_k)$.

Outline of *MakeClause* On the basis of the coding function E defined above, we implement the operation *MakeClause* as follows Figure 2.

First, a string tile representing a rule in P is assembled with tiles representing substitutions for the variables in the rule. The details of the string tiles are explained later. The resulting large tile is then denatured to produce a single stranded circular strand, which corresponds to the result of applying the substitutions to the rule. The circular strand is then annealed and cut at several positions in order to reveal the sticky ends corresponding to the ground literals in the rule. If we use restriction enzymes to cut the strand, we insert restriction sites in the seed tile. The restriction site for the head atom is at the 5'-end of the strand in the MID tile (see below), and that for an atom in the body is at the 3'-end of the encoded predicate.

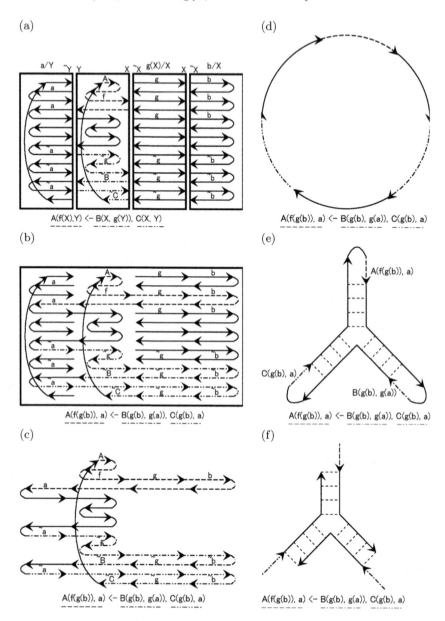

Fig. 2. An example of the *MakeClause* operation sequence. (a) A seed tile (representing the rule $A(f(X), Y) \leftarrow B(X, g(Y)), C(X, Y))$ and substitution tiles (representing the substitutions a for Y, $g(X)$ for X and b for X) are assembled. (b) Strands in tiles are connected with each other, and construct one circular strand. (c) Other strands are removed. (d) The single stranded circular strand has sequences corresponding to ground atoms. (e) The strand is self-annealed to make a 3-armed form. (f) The strand is then cut to reveal sticky ends corresponding to the ground atoms.

As for fact molecules in P, the operation does nothing since they are already ground.

Implementation of Substitutions The details of string tiles used in the implementation of *MakeClause* are as follows. We use two kinds of string tiles, *seed tiles* corresponding to a rule, whose arguments contain a variable, and *substitution tiles* corresponding to a substitution for a variable. The side of a tile at which it can connect with another tile is called a *port*. Since seed tiles and substitution tiles are linear string tiles (in the sense of [9]), they have at most two ports. A string in Σ^+, called a *binding label*, is assigned to each port, and only ports having reverse complementary binding labels can connect with each other.

The right binding label of a seed tile encodes X, while the left one encodes Y. One binding label of a substitution tile encodes X or Y, while the other encodes $\sim X$ or $\sim Y$, respectively.

A substitution occurs by a connection of tiles according to the binding labels. After a substitution tile representing a function is connects with a (seed) tile, a further substitution tile can connect with the resulting tile at the other port. Finally, the substitution for an argument is completed by a substitution tile representing a constant.

Structure of a Seed Tile The structure of a seed tile is expressed by a combination of four kinds of tiles as in Figure 3. Four kinds of tiles are (a) the HEAD tile corresponding to the head atom, (b) the BODY tiles corresponding to atoms in the body, (c) the MID tile inserted between the HEAD and the BODY tiles for making the size of the produced seed tile uniform, and (d) the SIDE tile for making the produced strand circular.

A seed tile is made by combining these tiles as in Figure 3(e). BODY1 and BODY2 denote the BODY tiles corresponding to the first and second atoms of the body, respectively.

3.3   Computational Power

In this subsection, we construct a restricted simple Horn program (an RSHP, for short) to simulate a nondeterministic Turing machine which computes a problem in NP. Through this simulation, we can give another proof that a polynomial number of operations using self-assembly of DNA molecules can compute any problem in NP.

Theorem 1. *For any decision problem in NP, there exists an RSHP which computes the problem in a polynomial number of steps.*

Proof. Let TM be a one-tape nondeterministic Turing machine which computes the problem in polynomial time. We will define an RSHP H which simulates TM in a polynomial number of steps.

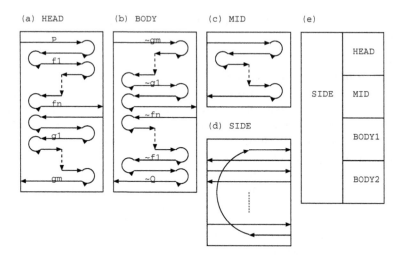

Fig. 3. The structure of a seed tile. (a) The HEAD tile corresponding to the head of a rule. This tile represents $P(f_1(\ldots f_n(X)\ldots), g_1(\ldots g_m(Y)\ldots))$. (b) The BODY tile corresponding to an atom of the body of a rule. This tile represents $Q(f_1(\ldots f_n(X)\ldots), g_1(\ldots g_m(Y)\ldots))$. (c) The MID tile to be inserted between the HEAD and BODY tiles. It is used to make the size of the tiles uniform. (d) The SIDE tile to make a circular strand. (e) The way for combining these tiles.

For each state s of TM, we prepare a binary predicate S_s for representing instantaneous descriptions of TM. The first and second arguments of S_s are used to store the tape symbols to the left and right of the head of TM, respectively. The second argument includes the symbol at the head as well.

For each tape symbol t, we prepare a unary function f_t. A state transition with the head moving to the left can be implemented as follows:

$$S_{s'}(X, f_{t_{-1}}(f_{t_0'}(Y))) \leftarrow S_s(f_{t_{-1}}(X), f_{t_0}(Y)),$$

where t_{-1} ranges over all tape symbols of TM. This rule corresponds to a transition rule such that the head in the state s reads t_0, writes t_0', changes the state to s', and moves to the left.

Similarly, a state transition with the head moving to the right can be implemented as follows:

$$S_{s'}(f_{t_0'}(X), Y) \leftarrow S_s(X, f_{t_0}(Y)).$$

This rule corresponds to a transition rule such that the head in the state s reads t_0, writes t_0', changes the state to s', and moves to the right.

We add to H a fact representing an initial configuration of the following form:

$$S_{p_0}(a_1, f_{t_1}(f_{t_2}(\ldots f_{t_n}(f_b(\ldots f_b(a_2)\ldots))\ldots))),$$

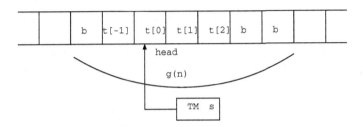

Fig. 4. An example of a configuration of a Turing machine. The machine is in the state s and the head is on the cell with the tape symbol t_0. The fact $S_s(f_{t_{-1}}(f_b(a_1)), f_{t_0}(f_{t_1}(f_{t_2}(f_b(f_b(a_2))))))$ represents the configuration of the Turing machine in this figure. The constants a_1 and a_2 represent the left and the right bounds of the tape, respectively.

where p_0 is an initial state, $w = t_1 t_2 \ldots t_n$ is the input string to TM, and b is a blank symbol. The constants a_1 and a_2 represent the left and the right bounds of the tape, respectively. Then, $H \cup \{S_{p_0}(a_1, f_{t_1}(f_{t_2}(\ldots f_{t_n}(f_b(\ldots f_b(a_2) \ldots)) \ldots)))\}$ precisely simulates TM with the input w.

To check whether TM accepts w or not, we add the following clause:

$$Goal \leftarrow S_{p_f}(X, Y),$$

where p_f is a final state of TM.

4   Discussion

4.1   3D Conformation of Branching Molecules

Proof trees consisting of fact molecules and rule molecules are complex DNA molecules having 3D conformation with high density. Here, we examine their conformation quantitatively.

If we regard a proof tree molecule as a binary tree with height n, then the tree has at most 2^n nodes. If the length of each edge in the tree is 1, the 2^n nodes are packed in a cube with volume n^3. Hence the volume each node can occupy is less than $2^n/n^3$. Therefore, the size of each node should become zero as n increases infinitely. Even if we satisfy the above condition, conformation with high density may block hybridization.

A solution to this problem is to introduce a new operation for simplifying proof trees. The operation "contraction" cuts self-annealed bulges, which are branches without sticky ends, in a proof tree molecule. This operation may be effective for reducing the density of conformation, unless the proof tree molecule has too many sticky ends.

This solution has two drawbacks. Firstly, the algorithm is required to repeat the additional operation "contraction." It is desirable that the operation also proceeds autonomously as proof trees are self-assembled. Secondly, the operation cannot cut branches with sticky ends.

Another solution is to make edges longer as trees get larger. This solution cannot be realized with the present set of DNA molecules, which are designed uniformly.

4.2   Bias and Errors in Hybridization

In our model, we assume that various forms of DNA molecules, which correspond to facts and rules derived from the initial Horn program, behave uniformly. In particular, the model ignores the difference between the sticky end sequences. Different DNA sequences have different rates of hybridization, which lead to bias in hybridization i.e., some pairs of sticky ends are hybridized more preferably than others. The analysis of the bias with respect to the time and space complexity of the model is also left as a future work.

Bias in hybridization is more critical when long terms are encoded in sticky ends for computing first-order logic. Sticky ends vary in their length as well as in their contents. The present encoding of terms has this fatal problem.

To make matters worse, the present encoding is error-prone in hybridization. The sticky ends $P(f(f(f(f(a)))))$ and $P(f(f(f(f(f(a))))))$ mis-hybridize easily, and they cannot be denatured easily. Hence it is unrealistic to assume that the sticky ends hybridize only if the sequences are completely complementary to each other.

We have no concrete solution to this problem now. But it is clear that a weak point of our algorithm is to hybridize long DNA sequences at once. So, an apparent solution is to split long sequences and repeat hybridization of short ones incrementally. Note that hybridization is also repeated in the phase of variable substitution. It is therefore desirable to have an algorithm in which variable substitution and deductive inference are executed in the same phase, and hybridization for deductive inference is also incremental. As described below, such an algorithm is also suitable for making substitution efficient.

4.3   Analysis of Chemical Reactions

As for the time complexity of our model, we claimed that the number of laboratory operations required for computing a Horn program is constant. However, we have not estimated the time necessary to complete each operation. It requires the examination of the reaction rate of DNA hybridization. The space complexity has a similar problem. More precise analysis of reactions from the aspect of thermodynamics and kinetics, as that of [3], is necessary to estimate the time and space complexity of the model and control the reactions as discussed in the final section.

4.4 Random Substitutions

The unification algorithm employed in the conventional Horn clause computation efficiently computes the most general unifier of two atoms. In our model, however, the generation of substitutions by connections of string tiles is completely random, and may cause combinatorial explosion of required string tiles.

In our present implementation, the phase of generating substitutions is separated from that of the deductive inference. Selective generation of substitutions as that of the unification algorithm requires the coupling of the two phases.

5 Conclusion

We proposed a DNA computing model based on Horn clause computation. It is implemented by self-assembly of DNA molecules. We extended the Horn clause computation to first-order logic by implementing variable substitutions using string tiles. This model has advantages over other models for DNA computing. First, because Horn clause computation is based on the high-level programming language PROLOG, it is easier to design algorithms in the model. Second, because the implementation employs self-assembly of DNA molecules, the number of operations to complete the computation is independent of the size of a problem instance.

It is, however, difficult to control self-assembly of molecules. At the microscopic control, we should carefully design DNA sequences or DNA structures. At the macroscopic level, the regulation of experimental conditions such as temperature and salt concentration is important.

In particular, the experimental conditions change the reaction rate and the error probability of molecular computation. Winfree [8] made one detailed analysis of the reaction rate and the error probability of a simple DNA tiling system using thermodynamics and showed their trade-off. This kind of analysis should also be done on our model.

Seeman et al. [5] emphasize the importance of chemical conditions for stabilizing special DNA nanostructures such as illustrated in Figure 2. In order to assemble large molecules in bulk enough to detect, we should also take chemical conditions into consideration from the aspect of scalability of autonomous computing.

In general, analysis of DNA computation as probabilistic algorithm will become more and more important for DNA computing to be practical.

References

1. Satoshi Kobayashi et al.: DNA Implementation of Simple Horn Clause Computation, *IEEE International Conference on Evolutionary Computation*, pp. 213–217, 1997.
2. Satoshi Kobayashi: Horn Clause Computation with DNA Molecules, *Journal of Combinatorial Optimization*, Vol. 3, pp. 277–299, 1999.
3. Stuart A. Kurtz et al.: Active Transport in Biological Computing, *DNA Based Computers II*, DIMACS Series, Vol. 44, pp. 171–179, 1998.

4. Thomas H. LaBean et al.: Experimental Progress in Computation by Self-Assembly of DNA Tilings, *Preliminary Proc. of Fifth International Meeting on DNA Based Computers*, pp. 121–138, 1999.
5. Nadrian C. Seeman et al.: The Perils of Polynucleotides: The Experimental Gap Between The Design and Assembly of Unusual DNA Structures, *DNA Based Computers II*, DIMACS Series, Vol. 44, pp. 215–233, 1998.
6. Piotr Wąsiewicz et al.: The Inference via DNA Computing, *Congress on Evolutionary Computation (CEC'99)*, pp. 988–993, 1999.
7. Piotr Wąsiewicz et al.: The Inference Based on Molecular Computing, *Int. Journal of Cybernetics and Systems*, 31/3, pp. 283–315, 2000.
8. Erik Winfree: Simulations of Computing by Self-Assembly, *Preliminary Proc. of Fourth International Meeting on DNA Based Computers*, pp. 213–239, 1998.
9. Erik Winfree et al.: String Tile Models for DNA Computing by Self-Assembly, *Preliminary Proc. of Sixth International Meeting on DNA Based Computers*, pp. 65–84, 2000.

DNA-based Parallel Computation of Simple Arithmetic

Hubert Hug[1] and Rainer Schuler[2]

[1] Universitäts-Kinderklinik Ulm, Prittwitzstrasse 43, D-89075 Ulm, Germany
[2] Abt. Theoretische Informatik, Universität Ulm, D-89069 Ulm, Germany

Abstract. We propose a model for representing and manipulating binary numbers on a DNA chip which allows parallel execution of simple arithmetic. As an example we describe how addition of large binary numbers can be done by using a DNA chip. The number of steps is independent of the size (bits) of the numbers. However, the time for some biochemical reactions is still large, and increases with the size of the sequences to be assembled.

1 Introduction

Biological macromolecules can be used for storing information in a new kind of computers and biochemical reactions, like nucleic acid hybridization and enzyme reactions, can be used to solve algorithmical problems [Adl94]. Since a vast number of biochemical reactions can take place at many molecules simultaneously, parallel computations involving millions of operations seem possible. Methods for solving several well known NP-complete problems have been proposed, and have been performed on small examples in many cases [Adl94, Adl98, LWF+00] [SGK+00, OKLL97, FCLL00]. Moreover, based on certain biochemical reactions, formal models capturing the power of DNA-computing have been developed [Lip95].

Many arithmetical operations, like addition and multiplication of binary numbers can be performed in parallel on classical hardware. In this paper we consider the implementation of simple arithmetic operations in parallel, using addressable DNA-chips.

The addition of (small) binary numbers with DNA-molecules has been done by Guarnieri et al. [GFB96]. However the procedure is sequential and does not use the power of DNA-computing for parallelization. Other approaches are based on selecting the correct result from a set of all possible values [Ata00, Fri00, QL98] [GPZ97, OR96, Rei97].

We propose a way of representing binary numbers on DNA-chips such that operations, which allow to initialize and manipulate numbers, can be performed with standard methods from gene technology. The methods used here have already been performed in practice by others. In particular, the use of a restriction enzyme to cut single stranded DNA which builds hairpin loops has been used in [SGK+00].

N. Jonoska and N.C. Seeman (Eds.): DNA7, LNCS 2340, pp. 321–328, 2002.
© Springer-Verlag Berlin Heidelberg 2002

We show that it is possible to read, write and manipulate numbers. For example, we describe how to

- store a number on the chip,
- read the number on the chip,
- add a (second) number to the number on the chip,

Each of the above operations can be performed in parallel (for each bit) with a constant number of steps for each bit of the numbers.

2 The Model

We use addressable DNA-chips for storing and manipulating large binary numbers. A standard tool used in DNA computing is hybridization of reverse complementary sequences. To minimize errors we assume that different sequences are chosen in such a way, that cross hybridization (i.e. the hybridization of sequences which are not reverse complementary) is minimal [FLT+97]. We use a restriction enzyme R_A to cut a double stranded DNA sequence. The restriction enzyme and the recognition site, denoted by A, is also chosen such that cross hybridization is minimal. Recall that a restriction site is a palindromic sequence of typically 4 or 6 bp length.

2.1 Representing Numbers

A number t is represented by a binary string $t_{n-1} \cdots t_1 t_0$, where $t_i \in \{0, 1\}$, such that $t = \sum_{i=0}^{n-1} t_i \cdot 2^i$. The number is stored bitwise, each bit is represented by specific DNA-sequences which are attached at a particular position (spot) on the chip via its 3' end (e.g. from ThermoInteractiva). The sequences for the i-th bit t_i consist of the sequences p_i, A, and t_i (see Figure 1). The sequence p_i indicates the position i and is specific for each i. Sequence A is the recognition site of a restriction enzyme R_A and will be cut in the presence of R_A if paired with its complementary sequence. The sequence t_i is either a sequence indicating value 0 or a sequence indicating value 1 (depending on whether the i-th bit t_i is 0 or 1). We note here that for all bits the same sequences are used to indicate the values 0 and 1. In order to minimize cross hybridization, sequence selection can be done as described by Frutos et al. [FLT+97].

2.2 Basic Operations

Let us first recall how to read a sequences from a chip. To do this we use reverse complementary oligonucleotides for 0 and 1 which are marked with different fluorescent dyes, say 0 with red (e.g. Cy3) and 1 with green (e.g. Cy5) [IER+99]. These sequences are put on the chip, where the reverse complementary sequences bind to the 0/1 sequences on the chip. Additional sequences are washed away, and the chip is read using a chip reader. All positions (bits) equal to 0 will be marked red and all positions (bits) equal to 1 will be marked green.

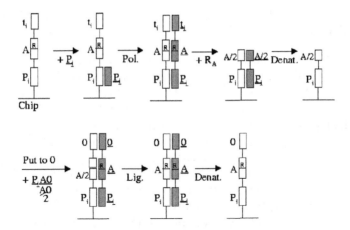

Fig. 1. Representing numbers with DNA on an addressable chip surface. p_i carries the positional information, A contains the recognition site for a restriction enzyme R_A, and t_i codes either 1 or 0. Reverse complementary sequences are underlined. Abbr.: Pol., DNA polymerase; Denat., denaturation; Lig., ligase.

We consider the addition of two numbers a and b using a small set of basic operations. In a first step, number a is stored on the chip. In a second step, number b is added bitwise to t. (That is bitwise addition modulo 2 of a_i and b_i is computed on the chip). In a third step the carry bits c are added to the number on the chip. We will further show that the carry bit can be computed in parallel using the method of Adleman [Adl94].

Let I_a denote the positions such that $a_i = 1$. Similarly let I_b and I_c denote the positions such that $b_i = 1$ and $c_i = 1$ respectively. The the addition of a and b can be done as follows.

For all $i \in I_a$ do select i and set $t_i = 1$
For all $i \notin I_a$ do select i and set $t_i = 0$
For all $i \in I_b$ do select i set $t_i = t_i \oplus 1$
For all $i \in I_c$ do select i set $t_i = t_i \oplus 1$

The first two steps store a number a on the chip. The third and fourth step perform bitwise addition modulo 2.

2.3   Writing Numbers on the Chip

Writing numbers on a chip is done in two parts. In the first part all bits which are equal to 0 are set, and in the second part all bits which are equal to 1 are set. To do this we prepare

- for all bits i which will be set to 0, reverse complementary sequences $\underline{p_i}$, $\underline{p_i A0}$ and sequences $0A/2$, and

- for all bits k which will be set to 1, reverse complementary sequences $\underline{p_k}$, $\underline{p_k}A1$ and sequences $1A/2$,

where $A/2$ is the sequence A cut by the restriction enzyme R_A. To set bits to 0, two steps are performed. In a first step the values of the respective positions are deleted, i.e. all positions $i \notin I_a$ are selected. In a second step all deleted positions are extended to 0, i.e. all selected positions are set to 0.

Select positions i: To delete the values at positions i we use reverse complementary sequences $\underline{p_i}$ to bind to the corresponding sequences p_i on the chip. Using DNA polymerase and free nucleotides the complementary sequence is extended to yield the double stranded DNA sequence $p_i At_i$, where t_i is 0 or 1 depending on the value at position i. Using the restriction enzyme R_A, the sequence is cut at position $p_i A$ (see Figure 1).

Set $t_i = 0$ **for all selected positions** i: To set the bit at the selected positions to 0, we hybridize oligonucleotides consisting of the reverse complementary sequences $\underline{p_i A0}$ and sequences $0A/2$. Sequence $0A/2$ is able to bind to its reverse complement, thus building a sequence $\underline{p_i A0}$ which is double-stranded in the $0A/2$ region. (We assume that $0A/2$ is in excess such that all $\underline{p_i A0}$ are double-stranded in the $0A/2$ region). The sequences are put on the chip and the single stranded part of the $\underline{p_i A0}$ sequence which is complementary to $p_i A/2$ will bind to the sequence $p_i A/2$ of a selected position. Using DNA ligase the free neighboring ends in the resulting double stranded DNA sequence can be closed (see Figure 1).

To set bits to 1 is done similarly by using the reverse complementary sequences $\underline{p_k}$, $\underline{p_k}A1$ and sequences $1A/2$.

2.4   Bitwise Addition Modulo 2

As second operation we consider addition modulo 2. Assume that a number t is stored on the surface of the chip and a number b will be added bitwise modulo 2. That is, for all i such that the i-th bit of b is equal to 1, the values of t_i have to be flipped. To achieve this we use the fact that reverse complementary sequences in a single stranded DNA molecule form hairpin loops.

The bit flipping is done in two parts. In the first part the bits t_i equal to 0 are set to 1, in the second part the bits t_i equal to 1 are set to 0 (see Figure 2). To do this we prepare

- for all bits k such that $b_k = 0$, reverse complementary sequences $\underline{p_i}$, and
- for all bits i such that $b_i = 1$, sequences $\underline{0}A$, $\underline{1}A$ and $\underline{A}$. Note that $\underline{0}A$ and $\underline{1}A$ consist of reverse complementary sequences of 0 or 1 followed by sequence A.

In general, reverse complementary sequences are underlined. To set t_i from 0 to 1 is done as follows.

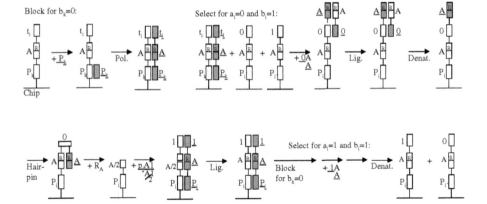

Fig. 2. Adding modulo 2 on a chip surface. Abbr. are as for Figure 1.

1. All positions k such that b_k is equal to 0 are blocked from further reactions.
2. All remaining positions i such that t_i is equal to 0 are selected. In this reaction the blocking of the positions k such that b_k is equal to 0 is released. Selected positions are set to 1 and blocked from further reactions.
3. All positions k such that b_k is equal to 0 are blocked from further reactions. All remaining positions i such that t_i is equal to 1 are selected.
4. Selected positions are set to 0.

We describe the procedure in detail. In the first step, reverse complementary oligonucleotides p_k, for all k such that $b_k = 0$, are added to the chip and will bind to the sequences $p_k At_k$. The 3' end of p_k will be elongated by adding DNA polymerase and free nucleotides (see Figure 2).

For the next step, partially reverse complementary oligonucleotides $\underline{0}A$ and $\underline{A}$ are hybridized and added to the chip. Now oligonucleotides $\underline{0}A$ which are double stranded on the A part, will hybridize to 0 at all positions $p_i A0$ which are not blocked by step 1. The DNA ligase is used to attach $\underline{A}$ to 0. Then $A0$ is removed by denaturation. Note that this also removes the blocking of positions p_k performed in the first step above.

Now $\underline{A}$ can hybridize intramolecularly with sequence A to form a hairpin loop. This double stranded part of the DNA sequence can be cut using restriction enzyme R_A. Now all positions p_i such that $b_i = 1$ and $t_i = 0$ are selected and can be set to 1 as described in paragraph "Set $t_i = 0$ for all selected positions i" above.

3 Computing the Carry Bits

Our construction follows the method of Adleman to build all possible paths in a graph. We prepare nodes and edges as follows. Each node specifies whether position i generates, propagates, or deletes a carry. An edge connects a node $i + 1$ to node i, if position i propagates a carry.

To build the nodes and edges we prepare

- for all bits i such that $a_i = 1$, reverse complementary sequences $\overline{p_{i+1}A1_i}$,
  for all bits i such that $a_i = 0$, reverse complementary sequences $\overline{p_{i+1}A0_i}$.
  The 5' ends of the oligonucleotides are biotinylated.
- for all bits i such that $b_i = 1$, sequences $0_i q_i$ and $1_i 1$ (i.e. sequences 3'-$0_i q_i$-5'
  and 3'-$1_i 1$-5'),
  for all bits i such that $b_i = 0$, sequences $0_i 0$ and $1_i q_i$ (i.e. sequences 3'-$0_i 0$-5'
  and 3'-$1_i q_i$-5').
- for all i, sequences $q_i p_i$ (i.e. sequences $3' - q_i p_i - 5'$).
- reverse complimentary sequences $\underline{p_0 A0}$. The 5' end is biotinylated.

First we compute the sequences for the nodes, indicating whether a carry is generated, propagated or deleted. For all i, we select the respective oligonucleotides for a_i and b_i as defined above. The oligonucleotides are mixed under conditions that complementary sequences hybridize. Double strands are extended by DNA polymerase, then the double strands are denatured. Single strands containing a biotin are coupled with paramagnetic streptavidine molecules (e.g. from Promega). They can now be separated by using a magnet from the rest of the DNA molecules in the solution [Adl94]. Observe, that for all positions i,

| if $a_i = b_i = 1$ | then the nodes are sequences $\overline{p_{i+1}A1_i 1}$ |
| if $a_i = b_i = 0$ | then the nodes are sequences $\overline{p_{i+1}A0_i 0}$ |
| if $a_i = 1$ and $b_i = 0$ then the nodes are sequences $\overline{p_{i+1}A1_i q_i}$ |
| if $a_i = 0$ and $b_i = 1$ then the nodes are sequences $\overline{p_{i+1}A0_i q_i}$ |

Now we add the sequences corresponding to the edges, i.e. sequences $q_i p_i$. (Note that some edges are connecting nodes that are not present). Furthermore sequences $\underline{p_0 A0}$ are added since at position 0 no carry has to be considered. Complementary sequences hybridize. Double strands are extended by DNA polymerase and then denatured. Single strands containing a biotin are separated as described above.

We observe that for every position i, the resulting DNA molecules contain sequences starting with p_{i+1} at the 5' end, and ending with a 0 or a 1 at the 3' end. Moreover, if the carry bit $c_{i+1} = 1$, then the sequences end with a 1. If the carry bit $c_{i+1} = 0$, then the sequences end with a 0.

We prepare a dextran matrix, to which the oligonucleotides 0 are attached (e.g. from ThermoHybaid). This selects two sets of DNA, one set containing sequences with p_i for all positions i such that the carry bit is 1, and one set containing sequences with p_k for all positions k such that the carry bit is equal to 0.

We add to each set the oligonucleotide A, hybridize and cut with restriction enzyme R_A. The double strands are denatured and separated by using a magnet. The second set contains $p_k A/2$ or $q_k p_k A/2$ for all k such that $b_k = 0$, and can be used to block positions $\overline{k}$ as described in subsection "Bitwise addition modulo 2."

4   Discussion

Standard methods of molecular biology are used to carry out all described calculations. Since all involved methods are feasible [ABK+01], the whole procedure seems also feasible. Single reactions might take a long time (e.g. ligation) and are error prone. Since we are using a constant number of steps for each position i, the overall error will be small enough to read out the result on the chip. Except cutting with restriction enzymes (e.g. BamH1, HindIII, EcoR1 etc.) the reactions are reversible. DNA double strands are more stable at lower temperature and higher salt concentration and the conditions for hybridization and denaturation have to be optimized. For the ligation reaction we can use T4 DNA ligase, for extending the 3' ends of p_i possible candidates are the Klenow fragment of DNA polymerase I, the large fragment of Bst-DNA polymerase or T4 DNA polymerase [ABK+01].

A crucial step involved is the computation of the carry bits. Here the length of the sequences depends on the number of bits used, slowing down the hybridization kinetics. A possible approach to keep sequences to an expected constant length could be to involve splicing mechanisms. (Splicing removes middle parts of the sequence, e.g. removes the $q_i p_i 0_i A$ part from $p_{i+1} A 1_i q_i p_i A 0_i 0$). For example, if we use RNA instead of DNA [FCLL00], we could exploit self splicing reactions. Currently there are limitations in exploiting splicing mechanisms: (1) RNA is less stable than DNA. (2) Splicing mechanisms are complex but generally irreversible. They depend on consensus sequences. Self splicing introns, which could be used, are RNA sequences with catalytic activity, so called ribozymes. Splicing of pre-mRNAs of higher eucaryotes is mediated by spliceosomes which are complexes of proteins and small RNAs [Sha94] and are therefore not applicable.

By removing biotin-streptavidine labeled oligonucleotides via a magnet [Adl94], or selecting oligonucleotides $\underline{0}$ losses may occur. These losses should not fall below a certain threshold level. The exact amounts have to be determined experimentally.

Currently we are investigating further biochemical reactions to simplify the procedure and to optimize the involved algorithm. The proposed model will be optimized to use as little as possible manual steps. An interesting question is, whether the calculation and addition of the carry bits is possible on the chip surface.

References

[ABK+01]   F.M. Ausubel, R. Brent, R.E. Kingston, D.D. Moore, J.G. Seidman, J.A. Smith, and K. Struhl, editors. *Current Protocols in Molecular Biology.* Wiley and Sons, 2001.

[Adl94]   L.M. Adleman. Molecular computation of solutions to combinatorial problems. *Science*, 266:1021–1024, November 11, 1994.

[Adl98]   L.M. Adleman. Computing with DNA. *Scientific American*, 279(2):54–61, August 1998.

328   Hubert Hug and Rainer Schuler

[Ata00]   A. Atanasiu. Arithmetic with membranes. In *Workshop on Multiset Processing, Curtea de Arges, Romania, August 2000*, pages 1–17. C. S. Calude and M. J. Dinneen and Gh. Păun, 2000.

[FCLL00]   D. Faulhammer, A. R. Cukras, R. J. Lipton, and L. F. Landweber. Molecular computation: RNA solutions to chess problems. *Proc. Natl. Acad. Sci. USA 97*, pages 1385–1389, 2000.

[FLT$^+$97]   A. G. Frutos, Q. Liu, A. J. Thiel, A. M. Sanner, A. E. Condon., L. M. Smith, and R. M. Corn. Demonstration of a word design strategy for DNA computing on surfaces. *Nucleic Acids Res*, 25(23):4748–4757, 1997.

[Fri00]   P. Frisco. Parallel arithmetic with splicing. *Romanian Journal of Information Science and Technology (ROMJIST)*, 2000. to appear.

[GFB96]   F. Guarnieri, M. Fliss, and C. Bancroft. Making DNA add. *Science*, 273(5272):220–223, July 12 1996.

[GPZ97]   V. Gupta, S. Parthasarathy, and M.J. Zaki. Arithmetic and logic operations with DNA. In *Proceedings of the 3rd DIMACS Workshop on DNA Based Computers, held at the University of Pennsylvania, June 23-25, 1997* [RW99], pages 212–220.

[IER$^+$99]   V.R. Iyer, M.B. Eisen, D.T. Ross, G. Schuler, T. Moore, J.C. Lee, J.M. Trent, L.M. Staudt, J. Hudson, M.S. Boguski, D. Lashkari, D. Shalon, D. Botstein, and P.O. Brown. The transcriptional program in the response of human fibroblasts to serum. *Science*, 283:83–87, 1999.

[Lip95]   R. J. Lipton. DNA solution of hard computational problems. *Science*, 268:542–545, April 28, 1995.

[LWF$^+$00]   Q. Liu, L. Wang, A. G. Frutos, A. E. Condon, R. M. Corn, and L. M. Smith. DNA computing on surfaces. *Nature*, 403:175–179, 2000.

[OKLL97]   Q. Ouyang, P. D. Kaplan, S. Liu, and A. Libchaber. DNA solution of the maximal clique problem. *Science*, 278:446–449, 1997.

[OR96]   M. Ogihara and A. Ray. Simulating boolean circuits on a DNA computer. Technical Report TR 631, University of Rochester, Computer Science Department, August 1996.

[QL98]   Z. F. Qiu and M. Lu. Arithmetic and logic operations for DNA computers. *Second IASTED International Conference on Parallel and Distributed Computing and Networks*, pages 481–486, December 1998.

[Rei97]   J. H. Reif. Local parallel biomolecular computing. In *Proceedings of the 3rd DIMACS Workshop on DNA Based Computers, held at the University of Pennsylvania, June 23-25, 1997* [RW99], pages 243–264.

[RW99]   H. Rubin and D. Wood, editors. *Proceedings of the 3rd DIMACS Workshop on DNA Based Computers, held at the University of Pennsylvania, June 23-25, 1997*, volume 48 of *DIMACS: Series in Discrete Mathematics and Theoretical Computer Science.*, Providence, RI, 1999. American Mathematical Society.

[SGK$^+$00]   K. Sakamoto, H. Gouzu, K. Komiya, D. Kiga, S. Yokoyama, T. Yokomori, and M. Hagiya. Molecular computation by DNA hairpin formation. *Science*, 288:1223–1226, 2000.

[Sha94]   P.A. Sharp. Nobel lecture: Split genes and RNA splicing. *Cell*, 77:805–815, 1994.

On P Systems with Global Rules

Andrei Păun

Department of Computer Science, University of Western Ontario
London, Ontario, Canada N6A 5B7
apaun@csd.uwo.ca

Abstract. We contribute to the vivid area of membrane computing (P systems) by considering the case when the same evolution rules are valid in all regions of a system. Such a P system is called *with global rules*. We consider the case of string-objects, with the evolution rules based on splicing and by rewriting. Universality results are proved for both types of systems. For splicing we also try to minimize the diameter of the used rules, while for rewriting systems we improve a result from the literature, proving that two membrane suffice for simulating Turing machines.

1 Introduction

Membrane computing is a computing framework inspired from the functioning of alive cells: in the regions defined by a membrane structure one places objects (in the present paper they are considered strings over a given alphabet), and rules for processing these objects. By using these rules in a nondeterministic parallel manner, we get transitions between system configurations, hence computations. Many variants were already considered in the literature, and many of them were proved to be computationally universal (equal in power to Turing machines), while many are able to solve NP-complete problems in polynomial time by making use of an exponential space created by membrane division and/or by object replication.

In the basic model, as introduced in [7], with each region one associates a (possibly) different set of rules. A natural variant is to consider the case when the rules are not localized, but a unique set of rules exists (a "global" set), which are used in all regions of the system. We deal here with this case, for P systems with string objects processed by rewriting (by means of context-free rules) and by splicing.

It is known (see [7], [9]) that in the case when each region has the set of rules, rewriting P systems (with three membranes [7]) and splicing P systems (with two membranes, [9], [2]) are computationally universal, they generate all recursively enumerable languages. At the first sight, working with a unique set of rules is a strong restriction, but as we will prove below, we still get universality. This is true for both rewriting and splicing P systems. Again, in both cases two membranes are sufficient. For splicing P systems we also minimise the diameter of the used rules (the length of strings involved in the splicing rules).

N. Jonoska and N.C. Seeman (Eds.): DNA7, LNCS 2340, pp. 329–339, 2002.
© Springer-Verlag Berlin Heidelberg 2002

We want to emphasize here a common feature of splicing and rewriting P systems (which is considered here for the first time for rewriting P systems): because we use a terminal alphabet for selecting among the strings sent out of the system those which belong to the language generated, one does not need to take into account account only halting computations and to introduce a (terminal) string in the language only if the computation halts. (The use of halting computation is essential in the case of symbol-objects, when we count the objects leaving the system, and a way to stop the computation is necessary, but not in the case of string-objects.). This slightly changes the proof techniques, because instead of introducing trap rules of the form $Z \to Z$, for running a "wrong" computation forever, we have to make sure that a nonterminal symbol is introduced in a sentential form so that it will never lead to a terminal string.

2   Basic Definitions

We will first remind the splicing operation introduced in [3] as a formal model of the DNA recombination under the influence of restriction enzymes and ligases.

A *splicing rule* (over an alphabet V) is a string $r = u_1\#u_2\$u_3\#u_4$, where $u_1, u_2, u_3, u_4 \in V^*$ and $\#, \$$ are two special symbols not in V. (V^* is the free monoid generated by the alphabet V under the operation of catenation; the empty string is denoted by λ; the length of $x \in V^*$ is denoted by $|x|$.) For $x, y, w, z \in V^*$ and r as above we write

$$(x, y) \vdash_r (w, z) \text{ iff } x = x_1 u_1 u_2 x_2, \ y = y_1 u_3 u_4 y_2,$$
$$w = x_1 u_1 u_4 y_2, \ z = y_1 u_3 u_2 x_2, \text{ for some } x_1, x_2, y_1, y_2 \in V^*.$$

We say that we splice x, y at the *sites* $u_1 u_2, \ u_3 u_4$. These sites encode the patterns recognized by restrictions enzymes able to cut the DNA sequences between u_1, u_2, respectively between u_3, u_4.

Let us now pass to splicing P systems. (The reader is assumed familiar with basic elements of membrane computing: membrane, membrane structure, region associated with a membrane, representation by means of strings of matching parentheses; see, e.g., [7].)

A *splicing P system with global rules* (of degree $m \geq 1$) is a construct $\Pi = (V, T, \mu, L_1, \ldots, L_m, R)$, where:

(i) V is an alphabet; its elements are called *objects*;

(ii) $T \subseteq V$ (the *output* alphabet);

(iv) μ is a membrane structure of m membranes (labeled with $1, 2, \ldots, m$);

(v) $L_i, 1 \leq i \leq m$, are languages over V associated with regions $1, 2, \ldots, m$ of μ;

(vi) R is a finite set of *evolution rules* given in the following form:
$(r = u_1\#u_2\$u_3\#u_4; tar_1, tar_2)$, where $r = u_1\#u_2\$u_3\#u_4$ is an usual splicing rule over V and $tar_1, tar_2 \in \{here, out, in\}$.

Note that, as usual in H systems, when a string is present in a region of our system, it is assumed to appear in arbitrarily many copies (any number of copies of a DNA molecule can be obtained by amplification).

Any m-tuple $(M_1, \ldots, M_m)$ of languages over V is called a *configuration* of Π. For two configurations $(M_1, \ldots, M_m), (M_1', \ldots, M_m')$ of Π we write
$$(M_1, \ldots, M_m) \Longrightarrow (M_1', \ldots, M_m')$$
if we can pass from $(M_1, \ldots, M_m)$ to $(M_1', \ldots, M_m')$ by applying the splicing rules from R, in parallel, to all possible strings from the corresponding regions, and following the target indications associated with the rules. More specifically, if $x, y \in M_i$ and $(r = u_1\#u_2\$u_3\#u_4, tar_1, tar_2) \in R$ such that we can have $(x, y) \vdash_r (w, z)$, then w and z will go to the regions indicated by tar_1, tar_2, respectively. If $tar_j = here$, then the string remains in M_i, if $tar_j = out$, then the string is moved to the region immediately outside the membrane i (maybe, in this way the string leaves the system), if $tar_j = in$, then the string is moved to any region placed directly inside membrane i; if i is an elementary membrane, then the rule cannot be applied. Note that the strings x, y are still available in region M_i, because we have supposed that they appear in arbitrarily many copies (an arbitrarily large number of them were spliced, arbitrarily many remain), but if a string w, z is sent out of region i, then no copy of it remains here.

A sequence of transitions between configurations of a given P system Π, starting from the initial configuration $(L_1, \ldots, L_m)$, is called a *computation* with respect to Π. The result of a computation consists of all strings over T which are sent out of the system at any time during the computation. We denote by $L(\Pi)$ the language of all strings of this type. We say that $L(\Pi)$ is *generated* by Π.

Note two important facts: if a string leaves the system but it is not terminal, or it remains in the system, then it does not contribute to the language $L(\Pi)$. It is also worth mentioning that we do not consider here halting computations. We leave the process to continue forever and we just observe it from outside and collect the terminal strings leaving it.

We denote by $SPLG_m$ the family of languages $L(\Pi)$ generated by splicing P systems as above, of degree at most $m \geq 1$, and with global rules.

We define the *diameter* of a splicing P system $\Pi = (V, T, \mu, L_1, \ldots, L_m, R)$, in a similar way as in the case of extended H systems ([5]), by
$$dia(\Pi) = (n_1, n_2, n_3, n_4), \text{ where } n_i = \max\{|u_i| \mid u_1\#u_2\$u_3\#u_4 \in R\}, 1 \leq i \leq 4.$$

We denote the family of languages generated by P systems of degree at most m, with global rules, and with diameter at most (n_1, n_2, n_3, n_4) by
$$SPLG_m(n_1, n_2, n_3, n_4).$$

For rewriting P systems we start by considering the general definition, when the rules are specific to regions. A *rewriting P system* is a construct:
$$\Pi = (V, T, \mu, L_1, \ldots, L_n, (R_1, \rho_1), \ldots, (R_n, \rho_n)), \text{ where } V \text{ is an alphabet, } T \subseteq$$
V is the terminal alphabet, μ is a membrane structure, $L_1, \ldots, L_n$ are finite languages over V, $R_1, \ldots, R_n$ are finite sets of context-free evolution rules, and $\rho_1, \ldots, \rho_n$ are partial order relations over $R_1, \ldots, R_n$, respectively.

The rules are provided with indications on the target membrane of the produced string, and always we use only context-free rules. Thus, the rules are of the form $X \to v(tar)$, where $tar \in \{here, out, in\}$, with the obvious meaning: the string produced by using this rule will go to the membrane indicated by tar.

The language generated by a system Π is denoted by $L(\Pi)$ and it is defined as follows: we start from an initial configuration of the system and proceed iteratively, by transition steps done by using the rules in parallel, to all strings which can be rewritten, obeying the priority relations relative to the membranes, and collecting the terminal strings sent out of the system during the computation.

Note that each string is processed by one rule only, the parallelism refers here to processing simultaneously all available strings by all applicable rules. So, even if we can apply more than one rule to a string, only one of the possible rules is applied. If we have priorities, then the high priority rule "forbids" the application of a low priority rule.

We say that such a system has global rules iff $R_1 = R_2 = \ldots = R_n = R$ and a partial order relation ρ over R is given; then we write the system in the following form: $\qquad \Gamma = (V, \mu, L_1, \ldots, L_n, R, \rho)$.

We denote by $RPL_n(Pri)$ the family of languages generated by rewriting P systems of degree at most $n, n \geq 1$, using priorities; and with $RPLG_n(Pri)$ the family of languages generated by rewriting P systems of degree at most n, with priorities and having global rules.

3   The Power of Splicing P Systems

We prove now that the splicing P systems with global rules and only two components are computationally universal:

Theorem 1. $SPLG_2 = RE$.

Proof. Let $G = (N, T, S, P)$ be a type-0 Chomsky grammar and let B be a new symbol. Assume that $N \cup T \cup \{B\} = \{\alpha_1, \ldots, \alpha_n\}$ and that P contains m rules, $u_i \to v_i, 1 \leq i \leq m$. Consider also the rules $u_{m+j} \to v_{m+j}, 1 \leq j \leq n$, for $u_{m+j} = v_{m+j} = \alpha_j$. Let us consider $n' = m + n$.

We construct the splicing P system (of degree 2):

$$\Pi = (V, T, \mu, L_1, L_2, R = R' \cup R''),$$
$$V = N \cup T \cup \{B, X, X', Y, Y', Z_1, Z_2, Z_1'\} \cup \{X_i \mid 0 \leq i \leq n'\}$$
$$\cup \{Y_i \mid 0 \leq i \leq n'\},$$
$$\mu = [_1[_2]_2]_1,$$
$$L_1 = \{Z_1', X'Z_1, Z_1Y'\} \cup \{Z_1Y_i \mid 0 \leq i \leq n'\} \cup \{X_iv_iZ_1 \mid 1 \leq i \leq n'\},$$
$$L_2 = \{XSBY, XZ_2, Z_2Y\} \cup \{Z_2Y_i \mid 1 \leq i \leq n'\} \cup \{X_iZ_2 \mid 0 \leq i \leq n'-1\},$$
$$R' = \{(X_iv_i\#Z_1\$X\#; \ in, \ out), (\alpha\#Y_i\$Z_1\#Y_{i-1}; \ in, \ out) \mid 1 \leq i \leq n',$$
$$\alpha \in N \cup T \cup \{B\}\}$$
$$\cup \{(\#Y_0\$Z_1\#Y'; \ in, \ out), (X_0\#\$X'\#Z_1; \ out, \ here) \mid \alpha \in N \cup T \cup \{B\}\}$$

$\cup \{(\#BY\$Z_1'\#;\ here,\ out),\ (X\#\$\#Z_1';\ out,\ out),$
$R'' = \{(\alpha\#u_iY\$Z_2\#Y_i; out, here)\}, (X_i\#a\$X_{i-1}\#Z_2; here, out) \mid 1 \leq i \leq n',$
$\quad \alpha \in N \cup T \cup \{B\}\} \cup \{(X'\#\$X\#Z_2; here, here), (\#Y'\$Z_2\#Y; out, here)\}$

The idea of this proof is the "rotate-and-simulate" procedure, as used in many proofs in the H systems theory. Here both the simulation of the rules in G and the circular permutation of strings are performed in Π in the same way: a suffix u of the current string is removed and the corresponding string, v, is added in the left end of the string. For a rule $u \to v$ from P, we simulate a derivation step in G; while for a symbol in $N \cup T \cup \{B\}$ we have one symbol "rotation" of the current string $(u = v)$.

The special marker B is used to ensure that the produced strings are in the right permutation since B marks the beginning of the word. For instance, if Xw_1Bw_2Y is produced in Π, this means that in G we have the word w_2w_1.

Let us see in more detail the work of the system.

The "main" axiom is $XSBY$; we will always process a string of the form Xw_1Bw_2Y (the axiom is of that form). We replace a suffix u_iY of this word with Y_i (in region 2) and the prefix X with X_jv_j (in region 1). Then we will decrease repeatedly the subscripts of Y_i and X_j by one (each time the string is sent to the other membrane). In the end we will replace Y_0 with Y and X_0 with X (this means that $i = j$; so we simulated the production $u_i \to v_i$). In this way we can simulate the productions from P (using the splicings that model $u_i \to v_i$, $1 \leq i \leq m$) and rotate the string (using the splicings that model $u_i \to v_i$, $m+1 \leq i \leq m+n$).

In the end, in membrane 1, we delete the markers B and Y together (to be sure that we have the right permutation of the word) and finally we delete the marker X and send the string out. Thus, we get $L(G) \subseteq L(\Pi)$.

Now we will prove the converse inclusion. We will start observing that the rules from the set R' can only be applied in membrane 1 and the rules from R'' can only be applied in membrane 2: To prove this, first observe that the rules with a target in can only be applied in the outer membrane (membrane 1). Because of this and because the "by products" of a splicing in membrane 1 are expelled from the system, the special characters Z_1, and Z_1' can never reach membrane 2. Because the rest of the rules from R' have always either Z_1 or Z_1' in their pattern, we obtain that the rules from the group R' cannot be applied in membrane 2 and also the rules from R'' cannot lead to terminal strings if they are used in membrane 1.

A simple discussion can show that we have to start in membrane 1 using a rule $(X_iv_i\#Z_1\$X\#;\ in,\ out)$, otherwise the computation halts in a few steps without producing any terminal strings.

So, we start by replacing X with X_iv_i in membrane 1 and then continue by cutting u_jY and replaceing it with Y_j in membrane 2. The string $X_iv_iwY_j$ gets in membrane 1, where the only possibility is to apply the rule $(\alpha\#Y_j\$Z_1\#Y_{j-1};\ in,\ out)$. The string $X_iv_iwY_{j-1}$ gets in membrane 2, where the only possibility is to apply the rule $(X_i\#a\$X_{i-1}\#Z_2;\ here,\ out)$, so the string $X_{i-1}v_iwY_{j-1}$ gets in membrane 1. We iterate the process until at least one of the subscripts of X or Y is 0. If we got X_0, then we decreased the subscript of X in membrane 2 and sent the string $X_0v_iY_{j-i}$ in membrane 1. Now we have two possibilities: $j \neq i$ or $j = i$.

If $j \neq i$ then $j - i \neq 0$, so we can decrease the subscript of Y and send the string in membrane 2. But here the string $X_0v_iwY_{j-i-1}$ can enter no further splicings. Before decreasing the subscript of Y, in membrane 1 we can also apply the splicing rule $(X_0\#\$X'\#Z_1;\ out,\ here)$; the string $X'v_iwY_{j-i}$ is sent to membrane 1 and we continue as before. In this case in membrane 2 we can replace X' by X using the rule $(X'\#\$X\#Z_2;\ here,\ here)$ and the string Xv_iY_{j-i-1} cannot enter any other splicings so it remains in membrane 2.

If $j = i$, then the only productions from region 1 that can be applied are these two: $(\alpha\#Y_0\$Z_1\#Y';\ in,\ out)$ and $(X_0\#\$X'\#Z_1;\ out,\ here)$. If we apply the first one, then the string X_0v_iwY' is sent to membrane 2, here we can only apply the rule that replaces Y' with Y, so the string X_0v_iwY gets in membrane 1. This string will never lead to a terminal string because we cannot delete the left marker (we can replace X_0 with X' but the string remains in this membrane and that marker cannot be deleted).

If we start with the rule $(X_0\#\$X'\#Z_1;\ out,\ here)$, then we get the string $X'v_iwY_0$. The only possibility to continue is to apply $(\alpha\#Y_0\$Z_1\#Y';\ in,\ out)$ and the string $X'v_iwY'$ gets in membrane 2. If we don't replace here X' with X, then again the string that gets into membrane 1 cannot lead to a terminal string. So first we replace X' with X (using the rule $X'\#\$X\#Z_2$) and then we replace Y' by Y by using the rule $(\#Y'\$Z_2\#Y;\ out,\ here)$. In this way we send the string Xv_iwY in membrane 1 and we can perform another step of rotating the word or simulating the rules from P.

Therefore, the computations in Π correctly simulate rules in G or circularly permute the string. In this way we get that $L(\Pi) \subseteq L(G)$.   □

In the following we will try to minimise the diameter of the used splicing P systems with global rules (the restriction sites of the restrictive enzymes have a limited length, so the problem is quite natural).

The following auxiliary result is easy to be proved.

Lemma 1. $SPLG_m(n_1, n_2, n_3, n_4) = SPLG_m(n_3, n_4, n_1, n_2)$, for all $m \geq 1$ and $n_i \geq 0$, $1 \leq i \leq 4$.

Theorem 2. $SPLG_3(1, 2, 1, 0) = SPLG_3(1, 0, 1, 2) = RE$.

Proof. We give only the construction for $SPL_3(1, 2, 1, 0) = RE$, the other equality follows from Lemma 1.

Let $G = (N, T, S, P)$ be a type-0 Chomsky grammar in the Kuroda normal form, (that is, P consists of context-free rules of the form $A \rightarrow x$, $A \in N$, $x \in (N \cup T)^*$, $|x| \leq 2$, and non-context-free rules of the form $AB \rightarrow CD$, $A, B, C, D \in N$) and let B be a new symbol. Assume that $N \cup T \cup \{B\} =$

$\{\alpha_1, \ldots, \alpha_n\}$ and that P contains m rules, $u_i \to v_i, 1 \le i \le m$. Consider also the rules $u_{m+j} \to v_{m+j}, 1 \le j \le n$, for $u_{m+j} = v_{m+j} = \alpha_j$.

We denote by P_1 the set of context-free rules considered above, and with P_2 the rest of the rules. One can see that the rules $u_{m+j} \to v_{m+j}, 1 \le j \le n$ are in P_1. We construct the splicing P system (of degree 3):

$$\Pi = (V, T, \mu, L_1, L_2, L_3, R = R' \cup R'' \cup R'''),$$
$$V = N \cup T \cup \{B, X, X', Y, Z_X, Z_{X'}, Z_Y, Z_\lambda, Z'_\lambda\} \cup \{Y_i, Z_{Y_i} \mid 0 \le i \le n+m\}$$
$$\cup \{X_i, Z_{X_i}, Z_i, Y'_i, Z_{Y'_i} \mid 1 \le i \le m+n\},$$
$$\mu = [_1[_2[_3]_3]_2]_1,$$
$$L_1 = \{XSBY, Z_\lambda, Z'_\lambda\} \cup \{Z_{Y_i}Y_i \mid 0 \le i \le n+m\} \cup \{Z_{Y'_i}Y'_i \mid 1 \le i \le n+m\},$$
$$L_2 = \{XZ_X, X'Z_{X'}\} \cup \{X_i v_i Z_i \mid 1 \le i \le m+n\}$$
$$\cup \{X_i Z_{X_i} \mid 1 \le i \le m+n-1\},$$
$$L_3 = \{Z_Y Y\},$$
$$R' = \{(\#u_i Y \$ Z_{Y_i} \#; in, out), \mid u_i \to v_i \in P_1\}$$
$$\cup \{(C\#DY\$Z_{Y'_i}\#; here, out), (\#CY'_i\$Z_{Y_i}\#; in, out) \mid u_i = CD \to v_i \in P_2\}$$
$$\cup \{(\alpha\#Y_i\$Z_{Y_{i-1}}\#; in, out) \mid 1 \le i \le n+m, \ \alpha \in N \cup T \cup \{B\}\}$$
$$\cup \{(\alpha\#BY\$Z_\lambda\#; here, out), (\#Z'_\lambda\$X\#; out, out) \mid \alpha \in T\},$$
$$R'' = \{(\alpha\#Z_i\$X\#; out, in) \mid 1 \le i \le n+m, \ \alpha \in N \cup T\}$$
$$\cup \{(X_{i-1}\#Z_{X_{i-1}}\$X_i\#; out, in) \mid 2 \le i \le n+m\}$$
$$\cup \{(X'\#Z_{X'}\$X_1\#; in, in), (X\#Z_X\$X'\#; out, in)\},$$
$$R''' = \{(\alpha\#Y_0\$Z_Y\#; out, out) \mid \alpha \in N \cup T \cup \{B\}\}.$$

This proof closely follows the proof of Theorem 1 from [9], with a special attention paid to the diameter of the splicing rules and also changing the rules to be global. We leave the details to the reader; the arguments used in the proof of Theorem 1 are also useful in proving the correctness of the given construction.   □

4   The Power of Rewriting P Systems

We now pass to the second model of P systems we discuss here, rewriting P systems. First we improve the main result from [7] about this variant, that is, we prove that two membranes are sufficient for universality.

Theorem 3. $RPL_2(Pri) = RE$

Proof. Let $G = (N, T, S, M, F)$ be a matrix grammar with appearance checking in the binary normal form [1], that is, with $N = N_1 \cup N_2 \cup \{S, \dagger\}$, with these three sets disjoint, and the matrices in M are of one of the following forms:

1. $(S \to XA)$, with $X \in N_1, A \in N_2$,
2. $(X \to Y, A \to x)$, with $X, Y \in N_1, A \in N_2, x \in (N_2 \cup T)^*$,

3. $(X \rightarrow Y, A \rightarrow \dagger)$, with $X, Y \in N_1, A \in N_2$,
4. $(X \rightarrow x_1, A \rightarrow x_2)$, with $X \in N_1, A \in N_2$, and $x_1, x_2 \in T^*$.

Moreover, there is only one matrix of type 1 and F consists exactly of all rules $A \rightarrow \dagger$ appearing in matrices of type 3. For each matrix of type 4 $(X \rightarrow x_1, A \rightarrow x_2)$, with $x_1, x_2 \in T^*$, we also introduce the matrix $(X \rightarrow X'x_1, A \rightarrow x_2)$, which is considered of type 4'; we also add the matrices $(X' \rightarrow \lambda)$; X' is a new symbol associated with X. Clearly, the generated language is not changed. We assume the matrices of the types 2, 3, 4' labeled in a one-to-one manner with $m_1, \ldots, m_k$. We construct the following rewriting P system:

$$\Pi = (V, T, \mu, L_1, L_2, (R_1, \rho_1), (R_2, \rho_2)),$$
$$V = N_1 \cup N_2 \cup \{E, Z, \dagger\} \cup T \cup \{X_i, X_i' \mid X \in N_1, 1 \le i \le k\},$$
$$\mu = [_1[_2\ ]_2]_1,$$
$$L_1 = \{XAE\}, \text{ where } (S \rightarrow SA) \text{ is the initial matrix of } G,$$
$$L_2 = \emptyset,$$
$$R_1 = \{r_\alpha : \alpha \rightarrow \alpha \mid \alpha \in V - T, \alpha \ne E\} \cup \{r_0 : E \rightarrow \lambda(out)\}$$
$$\cup \{t_i : X \rightarrow Y_i(in) \mid m_i : (X \rightarrow Y, A \rightarrow x) \text{ is a matrix of type 2}\}$$
$$\cup \{t_i : X \rightarrow Y_i(in) \mid m_i : (X \rightarrow Y, A \rightarrow \dagger) \text{ is a matrix of type 3}\}$$
$$\cup \{t_i : X \rightarrow X_i'x_1(in), \ X_i' \rightarrow \lambda \mid m_i : (X \rightarrow X'x_1, A \rightarrow x_2)$$
$$\text{is a matrix of type 4'}\} \cup \{Y_i \rightarrow Y, \ Y_i' \rightarrow Y \mid Y \in N_1, 1 \le i \le k\},$$
$$\rho_1 = \{r_\alpha > r_0 \mid \alpha \in V - T, \alpha \ne E\},$$
$$R_2 = \{r_i : Y_i \rightarrow Y_i, \ r_i' : A \rightarrow x(out) \mid m_i : (X \rightarrow Y, A \rightarrow x)$$
$$\text{is a matrix of types 2 or 4}\}$$
$$\cup \{r_i : X_i' \rightarrow X_i', \ r_i' : A \rightarrow x_2(out) \mid m_i : (X \rightarrow X'x_1, A \rightarrow x_2)$$
$$\text{is a matrix of type 4'}\}$$
$$\cup \{p_i : Y_i \rightarrow Y_i', \ p_i' : Y_i' \rightarrow Y_i, \ p_i'' : A \rightarrow \dagger(out) \mid m_i : (X \rightarrow Y, A \rightarrow \dagger)$$
$$\text{is a matrix of type 3}\} \cup \{p_0 : E \rightarrow E(out)\},$$
$$\rho_2 = \{r_i > r_j', \ r_i > p_j'', \ p_i > r_j', \ p_i' > r_j' \mid i \ne j, \text{ for all possible } i, j\}$$
$$\cup \{p_i'' > p_i, \ p_i > p_0, \ r_i > p_0 \mid \text{ for all possible } i\}.$$

We will now explain the work of the system. Observe first that the rules $\alpha \rightarrow \alpha$ from membrane 1 change nothing, can be used forever, and prevent the use of the rule $E \rightarrow \lambda(out)$, which sends the string out of the system. So, we can use the rule $E \rightarrow \lambda$ only after all nonterminal symbols have been rewritten into terminal ones.

Let us assume that in membrane 1 we have a string of the form XwE (initially, we have the string $\bar{X}\bar{A}E$). In membrane 1 one chooses the matrix to be simulated, m_i, and one simulates its first rule, $X \rightarrow Y$, by introducing Y_i (the subscript i keeps the information about what rule we are simulating); and then the string is sent to membrane 2.

Let us discuss now the case of matrices of type 2: In membrane 2 we can use the rule $r_i : Y_i \to Y_i$ forever. The only way to exit this membrane is by using the rule $A \to x$ appearing in the second position of a matrix of type 2 (we cannot use $p_0 : E \to E(out)$ because r_i has priority over p_0, and we cannot use a rule p_i'' because r_i has again priority). Due to the priority relation ρ_2, this matrix should be exactly m_i as specified by the subscript of Y_i (every other rule cannot be applied because then r_i have priority over all r_j' with $j \neq i$). Therefore, we can continue the computation only when the matrix is correctly simulated (we use the rule r_i').

The process is similar for matrices of type 3: The rules $Y_i \to Y_i', Y_i' \to Y_i$ can be used forever and we remain in membrane 2. We can quit this membrane either by using a rule $A \to \dagger(out)$ or by using the rule $E \to E(out)$, we cannot use a rule r_i' because $p_i > r_j'$ and $p_i' > r_j'$. In the first case the computation will never lead to a terminal string (we introduced the trap-symbol $\dagger$ that can never be removed). Because of the priority relation, such a rule must be used if the corresponding symbol A appears in the string. If this is not the case, then the rule $Y_i \to Y_i'$ can be used. If we now use the rule $Y_i' \to Y_i$, then we get nothing. If we use the rule $E \to E(out)$, and this is possible because Y_i is no longer present, so the higher priority rule p_i cannot be applied, then we send out a string of the form $Y_i'wE$.

In membrane 1 we replace Y_i or Y_i' by Y, and thus the process of simulating the use of matrices of types 2 and 3 can be iterated.

A slightly different procedure is followed for the matrices of type $4'$; they are of the form $m_i : (X \to X'x_1, A \to x_2)$. In membrane 1 we use $X \to X_i'x_1(in)$, which already introduces the string x_1, and the string arrives in membrane 2. Again the only way to leave this membrane is by using the associated rule $A \to x_2(out)$. In membrane 1 we have to apply $X_i' \to \lambda$. If no symbol different of E and the terminal symbols is present, then we can apply the rule $E \to \lambda(out)$. Thus, a terminal string is sent out of the system.

Therefore, in the language $L(\Pi)$ we collect exactly the terminal strings generated by the grammar G, that is $L(G) = L(\Pi)$. $\qquad\square$

By appropriately modifying the proof of Theorem 3, we can now show that rewriting P systems with global rules and only two components are computationally universal:

Theorem 4. $RPLG_2(Pri) = RE$

Proof. We use the notations from the previous proof. Starting from the system Π constructed in the proof of theorem 3, we construct the following rewriting P system with global rules:

$\Pi' = (V, T, \mu, L_1, L_2, R = R_1' \cup R_2', \rho_1' \cup \rho_2')$, where
$R_1' = R_1$, with the rules $Y_i \to Y$ and $Y_i' \to Y$ changed to
$\qquad s_i : Y_i \to Y(in)$ and $s_i' : Y_i' \to Y(in)$, respectively,
$\rho_1' = \rho_1 \cup \{t_i > r_j, \ t_i > r_j', \ t_i > p_j, \ t_i > p_j', \ t_i > p_j'', \ t_i > p_Y,$

$$s_i > r_j,\ s_i > r'_j,\ s_i > p_j,\ s_i > p'_j,\ s_i > p''_j,\ s_i > p_Y,$$
$$s'_i > r_j,\ s'_i > r'_j,\ s'_i > p_j,\ s'_i > p'_j,\ s'_i > p''_j,\ s'_i > p_Y \mid \text{for all } i, j, Y\},$$
$$R'_2 = R_2 \cup \{p_Y : Y \to Y(out)\},$$
$$\rho'_2 = \rho_2 \cup \{p_Y > r_i,\ p_Y > r'_i,\ p_Y > p''_i \mid i \geq 0\}.$$

The idea of the construction is the following: one can see that all the rules from R'_1, with the exception of $E \to \lambda(out)$ and $r_\alpha : \alpha \to \alpha$, have the target indication (in), so they cannot be applied in membrane 2. The additional priorities added to ρ'_1 make sure that no rule from R'_2 will be applied in membrane 1 (the rules from R'_1 have priority and always a rule from R'_1 can be applied: that is r_0) (the only exception to this is the rule p_0 that can be applied in membrane 1, but when it is applied it sends out a string containing E, so that string will not contribute to the language, thus the language is unchanged).

The only change in the rewriting rules was to introduce the target indication (in) to the rules $Y_i \to Y$ and $Y'_i \to Y$, but introducing this we had to add also the rule $p_Y : Y \to Y$ to R'_2. As we can see this doesn't change the language generated by the system, so because the rules R'_1 can only be applied in membrane 1 and the rules from R'_2 in membrane 2 using the previous proof we get that $L(\Pi') = L(G)$, which means that $RE \subseteq RPLG_2(Pri)$. □

5 Final Remarks

We have considered rewriting and splicing P systems with global rules and we have proved that this restriction does not decrease the generative power: characterizations of recursively enumerable languages are obtained also in this case, likewise to the case of systems with local rules (that is, rules associated with each membrane). This partially solves a problem formulated in [8]. The case of P systems of other types (for instance, using symbol-objects) remains to be investigated.

References

1. J. Dassow, Gh. Păun, *Regulated Rewriting in Formal Language Theory*, Springer-Verlag, Berlin, 1989.
2. P. Frisco, Membrane computing based on splicing: improvements, *Pre-proc. Workshop on Multiset Processing*, Curtea de Argeş, Romania, TR 140, CDMTCS, Univ. Auckland, 2000, 100–111.
3. T. Head, Formal language theory and DNA: an analysis of the generative capacity of specific recombinant behaviors, *Bull. Math. Biology*, **49** (1987), 737–759.
4. T. Head, Aqueous computations as membrane computations, submitted, 2001.
5. A. Păun, Controlled H systems of small radius, *Fundamenta Informaticae*, **31**, 2 (1997), 185 – 193.
6. A. Păun, M. Păun, On the membrane computing based on splicing, *Where Mathematics, Computer Science, Linguistics and Biology Meet* (C. Martin-Vide, V. Mitrana, Eds.), Kluwer, Dordrecht, 2001, 409–422.

7. Gh. Păun, Computing with membranes, *J. of Computer and System Sciences*, 61, 1 (2000), 108–143.

8. Gh. Păun, Computing with membranes (P systems): Twenty six research topics, *Auckland University, CDMTCS Report* No 119, 2000 (www.cs.auckland.ac.nz/ CDMTCS).

9. Gh. Păun, T. Yokomori, *Membrane Computing Based on Splicing, Preliminary Proc. of Fifth Intern. Meeting on DNA Based Computers* (E. Winfree, D. Gifford, eds.), MIT, June 1999, 213–227.

Computing with Membranes:
Variants with an Enhanced Membrane Handling

Maurice Margenstern[1], Carlos Martín-Vide[2], and Gheorghe Păun[3]

[1] Université de Metz, LITA, UFR MIM
Ile du Saulcy, 57045 Metz Cedex, France
`margens@lita.univ-metz.fr`
[2] Research Group in Mathematical Linguistics
Rovira i Virgili University
Pl. Imperial Tàrraco 1, 43005 Tarragona, Spain
`cmv@astor.urv.es`
[3] Institute of Mathematics of the Romanian Academy
PO Box 1-764, 70700 Bucureşti, Romania
`gpaun@imar.ro`

Abstract. Membrane computing is a recently introduced (very general) computing framework which abstracts from the way the living cells process chemical compounds in their compartmental structure. Many variants considered in the literature are computationally universal and/or able to solve NP-complete problems in polynomial (even linear) time — of course, by making use of an exponential working space created in a natural way (for instance, by membrane division).

In the present paper we propose a general class of membrane systems, where besides rules for *objects evolution* (the objects are described by strings over a finite alphabet), there are rules for *moving objects* from a compartment to another one (this is done conditionally, depending on the strings contents), and for *handling membranes*. Especially this latter feature is important (and new in many respects), because it makes possible to interpret several DNA computing experiments as membrane computations. Specifically, rules for *dividing membranes* (with the contents *replicated* or *separated* according to a given property of strings), *creating, merging,* or *dissolving* them are considered. Some of these variants generalize certain previous variants of membrane systems, for the new variants we investigate their power and computational efficiency (as expected, universality results, as well as polynomial solutions of NP-complete problems are found; the latter case is illustrated with the SAT problem).

Due to space restrictions, the paper is a preliminary, partially formalized one; more mathematical details are given in the appendices available at `http://bioinformatics.bio.disco.unimib.it/psystems`, where also current information about the membrane computing area can be found.

1 Introduction: The Basic Idea

The membrane systems — they are also called P systems — were introduced (in [11]) as a possible answer to the question whether or not the frequent state-

N. Jonoska and N.C. Seeman (Eds.): DNA7, LNCS 2340, pp. 340–349, 2002.
© Springer-Verlag Berlin Heidelberg 2002

ments (see, e.g., [3,10]) that the processes which take place in a living cell are "computations," that "the alive cells are computers," are just metaphors, or a formal computing device can be abstracted from the cell functioning. As we will see below, the answer turned out to be affirmative.

Three are the fundamental features of living cells which are basic to P systems: (1) the complex compartmentation by means of a **membrane structure**, where (2) **sets** (or **multisets**) of chemical compounds (we call them, in general, *objects*) evolve according to prescribed (3) **rules**.

A *membrane structure* is a hierarchical arrangement of membranes, all of them placed in a main membrane, called the *skin* membrane. This one delimits the system from its environment. The membranes should be understood as three-dimensional vesicles, but a suggestive pictorial representation is by means of planar Euler-Venn diagrams (see Figure 1). Each membrane is labeled and it precisely identifies a *region*, the space between it and all the directly inner membranes, if any exists. A membrane without any membrane inside is said to be *elementary*. Mathematically, a membrane structure can be represented by a string of labeled matching parentheses. For instance, the structure from Figure 1 is represented by $[_1[_2]_2[_3]_3[_4[_5]_5[_6[_8]_8[_9]_9]_6[_7]_7]_4]_1$.

In the regions of a membrane structure we place *objects*. In this paper, the objects are described by strings of symbols over a given finite alphabet and we do not count their multiplicities (we work with *sets*, not with *multisets*: as a possible interpretation, think about objects represented by DNA molecules, which can be present in any number of copies we need, produced by amplification).

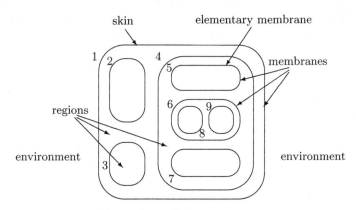

Fig. 1. A membrane structure.

The objects evolve by means of given *rules*, which are associated with the regions (the intuition is that the rules correspond to chemical reactions and each region has specific conditions, hence the rules from a region cannot necessarily act also elsewhere). These rules specify both object transformation (for instance, by rewriting symbols from the string-objects, by splicing them, etc) and object transfer from a region to another one. The passing of an object through a membrane is called *communication*. The communication of strings through membranes can also be controlled by the structure of the strings; for instance, a string can exit a given membrane only if it contains a specified substring. Such variants were considered in [2].

The rules are used *in a nondeterministic maximally parallel manner*: the objects to evolve and the rules to be applied to them are chosen in a nondeterministic way, but no object which can evolve at a given step may remain unchanged. Each string-object is processed by only one rule (the strings evolve in parallel, but the rewriting of each of them means using only one rule).

Other features can be introduced, such as the possibility to control the membrane thickness/permeability, or a priority relation among rules, but here we do not consider such features.

The membrane structure together with the objects and the evolution rules present in its regions constitute a *P system*. The membrane structure and the objects define a *configuration* of a given system. By using the rules as suggested above, we can define *transitions* among configurations. A sequence of transitions is called a *computation*. We accept as successful computations only the *halting* ones, those which reach a configuration where no further rule can be applied.

With a successful computation we can associate a *result*, for instance, by considering the objects which leave the system during the computation. More precisely, we can use a P system for solving three types of tasks: as a *generative* device (start from an initial configuration and collect all strings which describe the objects which have left the system during all successful computations), as a *computing* device (start with some input placed in an initial configuration and read the output at the end of a successful computation, by considering the objects which have left the system), and as a *decidability* device (introduce a problem in an initial configuration and wait for the answer in a specified number of steps). In all cases, instead of "reading" the result outside the system we can consider a given *output membrane* and consider as the result its contents at the end of a computation.

Among the numerous variants of P systems already considered in the literature, many are computationally universal (they can compute exactly what Turing machines can compute), while many can solve NP-complete problems in polynomial (often, linear) time, by making use of an exponential workspace.

2   Handling the Membranes

So far, two main ideas were used in order to generate an exponential workspace, both of them of a biochemical inspiration: dividing membranes [12] and replicating string-objects [5]. A combination of them is based on creating membranes from symbol-objects, as in [7], and providing rules only for increasing the number of objects.

Sugestively, but not completely formalized, a rule for dividing membranes is of the form $[_i \ ]_i \rightarrow [_i \ ]_i[_i \ ]_i$, with the meaning that the membrane with the label i is divided into two copies, each one inheriting the contents of the former membrane (both the objects and the membranes placed in the initial copy of membrane i are replicated in the two new copies of membrane i).

Similarly, a rule for creating a membrane [7] is of the form $a \rightarrow [_i b]_i$, with the meaning that the object a creates a membrane with label i, which contains the object b (because we know the label of the membrane, we also know the evolution rules which are associated with it).

A powerful, related, idea was recently considered in [6], where rules of the form $[_i a_1 a_2 \ldots a_n]_i \rightarrow [_i a_1]_i [_i a_2]_i \ldots [_i a_n]_i$ were used: membrane i is replicated in as many copies as many symbols exist in the string $a_1 a_2 \ldots a_n$.

Actually, Head's paper [6] is one of the main incentives for the present work: with the goal of formulating in terms of membrane computations several actual DNA computing experiments, Head has considered a variant of P systems with an enhanced membrane handling, including the possibility of creating, dividing, dissolving, and (this is an entirely new idea) *merging* membranes.

3   A General Class of P Systems

We consider here a general class of P systems, where rules of three types are used: for *evolving* the string-objects, for *communicating* objects through membranes, and for *handling membranes*. At each step of a computation, we first use the object evolution rules, in the nondeterministic, maximally parallel manner usual in P systems, then we use the communication rules (the objects which can be moved from a region to another one are moved), and then we process the membranes. The cycle is iterated.

The evolution rules will be context-free rewriting rules of the form $a \rightarrow u$: the symbol a is replaced by the string u. Because this might be of interest (and biologically motivated), we may impose to use a rule as above only for rewriting the symbol a in the leftmost or the rightmost position of a string, which is indicated by writing $a \rightarrow_{\text{pref}} u, a \rightarrow_{\text{suf}} u$, respectively. When no indication is given, no restriction on the place of the rewritten a is imposed.

For the communication rules we follow the idea of [2], namely we consider *predicates* associated with each membrane and telling when a string can exit the membrane or can go into an inner membrane. These predicates can be defined according to the structure of a string, taking into account its prefixes, suffixes, subwords, or the symbols appearing in it. We do not enter here into details, because [2] provides a systematic examination of such variants and of their influence on the power of P systems. We only mention the important fact that using rewriting rules without pref/suf restrictions and various combinations of communication predicates, we get computational universality, even for systems with a reduced number of membranes – see [2].

In what concerns the rules for membrane handling, the main topic of our paper, we can consider at least the following possibilities (when specifying the system, besides the alphabet, the initial membrane structure and the objects present in its regions, we also give sets of rules associated with all possible membranes; thus, the labels of new membranes precisely identify the rules corresponding to these membranes):

- *Merge:* the fact that two membranes, with labels i and j, can be merged into a single membrane, with label k, providing that they contain the sets of objects M_i, M_j, respectively, is indicated by $[_iM_i]_i[_jM_j]_j \to [_k \ ]_k$. The contents of the former membranes (objects and inner membranes included) are put together in the new membrane. Of course, in order to perform a merging operation, the two membranes must be adjacent.
- *Divide:* a membrane i, containing the set of objects M, can be divided into two new membranes by a rule $[_iM]_i \to [_j \ ]_j[_k \ ]_k$; the contents of the former membrane is reproduced in the two new membranes.
- *Separate:* if we want not to replicate the contents of a membrane, but to separate its objects according to a given property π, we can use a rule $[_i \ ]_i \to [_j\pi]_j[_k\neg\pi]_k$. The objects which have the property π are placed in membrane j, the objects which do not have property π are placed in membrane k.
- *Create:* we can also separate the objects of a membrane in such a way that the objects with the property π are put into a new membrane created inside it, by using rules of the form $[_i \ ]_i \to [_i[_j\pi]_j\neg\pi]_i$; the objects which do not have property π remain outside the new membrane.
- *Dissolve:* a membrane can be dissolved when it contains a given set of objects, by a rule of the form $[_iM]_i \to M$, or when it is empty. In the first case, the contents of the dissolved membrane remains free in the membrane which surrounds it. The skin membrane cannot be dissolved.

The previous rules were formulated for any type of membranes, but they are natural (and mathematically elegant) for elementary membranes. In particular, the skin membrane should be handled with care; for instance, by a division or a separation operation, we can duplicate the system itself. From a computational point of view, this does not make much sense (we have to change the definition of the computation, of its result, etc), but from other points of view, such as the replication of a system, this can be of a central interest.

In this paper, all operations are allowed only for elementary membranes.

In the same way as each string-object is processed at each step by only one rule (the strings which cannot be rewritten are passed unchanged to the next configuration), also the membranes are processed by at most one rule each: if there are some rules which can be applied to a membrane, then one of them is nondeterministically chosen and applied, but if no rule can be applied to a given membrane, then it remains unchanged.

4 The Computing Power

From the above (informal) discussion, the reader can realize that our class of P systems is rather general and powerful. First, it contains the variants considered in [2], where only rewriting and communication rules are used, and several universality results were obtained. Second, most of the operations considered in [6] are captured in our model, so that some of the experiments discussed by Head can also be formalized in the present context. (In this way, the aqueous computations performed by Head and his collaborators can be considered as a sort of membrane computations *avant la lettre*.) Third, even the test tube "programming language" of Lipton-Adleman, [8,1], is covered by the membrane handling rules considered here.

With respect to the last two statements, it is worth mentioning also the differences between the membrane formalism and the other mentioned ones. A membrane structure is more complex than a "test tube desk," because of the hierarchical structure of membranes, while the versatility of the model is much larger. Consider, for instance, the communication among compartments, which can be done at the level of each object. Then, the membrane formalism is specially designed to define "one macro-pot self-running" computations, in the Turing sense (an input-output function is computed), in a parallel, distributed manner. This makes possible the precise classification of variants of membrane systems and the mathematical comparison of their computing power with Turing machines and their variants.

We will examine here the power of some variants which were not considered before. Specifically, we will not use any communication rules, but only evolution rules and membrane handling rules. Because we cannot send strings out of the system, the result of a computation will be the set of terminal strings (that is, strings over a specified *terminal* alphabet) present at the end of a computation in a specified output membrane. The family of languages computed in this way by systems with at most n membranes present in any configuration, out of m possible types of membranes, using, for instance, the operations of creating and dissolving membranes, is denoted by $LP_{n/m}(create, dissolve, mode)$, where $mode$ is *any* when the rewriting can be done in any place, or *pref-suf* when only the first or the last symbol of a string can be rewritten. When m above is not bounded in advance, we replace it with $*$. Of course, the mathematical challenge (also of interest from a practical point of view) is to find combinations of operations as reduced as possible, but such that the generated family is as large as possible.

We mention here three results about such families. RE denotes the family of recursively enumerable languages, those recognized by Turing machines (the reader is assumed familiar with basic elements of formal language theory, as available in many monographs; in particular, the introductory chapters of [4] and [13] can be used).

Theorem 1. $RE = LP_{3/*}(create, dissolve, pref\text{-}suf) = LP_{4/*}(create, dissolve, any) = LP_{3/9}(create, separate, dissolve, any)$.

The proof of the first equality starts from a Chomsky type-0 grammar G in the Kuroda normal form and constructs a P system which simulates the derivations of G by using the "rotate-and-simulate" technique, much used in the H systems area, while the other two equalities are based on the characterization of recursively enumerable languages by means of the so-called matrix grammars with appearance checking (in the binary normal form).

In order to let the reader having an idea about these proofs and also having an example of a P system at work, we give here the proof for the last equality.

Consider a matrix grammar $G = (N, T, S, M, F)$, with $N = N_1 \cup N_2 \cup \{S, \#\}$ and the matrices in M of the form $m_i : (X \rightarrow \alpha, A \rightarrow u)$, for $X \in N_1, \alpha \in N_1 \cup \{\lambda\}, A \in N_2, u \in (N_2 \cup T)^*$, or of the form $(X \rightarrow Y, A \rightarrow \#)$, for $X, Y \in N_1, A \in N_2$; also an initial matrix $(S \rightarrow XA)$ is used. Assume that we have k matrices of the first type (with rules not used in the appearance checking mode). As recently proved, it is possible to assume that only two symbols, $B^{(1)}, B^{(2)}$, are used in appearance checking rules, $B^{(j)} \rightarrow \#, j = 1, 2$.

We construct the P system Π with the total alphabet

$$V = N \cup T \cup \{c_i \mid 1 \le i \le k\} \cup \{d_1, d_2, e, f, Z\},$$

the terminal alphabet T, one initial membrane, with label s (from "skin") and containing the strings X, eA, for $(S \rightarrow XA)$ being the initial matrix of G, and the following sets of rules (associated with membranes labeled with $s, 0, 1, 2, 3, 4, 5, 31, 32$):

$R_s : [_s \ ]_s \rightarrow [_s[_0 \lambda$ is a subword$]_0 \lambda$ is not a subword$]_s$,
$\quad\quad [_s \ ]_s \rightarrow [_s[_5 \lambda$ is a subword$]_5 \lambda$ is not a subword$]_s$;
$R_0 : X \rightarrow c_i Y$,
$\quad\quad A \rightarrow c_i u$, for each matrix $m_i : (X \rightarrow Y, A \rightarrow u) \in M$,
$\quad\quad X \rightarrow c_i f$,
$\quad\quad A \rightarrow c_i u$, for each matrix $m_i : (X \rightarrow \lambda, A \rightarrow u)$ with a terminal u,
$\quad\quad B \rightarrow Z$, for all $B \in N_2$,
$\quad\quad [_0 \ ]_0 \rightarrow [_1 c_i$ is a subword$]_1[_2 c_i$ is not a subword$]_2$,
$\quad\quad X \rightarrow d_j Y$, for $(X \rightarrow Y, B^{(j)} \rightarrow \#), j = 1, 2$,
$\quad\quad e \rightarrow d_j e$, for $j = 1, 2$,
$\quad\quad [_0 \ ]_0 \rightarrow [_{3j} d_j$ is a subword$]_{3j}[_4 d_j$ is not a subword$]_4, j = 1, 2$,
$\quad\quad [_0 f]_0 \rightarrow f$,
$\quad\quad e \rightarrow \lambda$;

$\quad\quad R_1 : c_i \rightarrow \lambda, 1 \le i \le k$,
$\quad\quad\quad\quad B \rightarrow Z$, for all $B \in N_2$,
$\quad\quad\quad\quad [_1 Y]_1 \rightarrow Y$, for all $Y \in N_1$;
$\quad\quad R_2 : c_i \rightarrow Z, 1 \le i \le k$,
$\quad\quad\quad\quad d_i \rightarrow Z, j = 1, 2$,
$\quad\quad\quad\quad B \rightarrow Z$, for all $B \in N_2$;
$\quad\quad R_{3j} : d_j \rightarrow \lambda$,
$\quad\quad\quad\quad [_{3j} \ ]_{3j} \rightarrow [_{3j}[_3 Y$ is a subword$]_3 Y$ is not a subword$]_{3j}$,
$\quad\quad\quad\quad [_{3j} Y]_{3j} \rightarrow Y$,

$$B^{(j)} \to Z, \text{ for } j = 1, 2;$$
$$R_3 : [_3 Y]_3 \to Y;$$
$$R_4 : c_i \to Z, \ 1 \le i \le k,$$
$$d_j \to Z, \ j = 1, 2,$$
$$B \to Z, \text{ for all } B \in N_2;$$
$$R_5 = \emptyset.$$

Membrane 5 is the output one. At any moment, we have exactly two strings in the system, one derived from the axiom X and one derived from the axiom eA, for $(S \to XA)$ the initial matrix of G. The strings derived from X are used for controlling the rewriting of the "main string," that derived from eA and which will lead to a terminal string of $L(G)$. Membranes 1 and 2 are used for simulating matrices m_i without appearance checking rules, while membranes with labels 31, 32, together with membranes 3 and 4, are used for simulating the matrices with appearance checking rules.

At each step, all strings of the skin membrane can be enclosed in a membrane with label 0. In this membrane, each string can be rewritten. Assume that X is replaced by $c_i Y$ corresponding to a matrix $m_i : (X \to Y, A \to u)$, but the string ew present at the same time in membrane 0 is not rewritten by the corresponding rule $A \to c_i u$. A separation is performed, according to the presence of c_i or d_j; all strings which arrive in membranes 2 or 4 will introduce the trap-symbol Z and remain here forever (these membranes dissolve if they contain no string). If both strings arrive in membrane 1, at the first step the symbols c_i are removed, then the membrane is dissolved, hence we return to the skin membrane with the strings Y, ew', which represent the correct simulation of the matrix m_i.

If we use the rules $X \to d_j Y, e \to d_j e$, then we can simulate the matrix $(X \to Y, B^{(j)} \to \#)$: the two strings should be both present in membrane $3j$, where both symbols d_j are removed; because of the string Y, we can create a membrane with label 3, where we send the string Y; at the next step, this membrane is dissolved and at the same time either no rule can be used in membrane $3j$, or the rule $B^{(j)} \to Z$ is used, providing that $B^{(j)}$ is present. At the next step, also membrane $3j$ is dissolved, because of the string Y. We return to the skin membrane with a correct simulation of the matrix.

The process can be iterated. At any moment, in membrane 0 we can also remove the symbol e, which makes impossible the simulation of appearance checking matrices. We can continue with simulating matrices without appearance checking. If the symbol f is produced, then no further simulation is possible. If we create again a membrane 0, it can only be dissolved by the rule $[_0 f]_0 \to f$. At any moment we can also create a membrane with label 5, where all strings are sent. If such strings are not terminal, then they are not accepted in the generated language. If they are terminal, then they correspond to strings generated by G. Consequently, Π exactly generated the strings in the language $L(G)$.

Because we use membranes of nine types, but at most three are simultaneously present in the system, we have the equality $LP_{3/9}(create, separate, dissolve, any) = RE$.

We do not know whether the results in Theorem 1 are optimal in the number of used membranes; in particular, we believe that the total number of membranes used in the first two equalities can be bounded, but we do not have a proof of this assertion. Also, further combinations of allowed rules must be considered.

5   Solving SAT in Linear Time

The computing power is of a clear mathematical interest (for instance, if we want to have *universal* — hence programmable — computing devices), but of a "practical" importance is the efficiency of computing models, in the sense of the time complexity of computations. Making use of the space provided by dividing membranes (and replicating their contents), P systems as above can solve NP-complete problems in a linear time. This is the case with SAT: consider m clauses C_i, involving n variables $a_1, \ldots, a_n$, and let us construct the P system with the initial membrane structure

$$[_s[_m[_{m-1} \cdots [_1[_0\ ]_0]_1 \cdots]_{m-1}]_m]_s,$$

with the strings a_1, d_1 placed in membrane 0; consider also the following rules:

$$[_0d_i]_0 \to [_{i1}\ ]_{i1}[_{i2}\ ]_{i2}, 1 \le i \le n,$$
$$a_i \to t_i a_{i+1},$$
$$d_i \to d_{i+1}, \text{ in membrane } i1, 1 \le i \le n,$$
$$a_i \to f_i a_{i+1},$$
$$d_i \to d_{i+1}, \text{ in membrane } i2, 1 \le i \le n,$$
$$[_{i1}d_{i+1}]_{i1}[_{i2}d_{i+1}]_{i2} \to [_0\ ]_0, 1 \le i \le n-1,$$
$$[_{n1}d_{n+1}]_{n1} \to d_{n+1},$$
$$[_{n2}d_{n+1}]_{n2} \to d_{n+1},$$
$$[_iw]_i \to [_i\ ]_iw, \text{ if } w \text{ satisfies } C_i, 1 \le i \le m,$$
$$[_sw]_s \to [_s\ ]_sw, \text{ for all } w.$$

By division (and replication) and making use of membranes $i1, i2$, for $1 \le i \le n$, in membrane 0 one generates all truth-assignments in the form of strings $w \in \{t_i, f_i \mid 1 \le i \le n\}^*$ of length n (this takes $3n$ steps); when this operation is completed and membranes $n1, n2$ are dissolved (the task of counting to n and to dissolve membranes $n1, n2$ is covered by the counters $d_i, 1 \le i \le n+1$), truth-assignments can pass through membranes providing that they satisfy the clauses associated with membranes (m more steps). In this way, in $3n + m + 1$ steps we get a string outside the system if and only if the set of clauses is satisfiable.

The communication rules can be avoided in the following way. Start from only one membrane inside the skin, membrane 0 as above, but with the objects ta_1, d_1 inside; use the previous rules for generating the truth-assignments. After dissolving membranes $n1, n2$ (hence membrane 0 is no longer created), we use the following rules:

$[_s]_s \rightarrow [_s[_{c1}w \text{ satisfies } C_1]_{c1}\text{otherwise}]_s,$

$[_{ci}]_{ci} \rightarrow [_s[_{c(i+1)}w \text{ satisfies } C_{i+1}]_{c(i+1)}\text{otherwise}]_{ci}, 1 \le i \le m - 1,$

$t_i \rightarrow \lambda,$ and

$f_i \rightarrow \lambda, 1 \le i \le n,$ in the membrane with the label $cm,$

$[_{ci}t]_{ci} \rightarrow t, 1 \le i \le m.$

The truth-assignments which satisfy the clauses are encapsulated in inner membranes which grow iteratively until producing a membrane with the label cm; in this membrane all symbols $t_i, f_i, 1 \le i \le n,$ are removed, which makes possible to dissolve all membranes $cj, 1 \le j \le m,$ under the influence of the object $t.$ Thus, we get t in the skin membrane (after $3n + 2m$ steps) if and only if the set of clauses is satisfiable.

Note. Work supported by NATO Project PST.CLG.976912, and, in the case of the third author, also by a grant of NATO Science Committee, Spain, 2000–2001.

References

1. L. M. Adleman, On constructing a molecular computer, in [9], 1–22.
2. P. Bottoni, A. Labella, C. Martin-Vide, and Gh. Păun, Rewriting P systems with conditional communication, submitted, 2000.
3. D. Bray, Protein molecules as computational elements in living cells, *Nature* 376 (1995), 307–312.
4. C. Calude and Gh. Păun, *Computing with Cells and Atoms*, Taylor and Francis, London, 2000.
5. J. Castellanos, A. Rodriguez-Paton, and Gh. Păun, Computing with membranes: P systems with worm-objects, *IEEE 7th. Intern. Conf. on String Processing and Information Retrieval, SPIRE 2000*, La Coruna, Spain, 64–74.
6. T. Head, Aqueous simulations of membrane computations, *Romanian J. of Information Science and Technology*, 4, 1–2 (2001).
7. M. Ito, C. Martin-Vide, and Gh. Păun, A characterization of Parikh sets of ET0L languages in terms of P systems, in vol. *Words, Semigroups, and Transductions* (M. Ito, Gh. Păun, and S. Yu, eds.), Word Scientific Publ., Singapore, 2001.
8. R. J. Lipton, Speeding up computations via molecular biology, in [9], 67–74.
9. R. J. Lipton and E. B. Baum, eds., *DNA Based Computers*, Proc. of a DIMACS Workshop, Princeton, 1995, Amer. Math. Soc., 1996.
10. W. R. Loewenstein, *The Touchstone of Life. Molecular Information, Cell Communication, and the Foundations of Life*, Oxford Univ. Press, New York, 1999.
11. Gh. Păun, Computing with membranes, *Journal of Computer and System Sciences*, 61, 1 (2000), 108–143.
12. Gh. Păun, P systems with active membranes: Attacking NP-complete problems, *J. Automata, Languages and Combinatorics*, 6, 1 (2001), 75–90.
13. Gh. Păun, G. Rozenberg, and A. Salomaa, *DNA Computing. New Computing Paradigms*, Springer-Verlag, Berlin, 1998.

Towards an Electronic Implementation of Membrane Computing: A Formal Description of Non-deterministic Evolution in Transition P Systems

Angel V. Baranda[1], Fernando Arroyo[2], Juan Castellanos[1], and Rafael Gonzalo[1]

[1] Dept. Inteligencia Artificial,
Facultad de Informática, Universidad Politécnica de Madrid,
Campus de Montegancedo, Boadilla del Monte, 28660 Madrid, Spain
{jcastellanos, rgonzalo}@fi.upm.es
http://www.dia.fi.upm.es
[2] Dept. Lenguajes, Proyectos y Sistemas Informáticos,
Escuela de Informática, Universidad Politécnica de Madrid,
Crta. de Valencia km. 7, 28031 Madrid, Spain
farroyo@eui.upm.es
http://www.lpsi.eui.upm.es

Abstract. This paper is part of a program of our research group, aiming to implement membrane computing on electronic computers. We here present Transition P systems, which is the preliminary steep of our approach. The formalisation we before have in mind the functional programming framework for developing the software modules. Part of these modules has already realised, in *Haskell*, and they are briefly described in the second section of the paper.

1 Introduction

Gheorghe Paun has introduced a new computability model, of a distributed parallel type, based on biological cells called Transition P systems. Such structure is recursively defined, and it is composed of several membranes enclosed by a unique skin membrane; membranes without any other membrane inside them are said to be elementary [1]. It is easy to give a plane representation for such a structure as Venn diagrams. Membranes delimit regions, and regions contain objects and rules partially ordered by a priority relation. Rules are responsible for making evolve objects of their regions, and moreover, they can change the membrane structure of the System dissolving the membrane of its region. The P system evolution is done in parallel, for all membranes in the system all objects that are able to evolve (that is, having access to a rule) should use it. The only restriction is the priority relation defined among rules in membranes.

The Membrane structure is the main component of a Transition P system. It is a hierarchised structure uniquely labelled that can be represented as a tree [2].

N. Jonoska and N.C. Seeman (Eds.): DNA7, LNCS 2340, pp. 350–359, 2002.
© Springer-Verlag Berlin Heidelberg 2002

From now on, we will describe the static structure of Transition P system using terms of abstract data types until giving a definition of a Transition P system in terms of them. The goal of this formalisation is to obtain a software product to implementing/simulating a Transition P system on a usual computer. Became we believe that a functional language (in particular, Haskell) is appropriate to this aim, we will look for a formalisation as close as possible to such a framework.

We have partially realised the software program none information about them and about its functioning is given in the second part of this paper.

We emphasise the fact that implementing P system on a digital computer is not a goal *per se*, but we want to explore in this way the possibility of devising a sort of cellular computing, starting from the way the living cells "compute." From this point-of-view, P system is one promising formalisation of the cell structure and functioning. It is also attractive from computer science point of view: just remember that many classes of P system are computationally universal and that NP-Complete problems can be solve in this framework in polynomial, or even lineal time, in a natural and theoretical easy manner.

2   Static Structure of Transition P Systems

The static structure of Transition P systems is related to everything concerning to membranes, and objects to be contained in membranes. So, it is related to multisets, evolution rules and regions. There are some relations among these elements: multisets will be necessary for defining regions and evolution rules; evolution rules and again multisets will define a region, and finally regions will give the static structure definition of a Transition P system.

Therefore, it is very important to determine a data structure for defining multisets, evolution rules and regions and the operations that can be done over such data structures [3].

2.1   Multisets

The natural way to represent object inside a region is the multiset. The region contains a collection of different objects, so they can be represented by a determined multiset. Let us to define more precisely a Multiset.

Let U an arbitrary set and let N the natural number set: a multiset M over U is a mapping from U to N.

$$M : U \to N \tag{1}$$
$$a \to Ma$$

Over a multiset is possible to define several operations.

Definitions: Let M, M_1 and M_2 multisets over U, then:

Empty Multiset: Is the 0 map, that is, $\forall a \in U$, $0a = 0$.

Multiset Inclusion: $M_1 \subset M_2$ if and only if $\forall a \in U$, $(M_1\, a) < (M_2\, a)$.

Multiset Union: $\forall a \in U$, $(M_1 \cup M_2)\, a = (M_1\, a) + (M_2\, a)$.

Multiset Difference: $\forall a \in U$, $(M_1 - M_2)\, a = (M_1\, a) - (M_2\, a)$, if $M_1 \subset M_2$.

Multiset Size: $Size\ M = \sum_{a \in U} M\, a$.

Following definition is related to an important subset of the multiset.

Support Multiset: $Supp\ M = \{a \in U | M\, a > 0\}$.

2.2 Evolution Rules

Once have formally been defined multiset, we can define evolution rules in terms of multisets.

Evolution Rules are responsible for Transition P system evolution they make evolve the region they are associated. An evolution rule is formed by one antecedent and by one consequent. Both of them could be represented by multiset over different sets. Moreover, an evolution rule can make disappear the external membrane of its region, which produces some consequences in the Transition P system. With these considerations, we can define an evolution rule as:

Let L a label set, let U an object set, and let $D = \{out, here\} \in \{in\ j | j \in L\}$ the set of regions which a rule can send objects. An evolution rule with label in L and objects in U is a tern (u, v, δ), where u is a multiset over U, v is a multiset over $U \times D$ and $\delta \in \{dissolve, not\ dissolve\}$.

With this evolution rules representation, it can be defined several operations over evolution rules.

Definitions: Let $r = (u, v, \delta)$, $r_1 = (u_1, v_1, \delta_1)$ and $r_2 = (u_2, v_2, \delta_2)$ evolution rules with labels in L and objects in U. Let $a \in U$ and n a natural number. Then:

Evolution rules add: $r_1 + r_2 = (u_1 \cup u_2, v_1 \cup v_2, d_1 \vee d_2)$.

Product of an evolution rule by a natural number: $n\ r = \sum_{i=1}^{n} r$.

Inclusion of evolution rules: $r_1 \subset_u r_2$ if and only if $u_1 \subset_u u_2$.

Input of an evolution rule: $Input\ r = u$.

Dissolve: $Dissolve\ r = \delta$.

Now it is necessary to define some function over evolution rules related to the rule consequent. These functions will provide important information over what will happen when the rule was applied.

Membrane labels set the rule is sending objects: $Ins\ r = \{j \in L | (\_, in\,j) \in Supp\ v\}$.

Evolution rule's outputs:

$$(OutputToOut\ r)a = v(a, out)$$
$$(OutputToHere\ r)a = v(a, here)$$
$$((OutputToIn\ j)r)a = v(a, in\ j)$$

These last functions $OutputToOut$, $OutputToHere$, $OutputToIn$, return the multiset that the rule is sending to his father ($ToOut$), to itself ($ToHere$) and to a determined region ($ToIn\cdots$). They will be very useful in order to define the dynamic structure of a Transition P system.

2.3 Regions

A region is the inter-membranes space. Therefore, a region is the enclosed area by a membrane and any inner membrane to the first one do not enclose it. Regions in the Transition P system membranes structure are uniquely labelled in some set L, contain objects and evolution rules with a priority relationship defining a partial order among rules of the region. In order to define more precisely a region:

Let L a label set and U an object set. A region with labels in L and objects in U is a tern $(l, \omega, (R, \rho))$ where $l \in L$, ω is a multiset over U, R is a set of evolution rules with labels in L and objects in U and r is a partial order relation over R.

3 Transition P System Static Structure

Before define the Transition P system static structure, we will give some more definitions.

Definitions:

A Tree over an arbitrary set U is a pair where the first component is an element in U and the second one is a set of trees over U.

In this definition, the second one pair element can be the empty set, in which case the element is a leaf.

Over trees it can be defined the multiplicity function which provides the occurrences number of a given element $a \in U$ in the tree.

Multiplicity:

Let $T = (u, S)$, where $u \in U$ and S is a set of trees over U, then

$$(Mult\ T)a = \begin{cases} 1 + \sum_{s \in S}(Mult\ s)a & \text{if } a = u \\ \sum_{s \in S}(Mult\ s)a & \text{if } a \neq u \end{cases} \tag{2}$$

Transition P system membrane structure define a tree in which the root of the tree is the "skin" and elemental membranes are the leaves of the tree being $U = \emptyset$.

Finally, for defining a Transition P system, it is necessary to provide content to membranes. With all given definitions, a Transition P system can be defined as a tree of regions uniquely labelled. In a formal way:

Let L a labels set, let U an objects set. A Transition P system with label in L and objects in U is Π, being Π a tree of regions with labels in L and objects in U provided:

$$\forall l \in L, (Mult\ \Pi)(l, \_, \_) < 2 \tag{3}$$

An alternative definition can be given: A Transition P system with labels in L and objects in U is a pair whose first element is a region with labels in L and objects in U and the second one is a set of Transition P systems with labels in L and object in U, and regions are uniquely labelled.

$$\Pi = ((l, \omega, (R, \rho)), \Pi\Pi) \text{ and } \forall l \in L, (Mult\ \Pi)(l, \_, \_) < 2 \tag{4}$$

Where ω is a multiset over U, R is a set of evolution rules over L and U and $\Pi\Pi$ is a set of Transition P system with labels in L and objects in U.

These algebraic representations define precisely the static structure of Transition P systems.

We have name it Static Structure because these representation do not take into account evolution in Transition P systems, only represent Transition P systems in a static manner.

4   Transition P System Dymanic Structure

With dynamic structure Transition P systems, we are naming everything related to evolution. During evolution, a Transition P system is changing its region contents in several ways. Objects are changed by evolution rules application. Evolution rules that can be applied in a determined configuration when the P system has evolved cannot be applied anymore, and even, a rule can dissolve the external membrane of the region and the region disappears in the static structure of the P system sending its objects to its father region and disappearing their evolution rules.

Transition P system dynamic has been described in terms of parallel and non-deterministic devices, "evolution is done in parallel for all objects able to evolve," "All these operations are done in parallel, for all possible applicable rule $u \rightarrow v$, for all occurrences of multiset u in the region associated with the rules, for all regions at the same time" [1].

If we carefully look to Transition P system evolution, it is possible to distinguish between at least two different parallelism types. The first one is related to what is happened inside a region, and the second one is related to what is happened in every region of the Transition P system. The first one is the regional local parallelism and the second one is the system global parallelism, both of them are involved in System evolution, they are not independent because regions can disappear from the system and regions send objects to other regions; but in some way they can be independently described.

4.1   Regional Local Parallelism

At this exposition point, we will try to show how it is possible to reduce parallel evolution rules application in a determined region to the application of only one rule, and how it is obtained such a rule.

A region has been defined as a tern:

$$Region : (label, multiset, (Rules, Partial\,Order\,Relation)) = (l, \omega, (R, \rho)) \quad (5)$$

Evolution inside the region involves objects consumption by rules, sending objects to other regions and some times membrane dissolution and consequently region vanishing from the System. All these things are made in parallel and the only constrain is that a rule with higher precedence than others inhibits the second ones.

Following definitions are very important in order to exactly define rules that can be applied in the region in order to make evolve the region.

Definitions:

Maximal Set: Let $(U, <)$ a partial order over U, then:

$$Maximal\,U = \{u \in U| \; \not\exists v \in U, u < v\} = \{u|(u \in U) \wedge (\neg(u < \_))\} \quad (6)$$

Useful Rules: An evolution rule in a rules set is useful over a given labels set L if and only if the $Ins\,r$ is included in L. In a more formal way:

Let R an evolution rule set with labels in L and objects in U. Let $L' \in L$.

$$(Useful\,R)L' = \{r \in R|Ins\,r \subseteq L'\} = \{r|(r \in R) \wedge (Ins\,r \subseteq L')\} \quad (7)$$

By Transition P system definition, objects sent by rules can pass through one membrane. Therefore, a rule can send objects only to regions separate only by one membrane. In order to calculate useful rules in a region, L' is the label set constituted by labels corresponding to adjacent regions to the first one.

Applicable Rules: A rule is applicable over a multiset ω if and only if its antecedent is included in ω. Let R an evolution rules set with labels in L and objects in U. Let ω a multiset over U.

$$(Applicable\,R)\omega = \{r \in R|Input\,r \subseteq \omega\} = \{r|(r \in R) \wedge (Input\,r \subseteq w)\} \quad (8)$$

Therefore, a rule is applicable in a region if and only if the rule antecedent is included in the multiset the region has. In addition, only applicable rules can be used in a determined evolution step of Transition P systems.

Active Rules: A rule r, in a rules set is active over a partial order relationship defined over R if and only if there is not any rule in R higher than r according to the partial order relationship. Therefore, actives rules are those belonging to the maximal set definition given above.

Let (R, ρ) a partial order relation defined over an evolution rules set R with labels in L and objects in U, then:

$$(Active\,R)\rho = (Maximal\,R)\rho \quad (9)$$

Adjacent regions to a region: Let reg a region of a Transition P system P with labels in L and object in U. Let $(reg, \Pi\Pi)$ the sub-tree of P with root in reg. Let us define adjacent region in P to reg as:

$$ADJACENT_\Pi\,reg = \{r|(r, \Pi\Pi') \in \Pi\Pi\} = \{r|(r, \_) \in \Pi\Pi\} \quad (10)$$

Rules able to make evolve the region: With these definitions, it is possible exactly determine evolution rules that can be applied in order to make evolve the region reg. By definition, they are those rules satisfying following equations:

$$L' = \{l \in L | (l, \_, \_) \in ADJACENT_\Pi \ reg\}$$
$$Active(Applicable((Useful \ R)L')\omega)\rho \tag{11}$$

Traditional description of Transition P system dynamic describes evolution in a region in terms of every rule satisfying (11) can be applied in parallel "for all occurrences of rule antecedent multiset, for all regions at the same time"[1]. What this sentence is telling us in terms of operations defined among rules is the following: We must search for a lineal combination of rules satisfying (11) being complete; that is no one more rule satisfying (11) can be added to it.

Application of two rules r_1, $r_2 \in R$ at the same evolution region step, in parallel, is equivalent to applying the rule $r_1 + r_2$. In the same sense, we can say that applying n times the same rule in parallel has the same effect that applying the rule defined by nr.

So, parallel rules application inside a region can be solved by a lineal combination of region evolution rules. Moreover, such a lineal combination of evolution rules can be represented as a multiset over R, the set of evolution rules of the region.

Finally, the region evolve in only one step by application of only one rule formed by a complete lineal combination of evolution rules in R satisfying (11).

5   Non-deterministic Evolution in a Transition P System

Until now, we have provide an exact definition of rules can be applied in order to make evolve a region in one evolution step of a Transition P system. Moreover, parallel rules execution inside region can be replaced by the execution of only one rule. Then, where is the non-deterministic evolution in the region? Of course, more than one complete multiset over R (region evolution rules set) can be obtained for making evolve the region.

Now several definitions over multiset of evolution rules will be given in order to obtain an expression for identifying complete multisets of evolution rules in the region.

Definitions: Let R an evolution rules set with labels in L and objects in U. Let MR, MR_1 and MR_2 multisets over R and let ω a multiset over U.

MultiAdd: This operation performs the transformation of a multiset of evolution rules in a lineal combination of evolution rules. From now on, we will keep in mind this equivalence between lineal combinations of evolution rules and multiset of evolution rules.

$$\oplus MR = \sum_{r \in R} (MRr)r \tag{12}$$

MultiInclusion$_u$: This binary relationship defines a partial order relation over multiset over R. It will be very useful in order to determine which multiset over R will be completes.

$$MR_1 \sqsubseteq_u MR_2 \equiv (\oplus MR_1) \subset_u (\oplus MR_2) \tag{13}$$

MultiApplicable: This is the set of multiset over R (evolution rules set) that can be applied over ω.

$$(MultiApplicable\, R)\omega = MR|(MR : R \to N) \wedge ((Input\,(\oplus MR)) \subseteq w) \tag{14}$$

MultiComplete:

$$(Multicomplete\, R)\omega = (Maximal((MultiApplicable\, R)\omega)) \sqsubseteq_u \tag{15}$$

The above expression describes the set of complete multisets over $(R, \sqsubseteq_u)$ and ω.

Finally, let P a Transition P system with labels in L and objects in U, let $reg = (l, \omega, (R, \rho))$ one region of P and let $(reg, \Pi\Pi)$ the sub-tree of P which root is reg. The set of complete rules multiset is defined by following equations:

$$LABELS = l \in L|((l, \_, \_), \_) \in \Pi\Pi \tag{16}$$

$$ACTIVES = Active(Applicable((Useful\, R)LABELS)\omega)\rho \tag{17}$$

$$COMPLETES = (Multicomplete\, ACTIVES)\omega \tag{18}$$

The $LABELS$ set defines the set of adjacent regions'labels to region reg.

The $ACTIVES$ set defines the set of evolution rules that can be applied in order to make the region evolves.

The $COMPLETES$ set defines the set of complete multisets over R (the set of the region evolution rules) capable of making evolve the region in one step by application of one and only one evolution rule. Therefore, there are as many different possibilities for the region evolution as elements $COMPLETES$ has. A non-deterministic election of a multiset in $COMPLETES$ is enough for having non-deterministic evolution in the region.

Once has been formally stated the local non-determinism, it is possible to describe the global non-deterministic parallelism in a Transition P system in terms of regional local non-deterministic parallelism. That is, lets evolve independently every Transition P system region applying one of the $COMPLETES$ multiset in each one and non-deterministic election of a $COMPLETES$ multiset in each region will provide parallel non-deterministic evolution for the Transition P system.

6 Towards a Software System

Based on this formalism, we are developing a set of software modules in order to implement Transition P systems in a digital computer.

The Software System has two different kinds of modules, the first one related to algebraic structures, and the second one related to Transition P systems architecture.

Modules included in the algebraic structures are tuples, set, multiset and relationships. Included in the P system architecture we can find modules of directions, rules, regions and transition P system.

Functional dependencies among modules are shown in Figure 1.

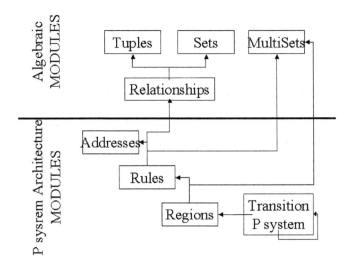

Fig. 1. One

7 Conclusions

This paper addresses the formalisation problem of Dynamic Structure of Transition P systems and the problem has been solved in a way chosen to be very close to functional programming style. We hope that these results can be translated to a functional programming language in order to develop a simulation of Transition P system in a digital computer. Part of this translation has started — some details are given in the end of the paper. Two continuation of this work are obvious, and we work on them:

- Complete the "Biosoftware" and to check its performances
- Add to the starting systems new features, such as membrane dissolution or membrane creation, in order to make possible to handle, theoretically, hard problems in a feasible time.

References

1. G. Paun, Computing with Membranes, Journal of Computer and System Sciences, 61, 1, (2000) 108–143.
2. G. Paun, P Systems with Active Membranes: Attacking NP Complete Problems, J. Automata, Languages, Combinatorics, 6,1 (2001), 75–90.
3. A. Baranda; Castellanos J.; Gonzalo Molina R.; Arroyo, F.; Mingo, L.F., Data Structure for Implementing Transition P System in Silico, Proceedings of the Workshop on Multiset Processing, (2000), 21–34.

Insertion-Deletion P Systems

Shankara Narayanan Krishna and Raghavan Rama

Department of Mathematics,
Indian Institute of Technology, Madras
Chennai-36, Tamilnadu, India
`ramar@iitm.ac.in`

Abstract. New computability models, called P systems, based on the evolution of objects in a membrane structure were recently introduced. This paper presents a new variant of P systems with string objects having insertion-deletion rules as the control structure. The use of this control structure is motivated from the DNA Computing area where insertions and deletions of small strands of DNA happen frequently in all types of cells and constitute also one of the methods used by some viri to infect a host. We investigate here the power of this type of systems with less than four membranes, in comparison with the families of CF, MAT and RE.

1 Introduction

P Systems were recently introduced as distributed computing models. In his seminal paper [4], Gh. Păun considers systems based on a hierarchically arranged finite cell structure consisting of several cell like membranes embedded in a main membrane called *skin*. A membrane structure corresponds to a string of correctly matching parantheses [,], with a matching pair at the ends. Each matching pair [,], of parantheses appearing in a membrane structure is called a *membrane*. A membrane having no membranes embedded within it is called an *elementary membrane*. The closed space delimited by any two membranes is called a *region*. The number of membranes in a membrane structure μ is called the degree of μ. The *depth* of a membrane structure μ denoted by $dep(\mu)$ is defined recurrently as follows:

1. If $\mu = [\ ]$, then, $dep(\mu) = 1$;
2. If $\mu = [\mu_1 \ldots \mu_n]$, then $dep(\mu) = \max\{dep(\mu_i) \mid 1 \leq i \leq n\} + 1$, where $\mu_1, \ldots, \mu_n$, are membrane structures embedded within μ.

The membranes delimit *regions*, where *objects*, elements of a finite alphabet V are placed. The objects evolve according to given evolution rules associated to a region; they contain symbols as $(a, here), (a, out)$ or (a, in_i) where a is an object. The meanings of the targets are : *here* indicates that the object remains in the membrane where it was produced; *out* means that the object is sent out of the membrane in which it was produced; in_i means that the object is sent to membrane i if it is reachable from the region where the rule is applied, if not

N. Jonoska and N.C. Seeman (Eds.): DNA7, LNCS 2340, pp. 360–370, 2002.
© Springer-Verlag Berlin Heidelberg 2002

the rule is not applied. An evolution rule involving the symbol δ can destroy the membrane in which it is. In this case, all the objects of the destroyed membrane pass to the immediately superior one and they evolve according to this membrane's evolution rules. The rules of the dissolved cell are lost. The skin membrane cannot be dissolved.

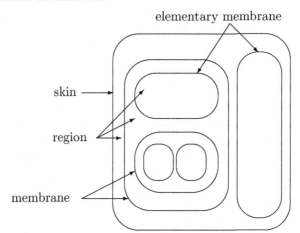

Figure 1: A membrane structure.

The membranes delimit *regions*, where *objects*, elements of a finite alphabet V are placed. The objects evolve according to given evolution rules associated to a region; they contain symbols as $(a, here), (a, out)$ or (a, in_i) where a is an object. The meanings of the targets are : *here* indicates that the object remains in the membrane where it was produced; *out* means that the object is sent out of the membrane in which it was produced; in_i means that the object is sent to membrane i if it is reachable from the region where the rule is applied, if not the rule is not applied. An evolution rule involving the symbol δ can destroy the membrane in which it is. In this case, all the objects of the destroyed membrane pass to the immediately superior one and they evolve according to this membrane's evolution rules. The rules of the dissolved cell are lost. The skin membrane cannot be dissolved.

Such a system evolves in parallel : at each step, all objects which can evolve should do so. A computation starts from an initial configuration of the system, defined by a cell structure with objects and evolution rules in each cell, and terminates when no further rule can be applied. The result of a computation consists of the multiplicity of objects present in a designated output membrane in a halting configuration.

Insertions and deletions of small linear DNA strands into long linear DNA strands are phenomena that happen frequently in nature and thus constitute an attractive paradigm for biomolecular computing. Insertion or deletion of short subsequences can be performed by using annealing of partially matching DNA strands. These operations with context dependence are also encountered in genetic area and were theoretically described in [5]. A model of DNA computation based on insertions and deletions that could be implemented in the laboratory by using a technique called PCR site-specific oligonucleotide mutagenesis has been proposed in [2,3]. In this paper, we relate the idea of computing with membranes with another important natural computing area, DNA Computing. Specifically, we consider P systems with objects in the form of strings and with evolution rules based on the insertion-deletion operation considered in [5]. Insertions and deletions of DNA strands depends on certain environmental conditions; to describe this action of nature, one has to provide different sets of insertion-deletion rules corresponding to different circumstances, with the assumption that at each moment these operations take place on different DNA strands. With this motivation, strings and insertion-deletion rules are localized to various regions; each of these strings are then rewritten by applying suitable insertion-deletion rules.

It is known that the insertion-deletion operation is powerful- see [2]. This observation is confirmed here : P systems based on insertion-deletion characterize the family of recursively enumerable languages. Moreover, very simple systems are enough : we need systems with only four membranes arranged in a four level structure. Insertion-deletion P systems with two membranes can generate CF; three membranes arranged in a two level structure are sufficient to generate non-context-free languages; four membranes arranged in a two level structure can generate MAT and four membranes arranged in a three level structure generate languages outside MAT.

2 Some Language Theory Prerequisites

In this section, we introduce some formal language theory notions which will be used in this paper; for further details, we refer to [6]. For an alphabet V, we denote by V^* the set of all strings over V, including the empty one, denoted by λ. By CF, RE we denote the families of context-free and recursively enumerable languages respectively, while MAT denotes the family of languages generated by matrix grammars without appearance checking.

A context-free grammar $G = (N, T, S, P)$ is said to be in the Chomsky normal form if the rules in P are of the form $X \to YZ, X \to x$, where $X, Y, Z \in N, x \in T$.

A type-0 grammar $G = (N, T, S, P)$ is said to be in Penttonen normal form if the rules in P are of the following three forms:

1. $X \to \alpha_1 \alpha_2$, for $\alpha_1, \alpha_2 \in N \cup T$ such that $X \neq \alpha_1, X \neq \alpha_2, \alpha_1 \neq \alpha_2$
2. $X \to \lambda$
3. $XY \to XZ$, for $X, Y, Z \in N$ such that $X \neq Y, X \neq Z, Y \neq Z$

It is known that for every context-free grammar G_1 and type-0 grammar G_2, there exist equivalent grammars G_1' and G_2' in the Chomsky normal form and Penttonen normal form respectively.

A matrix grammar without appearance checking is a construct $G = (N, T, S, M)$ where N, T are disjoint alphabets, $S \in N$ is the start symbol, M is a finite set of sequences of the form $(A_1 \longrightarrow x_1, \ldots, A_n \longrightarrow x_n), n \geq 1$, of context-free rules over $N \cup T$ with $A_i \in N, x_i \in (N \cup T)^*$. We say that N is the non terminal alphabet, T is the terminal alphabet, while the elements of M are called matrices. The rules of any matrix are applied in the order in which they appear. The family of languages generated by G is defined by $L(G) = \{w \in T^* \mid S \Longrightarrow^* w\}$.

A matrix grammar without appearance checking is said to be in *binary normal form* if $N = N_1 \cup N_2 \cup \{S\}$, with these three sets mutually disjoint, and the matrices in M are of the following forms:

1. $(S \longrightarrow XA)$,
2. $(X \longrightarrow Y, A \longrightarrow x)$,
3. $(X \longrightarrow \lambda, A \longrightarrow x')$

where $X, Y \in N_1$, $A \in N_2$, $x \in (N_2 \cup T)^*$, $x' \in T^*$ and $|x|, |x'| \leq 2$. So, $x = \alpha_1 \alpha_2$ for $\alpha_1, \alpha_2 \in N_2 \cup T \cup \{\lambda\}$ in matrices of types 2, and in matrices of type 3, x' is a string of length two or less over T^*. There is exactly one matrix of type 1 and a matrix of type 3 is used only once, at the last step of a derivation.

3   P Systems with Insertion-Deletion Rules

We refer to [1,7] for basic elements of P systems theory. Here, we introduce the variant of P systems we are going to investigate.

Definition 1. *An Insertion-deletion P system (Insdel P system) (of degree s) is a construct*

$$\Pi = (V, T, \mu, M_1, M_2, \ldots, M_s, R_1, R_2, \ldots, R_s, k)$$

where V is the total alphabet of the system; $T \subseteq V$ is the terminal or output alphabet; μ is a membrane structure consisting of s membranes; $M_i, 1 \leq i \leq s$ are finite languages over V, associated with regions i of μ; $R_1, R_2, \ldots, R_s$ are finite sets of developmental rules, associated with regions $1, \ldots, s$ of μ of the following form:

- *Insertion rules of the form $((u, \lambda/x, v), tar)$ where $u, x, v \in V^*$ and tar is one of $\{here, out, in_j\}$. This means that x can be inserted in between u and v. Then the produced string is moved to the membrane indicated by tar.*
- *Deletion rules of the form $((u, x/\lambda, v), tar)$ where $u, x, v \in V^*$. This means that x can be deleted from the context (u, v) and the resultant string can be moved to the membrane indicated by tar.*

k is a number between 1 and s which specifies the output membrane of Π.

We say that an Insertion-Deletion P system Π is of weight $(n, m; p, q)$ if

$$n = \max\{|\beta| \mid ((u, \lambda/\beta, v), tar) \in R\},$$
$$m = \max\{|u| \mid ((u, \lambda/\beta, v), tar) \in R \text{ or } ((v, \lambda/\beta, u), tar) \in R\}$$
$$p = \max\{|\alpha| \mid ((u, \alpha/\lambda, v), tar) \in R\}$$
$$q = \max\{|u| \mid ((u, \alpha/\lambda, v), tar) \in R \text{ or } ((v, \alpha/\lambda, u), tar) \in R\}$$

The work of such a system is defined as follows : in each time unit, each string which can be rewritten by applying rules of the above forms is rewritten; the strings and rules are localized to regions, so the strings in a given region can be rewritten only by rules in that region. A computation starts from an initial configuration defined by a cell structure with strings and insertion-deletion rules in each cell, and terminates when no more rules can be applied. Thus a sequence of transitions form a computation and we consider as successful computations the halting ones; (the computations which reach a configuration where no rule can be applied to any of the strings in any of the membranes); the result of a halting computation consists of all strings over T which are collected in the designated output membrane during the computation.

What is specific to our systems is the way of rewriting the strings. Specifically, each string is rewritten in the following way : take a string w in a region i; for any two words x and y which appear in w, and for which there are rules of the form $((x, \lambda/u, y), tar)$ or $((x, u/\lambda, y), tar)$ in the set R_i, we have to insert u or delete u between x and y. Then the string obtained by rewriting is sent to the membrane indicated by the target associated with the rule. If xy does not occur as a subword in w, no insertion can take place. Similarly, the (x, y) deletion of u from w is performed only if u occurs in w flanked by x on its leftside and by y on its right side. Note that in each time unit, exactly two words x and y of a string w are chosen nondeterministically and a string u is inserted or deleted between them (by an insertion or deletion rule nondeterministically chosen). In this way, we rewrite in parallel all strings in all membranes; each string being processed by a single rule.

We denote by $L(\Pi)$ the language generated by an Insdel P system Π; the family of languages of this type generated by systems of weight $(n', m'; p', q')$, degree s and depth t such that $n' \leq n, m' \leq m, p' \leq p, q' \leq q$ is denoted $INS_n^m DEL_p^q P_{s,t}$, $n, m, p, q, s, t \geq 0$; when one of the parameters n, m, p, q, s, t is not bounded, we replace it by $*$. Thus, the family of all insdel P languages is $INS_*^* DEL_*^* P_{*,*}$.

4   The Generative Power

We now investigate the computability power of P systems with insertion-deletion rules. We know that $INS_*^1 DEL_0^0 \subseteq CF$, [5]. In the following two theorems, we show that P systems with two membranes and weight (1,2;1,2) characterizes CF and systems having three membranes, depth two and weight (1,1;1,1) contain languages outside CF.

Theorem 1. $CF \subseteq INS_1^2 DEL_1^2 P_{2,2}$

Proof. Let $G = (N, T, S, P)$ be a context-free grammar in Chomsky normal form. Define the set $N' = \{B' \mid B \in N\}$. We construct the Insdel P system

$$\Pi = (V, T, [_1[_2 \ ]_2]_1, \{\#S\}, \lambda, R_1, R_2, 1),$$

with $V = N \cup N' \cup T \cup \{K_{YZ}, K_x \mid \exists$ rules in P of the form $X \to YZ, X \to x\} \cup \{r, \#\}$ and the following set of rules:

$$R_1 = \{((X, \lambda/K_x, \lambda), in_2) \mid X \to x \in P\} \cup \{((\lambda, X'/\lambda, \lambda), here) \mid X \in N\}$$
$$\cup \{((\lambda, r/\lambda, \lambda), here)\} \cup \{((\lambda, \#/\lambda, \lambda), here)\}$$
$$\cup \{((\#X, \lambda/K_{YZ}, \lambda), in_2) \mid X \to YZ \in P\}$$
$$\cup \{((\beta X, \lambda/K_{YZ}, \lambda), in_2) \mid X \to YZ \in P, \beta \in N \cup T\}$$
$$R_2 = \{((\beta\gamma, \lambda/r, XK_{YZ}), here) \mid \beta, \gamma \in N \cup T \cup \{\#, \lambda\}\}$$
$$\cup \{((\beta, \lambda/r, XK_x), here) \mid X \to x \in P, \beta \neq r\}$$
$$\cup \{((r, \lambda/Y', K_{YZ}), here) \mid Y, Z \in N\}$$
$$\cup \{((Y', \lambda/Z, K_{YZ}), here) \mid Y, Z \in N\}$$
$$\cup \{((rY', \lambda/Y, ZK_{YZ}), here) \mid Y, Z \in N\}$$
$$\cup \{((YZ, K_{YZ}/\lambda, \lambda), out) \mid Y, Z \in N\}$$
$$\cup \{((r, X/\lambda, K_{YZ}), here) \mid X \to YZ \in P\}$$
$$\cup \{((r, X/\lambda, K_x), here) \mid X \to x \in P\}$$
$$\cup \{((r, \lambda/x, K_x), here) \mid x \in T\}$$
$$\cup \{((\lambda, r/\lambda, xK_x), here) \mid x \in T\}$$
$$\cup \{((\beta x, K_x/\lambda, \lambda), out) \mid x \in T, \beta \neq r\}$$

The symbols K_{YZ} and K_x are markers : these symbols are inserted to the right of nonterminals X which correspond to rules $X \to YZ$ and $X \to x$. After this insertion operation, the string moves to membrane two. Here, a new symbol r is inserted to the left of XK_{YZ} and XK_x. If the string contains rXK_x, then we first delete X and then insert x between r and K_x. After this, r is deleted. Finally, we delete K_x and the string moves back to membrane one. In case we have rXK_{YZ} in the string, first X is deleted. After this, we insert Y' between r and K_{YZ}, Z between Y' and K_{YZ}, Y between rY' and ZK_{YZ} and delete K_{YZ} from the right of YZ. After the deletion, the string moves out. In membrane one, we delete Y' and r. It is clear that the terminal strings remaining in the skin at the end of a halting configuration belong to $L(G)$. $\square$

Theorem 2. $INS_1^1 DEL_1^1 P_{3,2} - CF \neq \emptyset$

In the next two theorems, we show that systems having four membranes, depth two and weight (1,2;1,2) generates the family MAT and systems with four membranes, depth three and weight (2,1;1,1) generate languages outside MAT.

Theorem 3. $MAT \subseteq INS_1^2 DEL_1^2 P_{4,2}$

Proof. Let $G = (N, T, S, M)$ be a matrix grammar in binary normal form generating the family MAT. Assume that there are n matrices labeled $m_1, m_2, \ldots, m_n$ in G. We construct the Insdel P system

$$\Pi = (V, T, [_1 [_2]_2 [_3]_3 [_4]_4]_1, M_1, \lambda, \lambda, \lambda, R_1, R_2, R_3, R_4, 1)$$

with

$$V = N_1 \cup N_2 \cup T \cup \{K_\alpha, K_{\alpha_1\alpha_2} \mid \exists \ rules \ A \to \alpha, A \to \alpha_1\alpha_2 \ in \ m_i, \ A \in N_2\}$$
$$\cup \{Y_i, A_i, \lambda_i \mid Y \in N_1, A \in N_2, 1 \le i \le n\}$$
$$\cup \{\kappa, \#_1, \#_2, \#', \#'', \epsilon, \epsilon', \alpha', \alpha'', \alpha''' \mid \alpha \in N_2 \cup T\}$$

$$M_1 = \{\#_1 X A \#_2 \mid (S \to XA) \ is \ the \ initial \ matrix\}$$

$$R_1 = \{((\#_1, \lambda/Y_i, X), here) \mid the \ matrix \ m_i \ contains \ the \ rule \ X \to Y\}$$
$$\cup \{((\#_1, \lambda/\lambda_i, X), here) \mid the \ matrix \ m_i \ contains \ the \ rule \ X \to \lambda\}$$
$$\cup \{((Y_i, X/\lambda, \beta), here), ((\lambda_i, X/\lambda, \beta), here) \mid X, Y \in N_1, \beta \in N_2 \cup T\}$$
$$\cup \{((A_i, \lambda/B_i, B), in_2), ((Y_i, \lambda/B_i, B), in_2) \mid Y \in N_1, \ A, B \in N_2 \cup T\}$$
$$\cup \{((\lambda_i, \lambda/B_i, B), in_2), ((B_i, \lambda/\#', \#_2), in_3) \mid B \in N_2 \cup T\}$$
$$\cup \{((\#', \lambda/\#'', \#_2), in_4)\} \cup \{((\lambda, \#'/\lambda, \#''), in_2)\}$$
$$\cup \{((\lambda, B'/\lambda, B), in_3), ((B', \#''/\lambda, \#_2), in_3) \mid B \in N_2 \cup T\}$$
$$\cup \{((B', B_i/\lambda, \#''), here), ((B', B_i/\lambda, \beta), in_2) \mid B \in N_2 \cup T, \beta \ne \#''\}$$
$$\cup \{((\alpha_1'', \lambda/\alpha_1', \lambda), in_2), ((\alpha_1'', \lambda/\alpha_1', \alpha_2' K_{\alpha_1\alpha_2}), here) \mid \alpha_1, \alpha_2 \in N_2 \cup T\}$$
$$\cup \{((\lambda, \alpha_1'''/\lambda, \alpha_1'\alpha_2'), in_2) \mid \alpha_1, \alpha_2 \in N_2 \cup T\}$$
$$\cup \{((\lambda, \epsilon/\lambda, \lambda), here)\} \cup \{((\lambda, \epsilon'/\lambda, \lambda), here)\}$$
$$R_2 = \{((A_i, A/\lambda, \beta), out) \mid A \in N_2 \cup T, \beta \in N_2 \cup T \cup \{\#_2\}\}$$
$$\cup \{((\kappa, \lambda/A', A_i\beta), out) \mid \beta \ne K_\alpha, K_{\alpha_1\alpha_2} \ and \ A, \alpha, \alpha_1, \alpha_2 \in N_2 \cup T\}$$
$$\cup \{((\gamma A', \lambda/B', B_i\beta), out) \mid \gamma = \kappa \ or \ C', \beta \ne K_\alpha, K_{\alpha_1\alpha_2}$$
$$and \ A, B, C, \alpha, \alpha_1, \alpha_2 \in N_2 \cup T\}$$
$$\cup \{((\kappa, \lambda/\alpha'', K_\alpha), here), ((A', \lambda/\alpha'', K_\alpha), here) \mid A, \alpha \in N_2 \cup T\}$$
$$\cup \{((A', B_i/\lambda, K_{\alpha_1}), here), ((A', B_i/\lambda, K_{\alpha_1\alpha_2}), here) \mid A, B, \alpha_1, \alpha_2 \in N_2 \cup T\}$$
$$\cup \{((\kappa, B_i/\lambda, K_{\alpha_1}), here), ((\kappa, B_i/\lambda, K_{\alpha_1\alpha_2}), here) \mid B, \alpha_1, \alpha_2 \in N_2 \cup T\}$$

$$\cup \{((\alpha'', K_\alpha/\lambda, \lambda), out), ((\lambda, \alpha''/\lambda, \alpha'\beta), here) \mid \alpha \in N_2 \cup T,$$
$$\beta \in \{A_i, A \in N_2 \cup T\}\}$$
$$\cup \{((\lambda, \alpha_1''/\lambda, \alpha_1'\#''), out), ((A', \lambda/\alpha_1''', K_{\alpha_1\alpha_2}), here) \mid A, \alpha_1, \alpha_2 \in N_2 \cup T\}$$
$$\cup \{((\kappa, \lambda/\alpha_1''', K_{\alpha_1\alpha_2}), here), ((\alpha_1''', \lambda/\alpha_2', K_{\alpha_1\alpha_2}), out) \mid \alpha_1, \alpha_2 \in N_2 \cup T\}$$
$$\cup \{((\alpha_1'\alpha_2', K_{\alpha_1\alpha_2}/\lambda, \beta), here) \mid \alpha_1, \alpha_2 \in N_2 \cup T, \beta \ne \#''\}$$
$$\cup \{((\alpha_1'\alpha_2', K_{\alpha_1\alpha_2}/\lambda, \#''), out) \mid \alpha_1, \alpha_2 \in N_2 \cup T\}$$
$$R_3 = \{((B_i, \lambda/K_\alpha, \beta), out) \mid m_i \ contains$$
$$B \to \alpha, B \in N_2, \beta \in \{A_i \mid A \in N_2 \cup T\} \cup \{\#'\}\}$$
$$\cup \{((B_i, \lambda/K_{\alpha_1\alpha_2}, \beta), out) \mid m_i \ contains$$
$$B \to \alpha_1\alpha_2, B \in N_2, \beta \in \{A_i \mid A \in N_2 \cup T\} \cup \{\#'\}\}$$
$$\cup \{((B', \lambda/B, \#_2), out), ((A', \lambda/A, B), out) \mid A, B \in N_2 \cup T\}$$
$$\cup \{((Y_i, \kappa/\lambda, A), here), ((\lambda_i, \kappa/\lambda, A), here) \mid Y \in N_1, A \in N_2 \cup T\}$$
$$\cup \{((\#_1, Y_i/\lambda, Y), out), ((Y_i, \lambda/Y, A), here) \mid Y \in N_1, A \in N_2 \cup T\}$$
$$\cup \{((\lambda_i, \lambda/\epsilon, A), here) \mid A \in N_2 \cup T\}$$

$\cup \{((\#_1, \lambda_i/\lambda, \epsilon), here), ((\#_1\epsilon, \lambda/\epsilon', \lambda), here),$
$\quad ((\lambda, \#_1/\lambda, \epsilon\epsilon'), here), ((\epsilon', \#_2/\lambda, \lambda), out)\}$
$\cup \{((\epsilon'\beta, \lambda/\epsilon', \gamma), here) \mid \beta \in N_2 \cup T, \gamma \in N_2 \cup T \cup \{\#_2\}\}$
$R_4 = \{((Y_i, \lambda/\kappa, A_i), out) \mid Y \in N_1, A \in N_2\} \cup \{((\lambda_i, \lambda/\kappa, A_i), out) \mid A \in N_2 \cup T\}$

The symbols K_α and $K_{\alpha_1\alpha_2}$ are markers : these symbols are inserted to the right of nonterminals $A \in N_2$ which correspond to rules $A \to \alpha$ and $A \to \alpha_1\alpha_2$ in matrices of type 2 or 4.

In the initial configuration, we have the string $\#_1 X A \#_2$ in the skin where $\#_1, \#_2$ are boundary markers and XA corresponds to the matrix of type one. To simulate the action of any matrix m_i, we first insert Y_i between the left boundary marker and X corresponding to the rule $X \to Y$. After this, X is deleted. Then, to apply the rule for the corresponding element of N_2, we proceed as follows: Using the rules $\{((Y_i, \lambda/B_i, B), in_2) \mid Y \in N_1, B \in N_2 \cup T\}, \{((A_i, \lambda/B_i, B), in_2) \mid A, B \in N_2 \cup T\}$, we index all the elements of $N_2 \cup T$ and the string moves to membrane two. By these rules, we insert to the left of every element in $N_2 \cup T$, the corresponding indexed element. In membrane two, we retain the indexed elements and delete the element to its right. This continues till all elements of N_2 in the string are indexed, after which the rule $\{((B_i, \lambda/\#', \#_2), in_3) \mid B \in N_2 \cup T\}$ is applied. In membrane three, K_α or $K_{\alpha_1\alpha_2}$ is inserted to the right of the indexed element $B_i, B \in N_2$ which occurs in the second position of the matrix $m_i : (X \to Y, B \to \alpha)$ or $(X \to Y, B \to \alpha_1\alpha_2)$. In the skin, $\#''$ is inserted between $\#'$ and $\#_2$ and the string moves to membrane 4. In membrane 4, we insert κ between Y_i and A_i, where $Y \in N_1, A \in N_2 \cup T$ and the string is sent out. Back in the skin, $\#'$ is deleted and the string is sent to membrane two.

Now we have to replace all the indexed elements by the original ones. For this purpose, we apply rules $((\kappa, \lambda/A', A_i\beta), out), ((\gamma A', \lambda/B', B_i\beta), out),$ $((A', B_i/\lambda, K_\alpha), here), ((A', B_i/\lambda, K_{\alpha_1\alpha_2}), here)$ in membrane two and $((B', B_i/\lambda, \beta), in_2)$ in the skin. Now we shall explain how K_α is rewritten by α, the procedure being similar for $K_{\alpha_1\alpha_2}$. We apply $((A', \lambda/\alpha'', K_\alpha), here)$ or $((\kappa, \lambda/\alpha'', K_\alpha), here)$ and then delete K_α by applying $((\alpha'', K_\alpha/\lambda, \lambda), out)$. In the skin membrane, α' is inserted to the right of α'' and the resultant string sent out to membrane two, where α'' is deleted. At the end of replacing all the indexed elements A_i by A', the string is sent to membrane three. All B''s are replaced by B by first applying rules $((B', \lambda/B, \#_2), out), ((B', \lambda/B, B), out)$ in membrane three; then the B''s are deleted in the skin membrane. After all B''s are replaced, κ is deleted, Y_i is replaced by Y and the string is sent back to the skin. Thus, the simulation of matrices of the type $(X \to Y, B \to \beta), |\beta| \le 2$ is done correctly. Simulation of matrices involving the rule $X \to \lambda, X \in N_1$ is also similar. Thus, the strings in $L(G)$ can be found in the skin membrane at the end of a halting configuration.   □

Theorem 4. $INS_2^1 DEL_1^1 P_{4,3} - MAT \ne \emptyset$.

It is known that $INS_1^2 DEL_1^1 = RE$ [5], whereas the size of the family $INS_1^1 DEL_1^1$ is not known. Now we show that a characterization of RE can be obtained by taking systems of weight $(1, 1; 1, 1)$, depth four and having four membranes.

Theorem 5. $RE \subseteq INS_1^1 DEL_1^1 P_{4,4}$

Proof. Let $G = (N, T, S, P)$ be a type-0 grammar in Penttonen normal form. We construct the insdel P system

$$\Pi = (V, T, [_1[_2[_3[_4 \ ]_4]_3]_2]_1, \{\$S\$\}, \lambda, \lambda, \lambda, R_1, R_2, R_3, R_4, 1)$$

with

$$
\begin{aligned}
V = \ & N \cup T \cup \{K_{\alpha_1\alpha_2}, r_Z \mid X \to \alpha_1\alpha_2, XY \to XZ \in P\} \\
& \cup \{\$, \#, \#'\} \cup \{\alpha' \mid \alpha \in N \cup T\} \\
R_1 = \ & \{((X, \lambda/r_Z, Y), in_2) \mid XY \to XZ \in P\} \\
& \cup \{((X, \lambda/K_{\alpha_1\alpha_2}, \lambda), in_2) \mid X \to \alpha_1\alpha_2 \in P\} \\
& \cup \{((\lambda, X/\lambda, \lambda), here) \mid X \to \lambda \in P\} \cup \{((\lambda, \$/\lambda, \lambda), here)\} \\
R_2 = \ & \{((r_Z, Y/\lambda, \lambda), in_3) \mid Y, Z \in N\} \cup \{((\lambda, \#/\lambda, \lambda), out)\} \\
& \cup \{((\lambda, X/\lambda, K_{\alpha_1\alpha_2}), in_3) \mid X \in N\} \cup \{((\lambda, \alpha'/\lambda, \lambda), out) \mid \alpha \in N \cup T\} \\
R_3 = \ & \{((r_Z, \lambda/Z, \lambda), in_4) \mid Z \in N\} \cup \{((\#, \lambda/\#', Z), here) \mid Z \in N\} \\
& \cup \{((\#, \#'/\lambda, \lambda), out)\} \cup \{((\lambda, \lambda/\alpha_1, K_{\alpha_1\alpha_2}), in_4) \mid \alpha_1, \alpha_2 \in N \cup T\} \\
& \cup \{((\alpha_1, \lambda/\alpha_2, \alpha_2'), out) \mid \alpha_1, \alpha_2 \in N \cup T\} \\
R_4 = \ & \{((X, \lambda/\#, r_Z), here) \mid X, Z \in N\} \cup \{((\#, r_Z/\lambda, \lambda), out) \mid Z \in N\} \\
& \cup \{((K_{\alpha_1\alpha_2}, \lambda/\alpha_2', \beta)), here) \mid \alpha_1, \alpha_2 \in N \cup T, \beta \in N \cup T \cup \{\$\}\} \\
& \cup \{((\lambda, K_{\alpha_1\alpha_2}/\lambda, \alpha_2'), out) \mid \alpha_1, \alpha_2 \in N \cup T\}
\end{aligned}
$$

The symbols r_Z and $K_{\alpha_1\alpha_2}$ are markers : these symbols are inserted to the right of nonterminals X which correspond to rules $XY \to XZ$ and $X \to \alpha_1\alpha_2$. To simulate $XY \to XZ$ and $X \to \alpha_1\alpha_2$, we insert r_Z in between X and Y and $K_{\alpha_1\alpha_2}$ to the right of X. The resultant string is then moved to membrane two. Here, we delete X from the left of $K_{\alpha_1\alpha_2}$, Y from the right of r_Z and move the string to membrane three. In membrane three, we insert Z to the right of r_Z, α_1 to the left of $K_{\alpha_1\alpha_2}$ and move the string to membrane four. Now, $\#$ is inserted to the left of r_Z and then r_Z is deleted; α_2' is inserted to the right of $K_{\alpha_1\alpha_2}$, then $K_{\alpha_1\alpha_2}$ is deleted and the string is sent out.

In membrane three, $\#'$ is inserted to the right of $\#$ and then $\#'$ is deleted; α_2 is inserted between α_1 and α_2' and the string is sent out. In membrane two, $\#$ and α_2' are deleted and the string sent out. The simulation of the rule $X \to \lambda$ is done by the rule $((\lambda, X/\lambda, \lambda), here)$. The boundary markers $\$$ can be deleted after completing all simulations. It is clear that the strings collected in the skin membrane at the end of a halting configuration belong to $L(G)$. □

5   Conclusion

We have investigated here, the P systems with string objects, which can evolve by using insertion-deletion operations. We have examined the power of systems with less than four membranes, comparing them with the families of CF, MAT and RE. The size of the family $INS_1^1 DEL_1^1$ was left as an open problem in [5]. Here, we show that systems of weight (1,1;1,1) having three membranes of depth two generate non-context-free languages while four membranes of depth four are sufficient for generating RE. Also, this variant resembles nature's action of insertion-deletion on different DNA strands simultaneously. Hence the implementation of Insdel P systems have considerable importance. However, the work reported in this paper is purely theoretical.

We synthesize the relations from the previous theorems in the following diagram; the arrows marked with a bullet correspond to inclusions, while the non-marked arrows indicate the comparison between two families.

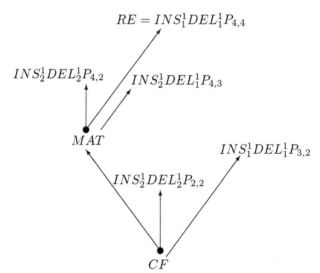

Figure 2: The INSDEL hierarchy

References

1. C.S. Calude and Gh. Păun, *Computing with Cells and Atoms–An Introduction to Quantum, DNA and Membrane Computing*, Taylor & Francis, London, 2001.
2. L. Kari, G. Paun, G. Thierrin, and S. Yu, *At the crossroads of DNA computing and formal languages: characterizing recursively enumerable languages by insertion-deletion systems*, DNA Based Computers III, H. Rubin and D. Wood, eds.

3. L.Kari and G.Thierrin, Contextual insertions/deletions and computability, *Information and Computation*, 131, pp. 47–61, 1996.
4. Gh. Păun, Computing with membranes, *Journal of Computer and System Sciences*, 61, 1 (2000), 108–143 (see also *Turku Center for Computer Science-TUCS Report No.* 208, 1998, www.tucs.fi).
5. Gh. Păun, G.Rozenberg, A.Salomaa, *DNA Computing- New Computing Paradigms*, Springer-Verlag, Berlin, 1998.
6. G. Rozenberg, A. Salomaa, eds., *Handbook of Formal Languages*, 3 volumes, Springer-Verlag, Berlin, 1997.
7. http://bioinformatics.bio.disco.unimib.it/psystems/

A Universal Time-Varying Distributed H System of Degree 1

Maurice Margenstern[1] and Yurii Rogozhin[2]

[1] Institut Universitaire de Technologie,
Université de Metz, Metz, France
Laboratoire d'Informatique Théorique et Appliquée
margens@lita.univ-metz.fr
[2] Institute of Mathematics and Computer Science of the
Academy of Sciences of Moldova
rogozhin@math.md

Abstract. Time-varying distributed H systems (TVDH systems shortly)
of degree n are a well known model of splicing computations which has
the following special feature: at different moments one uses different sets
of splicing rules (the number of these sets of splicing rules is called the
degree of the TVDH system). It is known that there is a universal TVDH
system of degree 2. Now we prove that there is a universal TVDH system of degree 1. It is a surprising result because we did not thought that
these systems are so powerful.
Recently both authors proved that TVDH systems of degree 1 can generate any recursively enumerable languages. We present here the short
description of the main idea of the proof that result.

1 Introduction

Time-varying distributed H systems are a well known theoretical model of biocomputing based on splicing operations [17, 18, 19]. We refer the reader to
[6, 7, 8, 9, 10, 11, 15, 23] for more details on the history of this model, its
extensions and connections with previous models based on splicing computations. That model introduces *components*, later see the formal definition, which
cannot all be used at the same time but one after another, periodically.

As a biochemical motivation, that model starts from the assumption that
the splicing rules are based on enzymes whose work essentially depends on the
environment conditions. Hence, at any moment, only a subset of the set of all
available rules is active. If the environment changes periodically, then the active
enzymes also change periodically.

Consider a model M of discrete computations. It consists in a family of
systems $\mathcal{S}_M$ that follow a certain paradigm: the one that is defined by the notion
of computation that is associated to M. Remember that a system U of M is called
universal when it is able to *simulate* any element of $\mathcal{S}_M$. Usually, the proof of the
simulation consists in the following: starting from an encoding $<G>$ of a formal
description of a member G of $\mathcal{S}_M$ and of an encoding $<\Delta>$ of some data Δ for

N. Jonoska and N.C. Seeman (Eds.): DNA7, LNCS 2340, pp. 371–380, 2002.
© Springer-Verlag Berlin Heidelberg 2002

G, to each step of computation s of G on Δ, there is a finite sequence $t_1, \ldots, t_k$ of computations of U on $(<G>,<\Delta>)$, such that the result of the application of $t_1, \ldots, t_k$ is an encoding of the result of the application of G on Δ at step s.

It is known, under the name of *Church's thesis*, that any model of this kind can be simulated by a Turing machine. Indeed, this can be formally proved for each model. As a result, in order to prove that a system U in M is universal, it is enough to prove that it is able to simulate any Turing machine or any system of computation that is known to be able to simulate any Turing machine.

In particular, one way to prove the universality property consists in proving that the system is able to generate any recursively enumerable language. Classical theorems of computer science prove that this is equivalent to the property of being able to simulate any Turing machine.

In [18] it is proved that 7 different *components* are enough in order to generate any recursively enumerable language.

In [16], the number of components was reduced down to 4. Recently, see [6, 7], both authors proved that two components are enough to construct a *universal time-varying distributed H system*, *i.e.* a time-varying distributed H system, which is capable of simulating the computation of any Turing machine.

Universality of computation and generating any recursively enumerable language are equivalent properties, but it is *a priori* not necessarily true, that the universality of some time-varying distributed H system with n components entails that there are time-varying distributed H systems which generate all recursively enumerable languages, with only n components. It is true for two components, because both authors proved that 2 different components are enough in order to generate any recursively enumerable language [11].

Now we prove that TVDH systems of degree one can carry out universal computations, i.e. that there is a TVDH system of degree one that is capable of simulating the computation of any Turing machine (for instance any universal Turing machine from [20, 21, 22]).

The question of whether or not a single component is enough for generating any recursively enumerable language is solved now, and we present here a short description of the main idea of the proof that TVDH systems of degree 1 can generate any recursively enumerable languages.

2   Basic Definitions

We recall some notions. An *alphabet* V is a finite, non-empty set whose elements are called *letters*. A *word* (over some alphabet V) is a finite (possibly empty) concatenation of letters (from V). The empty concatenation of letters is also called the *empty word* and is denoted by ε. The set of all words over V is denoted by V^*. A *language* (over V) is a set of words (over V).

An *(abstract) molecule* is simply a word over some alphabet. A *splicing rule* (over alphabet V), is a quadruple (u_1, u_2, u'_1, u'_2) of words $u_1, u_2, u'_1, u'_2 \in V^*$, which is often written in a two dimensional way as follows: $\dfrac{u_1 \big| u_2}{u'_1 \big| u'_2}$.

A splicing rule $r = (u_1, u_2, u_1', u_2')$ is applicable to two molecules m_1, m_2 if there are words $w_1, w_2, w_1', w_2' \in V^*$ with $m_1 = w_1 u_1 u_2 w_2$ and $m_2 = w_1' u_1' u_2' w_2'$, and produces two new molecules $m_1' = w_1 u_1 u_2' w_2'$ and $m_2' = w_1' u_1' u_2 w_2$. In this case, we also write

$$(w_1 u_1 | u_2 w_2, \ w_1' u_1' | u_2' w_2') \vdash_r (w_1 u_1 u_2' w_2', \ w_1' u_1' u_2 w_2)$$

or simply $(m_1, m_2) \vdash_r (m_1', m_2')$.

A pair $h = (V, R)$, where V is an alphabet and R is a finite set of splicing rules, is called a *splicing scheme* or an H *scheme*.

For an H scheme $h = (V, R)$ and a language $L \subseteq V^*$ we define:

$$\sigma_h(L) = \sigma_{(V,R)}(L) \stackrel{\text{def}}{=} \{w \in V^* | \exists w_1, w_2 \in L, \exists w' \in V^* :$$
$$\exists r \in R : (w_1, w_2) \vdash_r (w, w') \text{ or } (w_1, w_2) \vdash_r (w', w)\}.$$

A *Head-splicing-system* [1, 2], or H *system*, is a construct:

$$H = (h, A) = ((V, R), A),$$

which consists of an alphabet V, a set $A \subseteq V^*$ of initial molecules over V, the *axioms*, and a set $R \subseteq V^* \times V^* \times V^* \times V^*$ of splicing rules. H is called finite if A and R are finite sets.

For any H scheme h and any language $L \in V^*$ we define:

$$\sigma_h^0(L) = L,$$
$$\sigma_h^{i+1}(L) = \sigma_h^i(L) \cup \sigma_h(\sigma_h^i(L)),$$
$$\sigma_h^*(L) = \cup_{i \geq 0} \sigma_h^i(L).$$

The language generated by the H system H is defined as $L(H) \stackrel{\text{def}}{=} \sigma_h^*(A)$.

Thus, the language generated by the H system H is the set of all molecules that can be generated in H starting with A as initial molecules by iteratively applying splicing rules to copies of the molecules already generated.

A *time-varying distributed* H *system* (of degree n, $n \geq 1$), is a construct:

$$D = (V, T, A, R_1, R_2, \ldots, R_n),$$

where V is *an alphabet*, $T \subseteq V$ is *a terminal alphabet*, $A \subseteq V^*$ is *a finite set of axioms*, and *components* R_i are finite sets of splicing rules over V, $1 \leq i \leq n$.

At each moment $k = n \cdot j + i$, for $j \geq 0$, $1 \leq i \leq n$, only the component R_i is used for splicing the currently available strings. Specifically, we define

$$L_1 = A,$$
$$L_{k+1} = \sigma_{h_i}(L_k), \text{ for } i \equiv k(mod\ n), \ k \geq 1, 1 \leq i \leq n, \ h_i = (V, R_i).$$

Therefore, from a step k to the next step, $k + 1$, one passes only the result of splicing the strings in L_k according to the rules in R_i for $i \equiv k(mod\ n)$; the strings in L_k that cannot enter a splicing rule are removed.

The language generated by D is, by definition:

$$L(D) \stackrel{\text{def}}{=} (\cup_{k \geq 1} L_k) \cap T^*.$$

During the computations performed below to check our axioms and rules, we append the mark "↑" to any string that cannot enter a splicing rule and that is, therefore, ruled out from the later computation.

3   Main Result

Theorem 1. *There is a universal TVDH system of degree 1.*

In our proof, we use tag-systems which were introduced by E.Post. Cocke and Minsky proved that tag-systems are able to define universal computations [12, 13]. Tag-systems were used in the sixties for devising the smallest universal Turing machines, see a survey [14] and, more recently [5], leading to the smallest one known up to now constructed by the second author, see [20, 21, 22]. Tag-systems were also used for constructing small non-erasing Turing machines, see [4] and [5].

Let us now define tag-systems. For a given positive integer m and a given alphabet $\Omega = \{a_1, \ldots, a_n, a_{n+1}\}$, an m-tag-system on Ω transforms the word w on Ω as follows: we delete the first m letters of w and we append to the right of the result a word that depends on the first letter of w. This process is iterated until m letters cannot be deleted or the first letter is a_{n+1}, and then stops. Formally, we have the following definitions.

A *tag-system* is a three-tuple $\mathcal{T} = (m, \Omega, P)$, where m is a positive integer, $\Omega = \{a_1, \ldots, a_n, a_{n+1}\}$ is a finite alphabet, and P maps $\{a_1, \ldots, a_n\}$ into the set Ω^* of finite words on the alphabet Ω and a_{n+1} to $STOP$.

A tag-system $\mathcal{T} = (m, \Omega, P)$ is called m-tag-system when it is needed to stress on number m. Words $P_i = P(a_i) \in \Omega^*$ are called the *productions* of the tag-system $\mathcal{T}$. The letter a_{n+1} is the *halting symbol*. Productions are often displayed as follows:

$$a_i \rightarrow P_i, \quad i = 1, \ldots, n,$$
$$a_{n+1} \rightarrow STOP.$$

A *computation* of the tag-system $\mathcal{T} = (m, \Omega, P)$ on word $w \in \Omega^*$ is a sequence $w = v_0, v_1, \ldots$, of words on Ω such that, for all nonnegative integer k, a *current word* v_k is transformed into v_{k+1} by deleting the first m letters of v_k and appending the word P_i to the result if the first letter of v_k is a_i. The computation stops in k steps if the length of v_k is less than m or the first letter of v_k is a_{n+1}. In that latter case, the result is v_k.

Cocke-Minsky theorem more precisely states that there are universal 2-tag-systems ([13]), and therefore, we will deal only with 2-tag-systems. Due to the proof of Coke-Minsky theorem, we can restrict our attention to tag-systems which also have the following properties:

1. The computation of a tag-system stops only on a word beginning with the halting symbol a_{n+1}.

2. The productions P_i, $i = 1, \ldots, n$, are not empty.

3. The current word v_k, $k \geq 0$ contains at least 4 letters (it is always possible to require that for tag-systems that simulate two-register machines, see the details of Minsky's proof in his book [13]).

Henceforth, the tag-systems will be 2-tag-systems satisfying the above conditions.

Lemma 1. *For every tag-system T there is a time-varying distributed H system T' of degree 1 which simulates T.*

From the lemma, the proof of our theorem is straightforward.

Proof. Consider tag-system T: $\begin{array}{l} a_i \to P_i \quad (i = 1, \ldots, n), \\ a_{n+1} \to STOP. \end{array}$

Let be $\Omega = \{a_j\}$, $j = 1, \ldots, n+1$. Also denote the symbol a_{n+1} by h.

We construct the time-varying distributed H system T' as follows:
The terminal alphabet of T' is $T = \Omega$.
Let us introduce variables for $a_1, \ldots, a_{n+1}$, that we shall also denote by $\mathbf{a}, \mathbf{b}, \mathbf{c}, \mathbf{d}$.

The alphabet of T' is $V = \Omega \cup \{X, X', Y, Z, L, R, \overset{j}{\to}, \leftarrow\}$, $j = 1, \ldots, n+1$.

We denote $\overset{n+1}{\to}$ also by $\overset{h}{\to}$.

Axioms are given by $A = \{X'\mathbf{b} \overset{i}{\to} Z,\ L\mathbf{a} \overset{j}{\to} \mathbf{c}R,\ Z \leftarrow \mathbf{a}P_i Y,\ L\mathbf{a} \leftarrow \mathbf{b}R,\ X\mathbf{a}Z,$
$h \overset{h}{\to} Z,\ Z\mathbf{a}\}$, $\forall \mathbf{a}, \mathbf{b}, \mathbf{c}$ and $i = 1, \ldots, n$, $j = 1, \ldots, n+1$.

Block "START":

$$1.1 : \frac{X a_i \mathbf{ab} | \mathbf{c}}{X' \mathbf{b} \overset{i}{\to} | Z}, \quad \forall \mathbf{a}, \mathbf{b}, \mathbf{c},\ \forall i = 1, \ldots, n.$$

Block "RIGHT":

$$2.1 : \frac{\mathbf{b} \overset{j}{\to} \mathbf{ac}}{L | \mathbf{a} \overset{j}{\to} \mathbf{c}R}, \quad \forall \mathbf{a}, \mathbf{b}, \mathbf{c},\ \forall j = 1, \ldots, n+1;$$

$$2.2 : \frac{\mathbf{b} \overset{i}{\to} \mathbf{a}Y}{Z | \leftarrow \mathbf{a}P_i Y}, \quad \forall \mathbf{a}, \mathbf{b},\ \forall i = 1, \ldots, n;$$

$$2.3 : \frac{\mathbf{ba} \overset{j}{\to} \mathbf{c} | R}{L \overset{j}{\to} \mathbf{ac} | \mathbf{d}}, \quad \forall \mathbf{a}, \mathbf{b}, \mathbf{c}, \mathbf{d},\ \forall j = 1, \ldots, n+1;$$

$$2.4 : \frac{\mathbf{ba} \overset{j}{\to} \mathbf{c} | R}{L \overset{j}{\to} \mathbf{ac} | Y}, \quad \forall \mathbf{a}, \mathbf{b}, \mathbf{c},\ \forall j = 1, \ldots, n+1.$$

Block "LEFT":

$$3.1 : \frac{\mathbf{ab} \leftarrow \mathbf{c}}{L\mathbf{a} \leftarrow \mathbf{b}|R}, \quad \forall \mathbf{a}, \mathbf{b}, \mathbf{c};$$

$$3.2 : \frac{X'\mathbf{a} \leftarrow |\mathbf{b}}{X\mathbf{a}|Z}, \quad \forall \mathbf{a}, \mathbf{b};$$

$$3.3 : \frac{\mathbf{da}|\mathbf{b} \leftarrow R}{L\mathbf{a}| \leftarrow \mathbf{bc}}, \quad \forall \mathbf{a}, \mathbf{b}, \mathbf{c}, \mathbf{d};$$

$$3.4 : \frac{X'\mathbf{a}|\mathbf{b} \leftarrow R}{L\mathbf{a}| \leftarrow \mathbf{bc}}, \quad \forall \mathbf{a}, \mathbf{b}, \mathbf{c}.$$

Block "STOP":

$$4.1 : \frac{Xh|\mathbf{a}}{h \xrightarrow{h} |Z}, \quad \forall \mathbf{a};$$

$$4.2 : \frac{\mathbf{b}|\xrightarrow{h} \mathbf{a}Y}{Z|\mathbf{a}}, \quad \forall \mathbf{a}, \mathbf{b}.$$

Note. For each axiom $\alpha \in A$ there are rules $\frac{\alpha|\varepsilon}{\alpha|\varepsilon}$.

So, all the axioms $\alpha \in A$ are available at each step of operation under the TVDH system T'.

4   Details

The system T' works as follows.

Let the tag-system T start with a word $w = v_0 \in \Omega^*$. Then the system T' starts with a word XwY. That word, XwY, is added to the axioms of the system T' without adding a new rule $\frac{XwY|\varepsilon}{XwY|\varepsilon}$, as mentioned in Note above. So, the axiom XwY is available only at the first step of operation under the TVDH system T'.

At any time a current word $v_k \in \Omega^*$, $k \geq 0$ in the tag-system T is presented as a word Xv_kY, $X, Y \notin \Omega$ in the TVDH system T'.

We shall see that if the tag-system T stops on a word $hv' = v_k \in \Omega^*$, then the system T' produces the same word hv' and does not produce any other word over Ω^*.

Recall that $\mathbf{a}, \mathbf{b}, \mathbf{c}, \mathbf{d}$ are the variables for $a_1, \ldots, a_{n+1}$.

Let be $Xv_kY = Xa_i\mathbf{abc}w'Y$, $i \in \{1, 2 \ldots, n\}$.

At first the TVDH system T' erases the first two letters of the current word $a_i\mathbf{abc}w'$ $(Xa_i\mathbf{abc}w'Y)$. Simultaneously the tag-system T' sends a signal $\xrightarrow{i}$ to the right end of the current word, attached to the first erased letter a_i (Block START, rule 1.1):

$$(Xa_i\mathbf{ab}|cw'Y, \ X'\mathbf{b} \xrightarrow{i} |Z) \vdash_{1.1} (Xa_i\mathbf{ab}Z \uparrow, \text{ and } X'\mathbf{b} \xrightarrow{i} Z \text{ is an axiom.}$$

So, the word $X'\mathbf{b} \xrightarrow{i} \mathbf{c}w'Y$ passes to the next step of computation and the word $Xa_i\mathbf{ab}Z$ is ruled out, because it cannot enter any splicing rule.

That signal (symbol $\xrightarrow{i}$) crosses over the word (Block RIGHT, rules 2.1, 2.3 and 2.4) until Y is met:

$$(X'w'\mathbf{b}| \xrightarrow{i} \mathbf{a}cw''Y, \ L|\mathbf{a} \xrightarrow{i} \mathbf{c}R) \vdash_{2.1} (X'w'\mathbf{ba} \xrightarrow{i} \mathbf{c}R, \ L \xrightarrow{i} \mathbf{a}cw''Y),$$

$$(X'w'\mathbf{ba} \xrightarrow{i} \mathbf{c}|R, \ L \xrightarrow{i} \mathbf{a}c|dw'''Y) \vdash_{2.3} (X'w'\mathbf{ba} \xrightarrow{i} \mathbf{c}dw'''Y, \ L \xrightarrow{i} \mathbf{a}cR \uparrow),$$

where $w', w'', w''' \in \Omega^*$, $w'' = dw''''$ and $L\mathbf{a} \xrightarrow{i} \mathbf{c}R$ is an axiom.

If $w'' = \varepsilon$ then:

$$(X'w'\mathbf{ba} \xrightarrow{i} \mathbf{c}|R, \ L \xrightarrow{i} \mathbf{a}c|Y) \vdash_{2.4} (X'w'\mathbf{ba} \xrightarrow{i} \mathbf{c}Y, \ L \xrightarrow{i} \mathbf{a}cR \uparrow).$$

So, the word $X'w'\mathbf{ba} \xrightarrow{i} \mathbf{c}dw'''Y$ (or $X'w'\mathbf{ba} \xrightarrow{i} \mathbf{c}Y$) passes to the next step of computation and the word $L \xrightarrow{i} \mathbf{a}cR$ is ruled out, because it cannot enter any splicing rule.

The moment when Y is met indicates that the production P_i must be added to the right side of the current word and at that very moment, a resuming signal (a symbol $\leftarrow$) is sent backward to the beginning of the word in order to start again the process of performing one step of tag-system (Block RIGHT, rule 2.2):

$$(X'w'\mathbf{b}| \xrightarrow{i} \mathbf{a}Y, \ Z| \leftarrow \mathbf{a}P_iY) \vdash_{2.2} (X'w'\mathbf{b} \leftarrow \mathbf{a}P_iY, \ Z \xrightarrow{i} \mathbf{a}Y \uparrow).$$

So, the word $X'w'\mathbf{b} \leftarrow \mathbf{a}P_iY$ passes to the next step of computation and the word $Z \xrightarrow{i} \mathbf{a}Y$ is ruled out, because it cannot enter any splicing rule.

After that signal $\leftarrow$ crosses over the word to left (Block LEFT, rules 3.1, 3.3 and 3.4) until X' is met:

$$(X'w'\mathbf{ab} \leftarrow |\mathbf{c}w''Y, \ L\mathbf{a} \leftarrow \mathbf{b}|R) \vdash_{3.1} (X'w'\mathbf{ab} \leftarrow R, \ L\mathbf{a} \leftarrow \mathbf{b}cw''Y),$$

$$(X'w'''\mathbf{da}|\mathbf{b} \leftarrow R, \ L\mathbf{a}| \leftarrow \mathbf{b}cw''Y) \vdash_{3.3} (X'w'''\mathbf{da} \leftarrow \mathbf{b}cw''Y, \ L\mathbf{ab} \leftarrow R \uparrow),$$

where $w', w'', w''' \in \Omega^*$, $w' = w'''\mathbf{d}$ and $L\mathbf{a} \leftarrow \mathbf{b}R$ is an axiom.

If $w' = \varepsilon$ then:

$$(X'\mathbf{a}|\mathbf{b} \leftarrow R, \ L\mathbf{a}| \leftarrow \mathbf{b}cw''Y) \vdash_{3.4} (X'\mathbf{a} \leftarrow \mathbf{b}cw''Y, \ L\mathbf{ab} \leftarrow R \uparrow).$$

So, the word $X'w'''\mathbf{da} \leftarrow \mathbf{b}cw''Y$ (or $X'\mathbf{a} \leftarrow \mathbf{b}cw''Y$) passes to the next step of computation and the word $L\mathbf{ab} \leftarrow R$ is ruled out, because it cannot enter any splicing rule.

The moment when X' is met indicates that the circle of modelling is over: the signal $\leftarrow$ vanishes and a new current word v_{j+1} ($Xv_{j+1}Y$) is obtained (Block LEFT, rule 3.2) for a new step of the tag-system computation:

$$(X'\mathbf{a} \leftarrow |\mathbf{b}w''Y, \ X\mathbf{a}|Z) \vdash_{3.2} (X'\mathbf{a} \leftarrow Z \uparrow, \ X\mathbf{ab}w''Y).$$

If in the process of computation the first letter of the current word v_k is the halting symbol $h = a_{n+1}$ then the system T' deletes the symbol X from the beginning of the word Xv_kY and sends a special signal, for instance letter $\xrightarrow{h}$, to the right end of the word (Block STOP, rule 4.1).

$$(Xh|\mathbf{a}w''Y, \ h \xrightarrow{h} |Z) \vdash_{4.1} (XhZ \uparrow, \ h \xrightarrow{h} \mathbf{a}w''Y).$$

After that the signal $\xrightarrow{h}$ crosses over the word to the right (Block RIGHT, rules 3.1, 3.3 and 3.4) until Y is met. That moment indicates that the whole circle of modelling is over: system $\mathcal{T}'$ erases the signal $\xrightarrow{h}$ and the symbol Y, and so a result $v_{k+1} \in \Omega^*$ is obtained (Block STOP, rule 4.2).

$$(w'\mathbf{b}| \xrightarrow{h} \mathbf{a}Y, \ Z|\mathbf{a}) \vdash_{4.2} (w'\mathbf{ba}\uparrow, \ Z \xrightarrow{h} \mathbf{a}Y \uparrow).$$

It is easy to see that no other words than the result v_{k+1} (over the terminal alphabet Ω) can be obtained during computations of system $\mathcal{T}'$.

5   The Last New Result

Denote by RE the set of all recursively enumerable languages, by VDH_n, $n \geq 1$, the family of languages generated by time-varying distributed H systems of degree at most n, and by VDH_* the family of all languages of this type.

Theorem 2. $VDH_1 = VDH_* = RE$.

A short description of the main idea of proof.

We shall consider recursively enumerable sets of natural numbers instead of recursively enumerable languages. There are many well known methods to encode any formal languages into subsets of natural numbers. It will make our proof simpler.

It is a well known fact that from theoretical computer science that for every recursively enumerable set $\mathcal{L}$ of natural numbers, there is a Turing machine $T_{\mathcal{L}}$ which generates $\mathcal{L}$ as follows. Let $f_{\mathcal{L}}$ be a total recursive function enumerating the elements of $\mathcal{L}$, i.e. $\mathcal{L} = \{f_{\mathcal{L}}(x) \ ; \ x \in I\!\!N\}$. It is well known, see [3], chap.XIII, §68, Theorem XXVIII, how to construct a Turing machine $T_{\mathcal{L}}$ from the definition of $f_{\mathcal{L}}$ which would compute what f does, i.e. for every natural number x, machine $T_{\mathcal{L}}$ transforms any initial configuration $q_1 01^{x+1}$ into the final configuration $q_0 01^{f_{\mathcal{L}}(x)+1}0 \ldots 0$. And more, machine $T_{\mathcal{L}}$ never visits cells on the left at the cell with 0 of the initial configuration (i.e. machine $T_{\mathcal{L}}$ has a tape that is semi-infinite to the right) and no instruction of machine $T_{\mathcal{L}}$ is stationary, i.e. the head moves on every steps of the computation.

Let us assume for a while that we succeeded to simulate machine $T_{\mathcal{L}}$. It is not yet enough to prove the theorem. Indeed, starting from one word, say $q_1 01^{x+1}$ for some x, we would obtain an encoding of $f_{\mathcal{L}}(x)$ as $q_0 01^{f_{\mathcal{L}}(x)+1}$. But this is not what we need, because this gives us a single word.

It is not difficult to see that starting from machine $T_{\mathcal{L}}$, there is a Turing machine $T'_{\mathcal{L}}$ which computes $01^{x+2}q_0 01^{f_{\mathcal{L}}(x)+1}0 \ldots 0$, starting from $q_1 01^{x+1}$. It will be possible for us to device a time-varying distributed H system of degree 1 Γ which, arriving to that stage of the computation, will split the obtained word into two parts: $01^{f_{\mathcal{L}}(x)+1}$ as a *generated* word and $q_1 01^{x+2}$ as a new starting configuration. This will allow us to obtain exactly $\mathcal{L}$ as $\mathcal{L}(\Gamma)$.

Acknowledgement. The authors acknowledge the very helpful contribution of *INTAS project* 97-1259 and *NATO project* PST.CLG.976912 for enhancing their cooperation, giving the best conditions for producing the present result. For the same reasons they also acknowledge the help of the University of Metz (France), INRIA Lorraine and the French Ministry of Education and Research.

References

[1] Head, T.: Formal language theory and DNA: an analysis of the generative capacity of recombinant behaviors. Bulletin of Mathematical Biology **49** (1987), 737–759.

[2] Head, T., Păun, Gh., Pixton, D.: Language theory and molecular genetics. Generative mechanisms suggested by DNA recombination. Chapter 7 in vol.2 of G. Rozenberg and A. Salomaa, eds., *Handbook of Formal Languages*, 3 volumes, Springer-Verlag, Heidelberg, 1997.

[3] Kleene, S.: Introduction to Metamathematics. Van Nostrand Comp. Inc., New-York, 1952.

[4] Margenstern, M.: Non-erasing Turing machines: a new frontier between a decidable halting problem and universality. Lecture Notes in Computer Science **911**, in *Proceedings of LATIN'95* (1995), 386–397.

[5] Margenstern, M.: Frontier between decidability and undecidability: a survey. TCS **231-2** (2000), 217–251.

[6] Margenstern, M., Rogozhin, Yu.: A universal time-varying distributed *H*-system of degree 2. In Preliminary proceedings, Fourth International Meeting on DNA Based Computers, June 15–19, 1998, University of Pennsylvania (1998), 83–84.

[7] Margenstern, M., Rogozhin, Yu.: A universal time-varying distributed *H*-system of degree 2. *Biosystems*, **52** (1999), 73–80.

[8] Margenstern, M., Rogozhin, Yu.: Generating all recursively enumerable languages with a time-varying distributed *H*-system of degree 2, Technical Report in *Publications du G.I.F.M.*, Institut Universitaire de Technologie de Metz, ISBN 2-9511539-5-3, 1999.

[9] Margenstern, M., Rogozhin, Yu.: About Time-Varying Distributed H systems. In Proceedings of DNA Based Computers Conference, 6, Leiden, The Netherlands (2000), 55–64.

[10] Margenstern, M., Rogozhin, Yu.: About Time-Varying Distributed H systems. DNA Computing, 6th International Workshop on DNA-Based Computers, DNA 2000, Leiden, The Netherlands, June 13-17, 2000, LNCS **2054** (2001), 53–62.

[11] Margenstern, M., Rogozhin,Yu.: Time-Varying Distributed H-systems of Degree 2 Generate All Recursively Enumerable Languages. Where Mathematics, Computer Science, Linguistics and Biology Meet. Kluwer Academic Publishers (2001), 399–407.

[12] Cocke, J., Minsky, M.: Universality of tag systems with $P = 2$. J. Assoc. Comput. Mach. **11** (1964), 15–20.

[13] Minsky, M.L.: Computations: Finite and Infinite Machines. Prentice Hall, Englewood Cliffs, NJ, 1967

[14] Priese, L.: Towards a precise characterization of the complexity of universal and nonuniversal Turing machines. SIAM Journal of Computation, **8**, 4 (1979) 508–523

[15] Priese, L., Rogozhin, Yu., Margenstern, M.: Finite H-Systems with 3 Test Tubes are not Predictable. In Proceedings of Pacific Symposium on Biocomputing, Kapalua, Maui, January 1998, R.B. Altman, A.K. Dunker, L. Hunter, and T.E. Klein, (eds.), World Sci. Publ., Singapore (1998), 545–556.

[16] Păun, A.: On Time-Varying H Systems. *Bulletin of EATCS*, No. 67, (February 1999), 157–164.

[17] Păun, G.: DNA computing: distributed splicing systems. in *Structures in Logic and Computer Science. A Selection of Essays in honor of A. Ehrenfeucht*, LNCS **1261** (1997), 353–370.

[18] Păun, G.: DNA Computing Based on Splicing: Universality Results. TCS **231-2** (2000), 275–296.

[19] Păun, G., Rozenberg, G., Salomaa, A.: DNA Computing: New Computing Paradigms. Springer, 1998.

[20] Rogozhin, Yu.: Seven universal Turing machines. Mathematical Studies, Kishinev, Academy of Sciences, **69** (1982), 76–90. (In Russian).

[21] Rogozhin, Yu.: Small universal Turing machines. TCS **168-2** (1996) 215–240.

[22] Rogozhin, Yu.: A Universal Turing Machine with 22 States and 2 symbols. Romanian Journal of Information Science and Technology, **1**, 3 (1998), 259–265.

[23] Verlan, S.: On Extended Time-Varying Distributed H Systems. In Proceedings of DNA Based Computers Conference, 6, Leiden, The Netherlands (2000) 281.

A Note on Graph Splicing Languages

N. Gnanamalar David*, K.G. Subramanian, and D. Gnanaraj Thomas

Department of Mathematics,
Madras Christian College,
Chennai - 600059, India

Abstract. In this paper, we examine the relationship between graph splicing languages of Freund (1995) and Hyperedge replacement graph languages. We also extend the notion of self cross-over introduced by Dassow and Mitrana (1998) to graph languages and compare them with graph splicing languages.

1 Introduction

In order to model the recombinant behaviour of DNA molecules under the effects of restriction enzymes and ligases, Head [5] introduced in his pioneering paper, a new class of generative systems. These are called Splicing Systems and have been conceived as a generative formalism with the aim of analyzing the generative capacity of the recombinant behaviour of DNA molecules, in terms of formal languages. The splicing systems make use of a new operation, called splicing, on strings of symbols [5]. But DNA sequences are three-dimensional objects in a three-dimensional space. In fact, Jonoska [7] has proposed use of three dimensional graph structures solving computational problems with DNA molecules. Hence graphs seem to be more suitable objects for describing complex three-dimensional objects independently from their actual position in the three-dimensional space. One of the recent developments in the field of splicing systems is an interesting generalization of the concept of splicing to graphs proposed by Freund [3].

Graph grammars and graph languages have been studied since the late sixties, motivated by various application areas. Hyperedge replacement is one of the easiest and best studied types of graph rewriting (see for example Habel [4]). Hyperedge replacement graph grammars have an attractive generative power with interesting structural and decidability properties. To some extent, these nice properties result from the fact that hyperedge replacement is context-free in the sense of formal language theory. Extensive work has been done on hyperedge replacement grammars in the literature.

We examine the relationship between graph splicing languages [3] and Hyperedge replacement graph languages. Dassow and Mitrana [2] while modeling cross-over between a DNA molecule and its replicated version, have introduced and investigated a very simple and natural restriction on the splicing operation,

---

* ngdmcc@yahoo.com

N. Jonoska and N.C. Seeman (Eds.): DNA7, LNCS 2340, pp. 381–390, 2002.
© Springer-Verlag Berlin Heidelberg 2002

namely cross-over of identical strings, resulting in self cross-over systems. Here we extend the notion of self cross-over to graph languages in the context of graph splicing.

2   Preliminaries

We recall certain notions relating to grammars and languages. Let A be a finite set of symbols, called alphabet. A^* is the set of all finite strings or words over A. The length of a string w is the number of letters of the alphabet in w. λ denotes the string of length zero. $L \subseteq A^*$ is called a language over A.

In the literature, the notions of string grammars and languages have been exended to generate higher dimensional structures such as graphs. We briefly recall and review (mostly informally) the notion of hyperedge replacement with rendezvous [1] and certain basic results pertaining to these grammars and their graph languages.

Directed graph and decorated graph:

A directed graph is a pair (V,E) where V is a finite set of nodes (or vertices) and E is a subset of $V \times V$, called edges.

Let N be a set of nonterminal labels, each element A of N is associated with a natural number, its type, denoted by type(A). An (undirected) graph can be thought of as a directed graph by replacing each undirected edge (u,v) by two directed edges (u,v) and (v,u).

Hyperedge and Hypergraph:

A hyperedge is an atomic item with an ordered set of incoming tentacles and an ordered set of outgoing tentacles. Incoming tentacles are attached to nodes through a source function. Outgoing tentacles are attached to nodes through a target function.

Let C be an arbitrary, but fixed set, called set of labels (or color). A (directed, hyperedge-labeled) hypergraph over C is a system (V,E,s,t,ℓ) where V is a finite set of nodes (or vertices), E is a finite set of hyperedges, $s : E \to V^*$ and $t : E \to V^*$ are two mappings assigning a sequence of sources s(e) and a sequence of targets t(e) to each $e \in E$, and $l : E \to C$ is a mapping labeling each hyperedge.

Parallel hyperedge replacement [1] :

Parallel hyperedge replacement is a simple construction where the hyperedges of a decorated graph are removed, the associated decorated graphs are added disjointly and their external nodes are fused with the corresponding nodes formerly attached to the replaced hyperedges.

Hyperedge-replacement grammars with rendezvous [1] :

1. A hybrid table-controlled OL hyperedge replacement grammar with rendezvous (an RHTOL HR grammar) is a system $G = (N, T, \mathcal{R}, Z)$ where N is a typed set of nonterminals, T is a finite table set of productions, $\mathcal{R}$ is a rendezvous specification for T giving the connection relation for merging of nodes in the replaced graphs and $Z \in \mathcal{G}_0(N)$, called the axiom. Here $\mathcal{G}_0(N)$ is the set of all graphs with node labels from N. If $Z = S^\bullet$, the hyperedge with label S, for some $S \in N$, we may denote G by $(N, T, \mathcal{R}, S)$.

2. Given such a grammar, the generated graph language consists of all graphs derivable from Z, i.e., $L(G) = \{H \in \mathcal{G}_0 | Z \Rightarrow^*_{T,\mathcal{R}} H\}$.

Example 1 (Generation of all complete graphs) :

In order to generate the set of all complete graphs the RHTOL HR grammar K given by $K = (\{A\}, \{P_1, P_2\}, \mathcal{R}, A)$ can be used. Its components are defined as follows : $P_1 = \{p_{new}\}, P_2 = \{p_{del}\}$ where $p_{new} = (A, new)$ and $p_{del} = (A, del)$ with

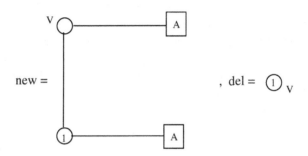

and the rendezvous $\mathcal{R}$ for table P_1 merges vertices (v,v) and the rendezvous for table P_2 is null.

This grammar generates the set of all complete graphs in the following way. In every intermediate graph, all nodes carry at least one A-labelled hyperedge. Each such hyperedge generates a new node v which is connected with the old one by an edge. By the rendezvous specification, all the new nodes are merged into one, which is thus connected with every old one.

3   Splicing on Graphs

The concept of splicing was introduced by Head [5], as a new operation on strings while studying the recombinant behaviour of DNA molecules. Freund [3] introduced an interesting extension of the notion of splicing to graphs. We now recall notions pertaining to graph splicing [3]. For completeness we give the full details.

A (directed, labelled) graph over V is a triple (N, E, L), where N is the set of nodes, E is the set of (directed) edges of the form (n,m) with $n, m \in N$, and $n \neq m$ (i.e., we do note allow loops), and L is a function from N to V assigning a label from V to each node from N. The set of all graphs over V is denoted by $\gamma(V)$. A graph $g \in \gamma(V)$ is called weakly connected if $g = (N, E, L)$ such that for all $n, m \in N, n \neq m$, there is a sequence $n = x[0], x[1], \ldots, x[k] = m$ of nodes from N such that for all i with $0 \leq i < k$ either $(x[i], x[i + 1]) \in E$ or $(x[i + 1], x[i]) \in E$. The set of all weakly connected graphs over V is denoted by $\gamma_C(V)$.

Graph splicing scheme [3] :

A *graph splicing scheme* σ is a pair (V, P), where V is an alphabet and P is a finite set of graph splicing rules of the form ((h[1], E'[1]), ..., (h[k], E'[k]); E), where $k \geq 1$ and for all i with $1 \leq i \leq k$,

- $h[i] \in \gamma(V), h[i] = (N[i], E[i], L[i])$,

- $E'[i] \subseteq E[i]$,

- the node sets $N[i]$ are mutually disjoint,

and E must obey the following rules :

1. Each edge $(n, m) \in E'[i]$ is divided into two parts, i.e., the start part $(n, m]$ and the end part $[n, m)$.
2. The elements of E are of the form $((n, m], [n', m'))$, where (n, m) and (n', m') are edges from $\cup_{1 \leq i \leq k} E'[i]$.
3. Every element from $\{(n, m], [n, m)/(n, m) \in \cup_{1 \leq i \leq k} E'[i]\}$ must appear exactly once in a pair of E.

Graph splicing rule:

Let $p = ((h[1], E'[1]), \ldots, (h[k], E'[k]); E)$ be a graph splicing rule, and let $r \in \gamma_C(V)$. If we can select k different subgraphs $g[1], \ldots, g[k]$ of r, then we can apply p to r, which yields some $s \in \gamma_C(V)$ in the following way:

1. For all i with $1 \leq i \leq k$, h[i] is a subgraph of g[i], where f[i] establishes the injective node embedding of h[i] into g[i].
2. The union of g[1], ..., g[k] can be looked at as a single graph g in $\gamma(V)$ and the union of the functions f[i] as a single function f embedding the h[i] into g. From g we eliminate all edges from $\cup_{1 \leq i \leq k} f(E'[i])$, but add all edges $(f(n), f(m'))$ such that $((n, m], [n', m')) \in E$ which yields the uniquely determined union of k' weakly connected graphs $g'[1], \ldots, g'[k']$.

We assume that any number of copies of a graph is available, if required for splicing.

Graph splicing system :

Let $\sigma = (V, P)$ be a graph splicing scheme and let $I \in \gamma_c(V), \sigma\{(I)\}$ is the set of all $I' \in \gamma_c(V)$ obtained by applying one graph splicing rule of P to I. Iteratively, for every $n \geq 2, \sigma^n\{(I)\}$ is defined by $\sigma^n(\{I\}) = \sigma(\sigma^{n-1}(\{I\})$ and $\sigma^0(\{I\}) = \{I\}$. The triple $S = (V, P, I)$ where $V = N \cup T$ is called an extended graph splicing system, (in an ordinary graph splicing system, no distinction between terminals and nonterminals is made).

$$\sigma^*(\{I\}) = \cup_{n \in N}\sigma^n(\{I\}) \text{ and } L(S) = \{g \in \gamma_c(V)/f(n) \in T \text{ for all } n \in V(g)\}.$$

Example 2 [3] : Let $\sigma = (\{A, T, C, G\}, P)$, where A, T, C and G denote the four deoxyribonucleotides that incorporate adenine, cytocine, guanine, thymine and $P = \{(g_1; \{((1, 2], [1', 2')), ((2, 6], [2', 6')), ((3, 7], [3', 7')), ((7, 8][7', 8')),$ $((1', 2'][1, 2)), ((2', 6'], [2, 6)), ((3', 7'], [3, 7)), ((7', 8'][7, 8))\})\}$.

As we are only interested in the connection points $(1,2][1,2),(3,4]$ etc., we can assign labels (e.g. numbers) only to these points and then represent P in a shorter way as

$$P = \{(g_2; \{(1, 2'), (3', 4), (5', 6), (7, 8'), (1', 2), (3, 4'), (5, 6'), (7', 8)\})\}.$$

where g_1 and g_2 are as in figure 1 and 2. Therefore, from two DNA molecules given in figure 3 (g_3) we obtain the two recombined DNA molecules given in figure 4 (g_4).

Every DNA splicing scheme can be expressed by a suitable graph splicing scheme.

We now prove that graphs that can be obtained by graph splicing operation of Freund [3], can be derived by hyperedge replacement graph grammar with rendezvous.

Theorem 1. *The class of all graph languages generated by graph splicing system is a subclass of the set of all languages generated by RHTOL HR grammars.*

Proof : Let $L = L(S)$ for some graph splicing system $S = (V, P, I)$. We construct an RHTOL grammar $H = (N, T, R, I)$ generating L as follows.

For each production $p = ((h[1], E'[1]), ..., ([h[k], E'[k])); E)$ in P, we construct two tables of productions - T_p and T_{pt}. Each edge $(n, m) \in E'[i], 1 \leq i \leq k$, is replaced by a hyperedge of type 2 with label (n,m)

$$[\text{ i.e., } \bigcirc \!\!\!\!\frac{\quad 1 \quad}{} \boxed{(n,m)} \frac{\quad 2 \quad}{} \!\!\!\!\bigcirc]$$

and add the production

to T_p and a terminating production to T_{pt}. Construct the rendezvous R_{T_p} as $R_{T_p} = \{((n,m],[n',m'))/((n,m],[n',m')) \in E\}$.

Since R_{T_p} contains a pair corresponding to each pair in E, the result of applying T_p is same as applying the splicing rule p, except that the edges remain as non-terminals. These non-terminals are then replaced by terminal edges by applying table T_{pt}. Hence $L = L(H)$. □

4   Self Cross-Over Graph Systems

We refer to [2] the notion of self cross-over systems on strings of symbols.

Informally, given a starting finite set of strings and a finite set of cross-over rules $(\alpha, \beta, \gamma, \delta)$, it is assumed that every starting string is replicated so that, we have two identical copies for every initial string. The first copy is cut between the segments α and β and the other one is cut between γ and δ. Then the last segment of the second string adheres to the first segment of the first string, and a new string is obtained. More generally, another string is also generated, by linking the first segment of the second string with the last segment of the first string. Iterating the procedure, we get a language.

The notion of self cross-over on strings [2], is considered here for graphs resulting in self cross-over graph systems. Two identical copies of a graph are taken and self cross-over of these graphs is done with respect to the splicing operation on graphs based on Freund's model [3]. The graph language obtained is called self cross-over graph language.

We illustrate self cross-over graph systems with examples.

Example 3 : Consider the self cross-over graph system $G_2 = (\{a\}, \{p\}, I)$ where and if $I = C_4$, cycle of length 4, then the language generated by G_2 is $L(G_2) =$

$\{C_n/n = 2^k, k \geq 2\}$, and if $I = C_3$, cycle of length 3, then $L(G_2) = \{C_n/n = 3 \times 2^k, k \geq 0\}$.

The self cross-over splicing on C_4 is illustrated in figure 5.

Example 4 : Consider the linear self cross-over graph system $G_3 = (\{a, b\}, \{p\}, I)$ where P and I as given in figure 6.

The linear graph language is given by the string language
$\{ba^{2^n}b/n \geq 1\} \cup \{bb\}$. $ba^{2^2}b$ is the linear graph

Example 5 : The linear self cross-over graph system $G_4 = (\{a, b\}, \{p_1, p_2\}, I)$ where

$$p_1 = (\; \text{(a)} \overset{1}{\bullet} \overset{2}{\bullet} \text{(b)} \quad \text{(a)} \overset{3}{\bullet} \overset{4}{\bullet} \text{(a)} \overset{5}{\bullet} \overset{6}{\bullet} \text{(b)} \quad \text{(a)} \; ;$$

$$E = \{\; (1,4), (5,2), (3,6)\;\} \;) \text{ and}$$

$$p_2 = (\; \text{(a)} \overset{1}{\bullet} \overset{2}{\bullet} \text{(b)} \quad \text{(a)} \overset{3}{\bullet} \overset{4}{\bullet} \text{(b)} \overset{5}{\bullet} \overset{6}{\bullet} \text{(b)} \quad \text{(a)} \; ; E)$$

and

$$I = \; \text{(a)} \text{---} \text{(b)} \text{---} \text{(a)} \text{---} \text{(b)}$$

generates the linear graph language given by the string language
$\{a^n b^m a^p b^q / n, m, p, q \geq 1\}$.

Example 6 : $G_5 = (\{a\}, \{p\}, I)$ where

$$p = (\; \text{(a)} \overset{1}{\bullet} \overset{2}{\bullet} \text{(a)} \;,\; \text{(a)} \overset{3}{\bullet} \overset{4}{\bullet} \text{(a)} \; ; \{\;(1,4), (3,2)\;\}) $$

and

$$I = \; \text{(a)} \text{---} \text{(a)} \text{---} \text{(a)}$$

generates the linear graph language given by the string language $\{a^n/n \geq 2\}$.
G_5 is both a graph splicing system and a self cross-over graph system.

Remark : It can be seen every self cross-over linear graph language over $\{a\}$, generated by self cross-over graph systems is either a finite set F or $F \cup \{a^n/n \geq$

$k\}$ for a finite set of F of linear graphs and some $k > 0$ where a^n represents the linear graph (n nodes).

Proposition 1 : There is a linear graph language generated by self cross-over graph system, but not generated by any graph splicing system.

Proof : The proof follows by noting that the self cross-over graph system of example 2 generates the linear graph language represented by the string language $\{ba^{2^n}b/n \geq 0\} \cup \{bb\}$. It is clear that there cannot be any graph splicing rule that gives rise to only linear graphs of the form $ba^{2^n}b$, since any such rule has to increase the number of a-labelled nodes but cannot produce the exponential increase of $2^n (n \geq 1)$ a-labelled nodes.

Remark : We note that the linear graph language given by the string language $\{a^{2^n}/n \geq 0\}$ is generated neither by a self cross-over graph system, nor by a graph splicing system.

Proposition 2 : There exist linear graph languages generated by graph splicing systems but not by any self cross-over system.

Proof : The proposition follows by observing that the linear graph language $\{a^n b^m a^p b^q \mid n, m, p, q \geq 1\}$ is not a self cross-over graph language as can be seen by an argument similar to the one in [[2], Lemma 3.1], but is generated by graph splicing system of Example 3.

Theorem 2. *The family of self cross-over languages and graph splicing languages are incomparable, but not disjoint.*

Proof : The result follows from propositions 1 and 2 and example 5.   □

References

1. N.G. David, F. Drewes, and H.J. Kreowski, Hyperedge Replacement with Rendezvous, Lecture Notes in Computer Science – 688, 1993, 169–181.
2. J. Dassow and V. Mitrana, Self Cross-over Systems, In 'Discrete Mathematics and Theoretical Computer Science', Springer Series, G. Paun (Ed.), 1998, 283–294.
3. R. Freund, Splicing Systems on Graphs, Proc. Intelligence in Neural and Biological Systems, IEEE Press, 1995, 189–194.
4. A. Habel, Hyperedge Replacement: Grammars and Languages, Lecture Notes in Computer Science, Springer-Verlag, Berlin Heidelberg, 1992.
5. T.Head, Formal Language Theory and DNA: An analysis of the generative capacity of recombinant behaviours, Bulletin of Mathematical Biology, 49, 1987, 735–759.

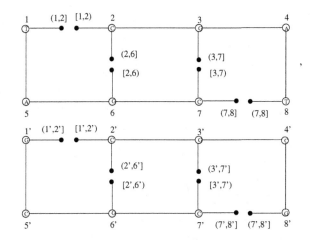

Fig. 1. g_1 of Example 2

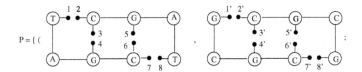

Fig. 2. g_2 of Example 2

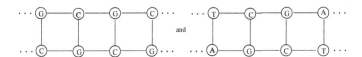

Fig. 3. g_3 of Example 2

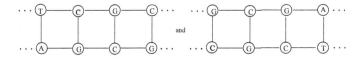

Fig. 4. g_4 of Example 2

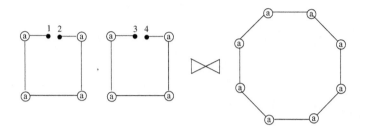

Fig. 5. Self Cross-over Splicing on C_4

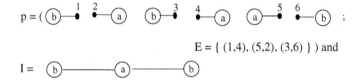

$$E = \{ (1,4), (5,2), (3,6) \}) \text{ and}$$

Fig. 6. P and I of example 4

6. J.E. Hopcroft and J.D. Ullman, Introduction to Automata Theory, Languages and Computation, Addison-Wesley Pub. Co., USA, 1979.
7. N. Jonoska, 3D DNA Patterns and Computation, Proc. Molecular Computing, India, 1998, 20–32.

Author Index

Akihiro, H., 182
Arroyo, F., 350

Balan, M.S., 290
Banzhaf, W., 23
Baranda, A.V., 350
Bi, H., 92

Castellanos, J., 350
Chen, J., 92
Chen, X., 191

David, N.G., 381
Deaton, R.J., 104, 138

Ehrenfeucht, A., 279

Feldkamp, U., 23
Freeland, S.J., 150

Gal, S., 191
Garzon, M.H., 117
Gast, F.-U., 33
Gifford, D.K., 300
Gonzalo, R., 350
Guo, B., 231

Hagiya, M., 104, 138, 308
Harju, T., 279
Hatnik, U., 1
Head, T., 191
Hinze, T., 1
Hiroto, Y., 213
Hug, H., 321
Hussini, S., 57

Jonoska, N., 70

Kari, L., 57
Kashiwamura, S., 14
Kawazoe, Y., 203
Khodor, J., 223, 300
Khodor, Y., 223
Kimbrough, S.O., 92
Kingsford, C., 231
Knight, T.F. Jr., 223

Kobayashi, S., 308
Konstantinidis, S., 57
Krishna, S.N., 360
Krithivasan, K., 290

LaBean, T.H., 231
Landweber, L.F., 150

Manca, V., 172
Margenstern, M., 340, 371
Martín-Vide, C., 340
Matoba, T., 213
Matsuura, N., 203
McCaskill. J.S., 33, 46
Mitchell, J.C., 258

Nakatsugawa, M., 14, 129
Nichols, M.J., 191
Noort, D. van, 33

Oehmen, C., 117
Ohuchi, A., 14, 129, 203

Păun, A., 329
Păun, G., 340
Penchovsky, R., 46
Pérez-Jiménez, M.J., 161
Petre, I., 279
Pirrung, M., 231

Rama, R., 360
Rana, V.S., 231
Rauhe, H., 23
Reif, J.H., 231
Rogozhin, Y., 371
Rose, J.A., 104, 138
Rozenberg, G., 279
Ruben, A.J., 150

Saghafi, S., 23
Saito, M., 70
Sakakibara, Y., 82
Sancho-Caparrini, F., 161
Schuler, R., 321
Shiba, T., 14, 129, 203
Simmel, F.C., 248
Sivasubramanyam, Y., 290

Sturm, M., 1
Subramanian, K.G., 381
Suyama, A., 104, 138

Tanaka, F., 129
Thomas, D.G., 381

Uejima, H., 308

Wakabayashi, K., 269
Wickham, G.S., 231

Wood, D.H., 92
Wu, D.-J., 92

Yamamoto, M., 14, 129, 203
Yamamura, M., 191, 213, 269
Yoichi, T., 182
Yurke, B., 248, 258

Zandron, C., 172

Lecture Notes in Computer Science

For information about Vols. 1–2275
please contact your bookseller or Springer-Verlag

Vol. 2276: A. Gelbukh (Ed.), Computational Linguistics and Intelligent Text Processing. Proceedings, 2002. XIII, 444 pages. 2002.

Vol. 2277: P. Callaghan, Z. Luo, J. McKinna, R. Pollack (Eds.), Types for Proofs and Programs. Proceedings, 2000. VIII, 243 pages. 2002.

Vol. 2278: J.A. Foster, E. Lutton, J. Miller, C. Ryan, A.G.B. Tettamanzi (Eds.), Genetic Programming. Proceedings, 2002. XI, 337 pages. 2002.

Vol. 2279: S. Cagnoni, J. Gottlieb, E. Hart, M. Middendorf, G.R. Raidl (Eds.), Applications of Evolutionary Computing. Proceedings, 2002. XIII, 344 pages. 2002.

Vol. 2280: J.P. Katoen, P. Stevens (Eds.), Tools and Algorithms for the Construction and Analysis of Systems. Proceedings, 2002. XIII, 482 pages. 2002.

Vol. 2281: S. Arikawa, A. Shinohara (Eds.), Progress in Discovery Science. XIV, 684 pages. 2002. (Subseries LNAI).

Vol. 2282: D. Ursino, Extraction and Exploitation of Intensional Knowledge from Heterogeneous Information Sources. XXVI, 289 pages. 2002.

Vol. 2283: T. Nipkow, L.C. Paulson, M. Wenzel, Isabelle/HOL. XIII, 218 pages. 2002.

Vol. 2284: T. Eiter, K.-D. Schewe (Eds.), Foundations of Information and Knowledge Systems. Proceedings, 2002. X, 289 pages. 2002.

Vol. 2285: H. Alt, A. Ferreira (Eds.), STACS 2002. Proceedings, 2002. XIV, 660 pages. 2002.

Vol. 2286: S. Rajsbaum (Ed.), LATIN 2002: Theoretical Informatics. Proceedings, 2002. XIII, 630 pages. 2002.

Vol. 2287: C.S. Jensen, K.G. Jeffery, J. Pokorny, Saltenis, E. Bertino, K. Böhm, M. Jarke (Eds.), Advances in Database Technology – EDBT 2002. Proceedings, 2002. XVI, 776 pages. 2002.

Vol. 2288: K. Kim (Ed.), Information Security and Cryptology – ICISC 2001. Proceedings, 2001. XIII, 457 pages. 2002.

Vol. 2289: C.J. Tomlin, M.R. Greenstreet (Eds.), Hybrid Systems: Computation and Control. Proceedings, 2002. XIII, 480 pages. 2002.

Vol. 2290: F. van der Linden (Ed.), Software Product-Family Engineering. Proceedings, 2001. X, 417 pages. 2002.

Vol. 2291: F. Crestani, M. Girolami, C.J. van Rijsbergen (Eds.), Advances in Information Retrieval. Proceedings, 2002. XIII, 363 pages. 2002.

Vol. 2292: G.B. Khosrovshahi, A. Shokoufandeh, A. Shokrollahi (Eds.), Theoretical Aspects of Computer Science. IX, 221 pages. 2002.

Vol. 2293: J. Renz, Qualitative Spatial Reasoning with Topological Information. XVI, 207 pages. 2002. (Subseries LNAI).

Vol. 2294: A. Cortesi (Ed.), Verification, Model Checking, and Abstract Interpretation. Proceedings, 2002. VIII, 331 pages. 2002.

Vol. 2295: W. Kuich, G. Rozenberg, A. Salomaa (Eds.), Developments in Language Theory. Proceedings, 2001. IX, 389 pages. 2002.

Vol. 2296: B. Dunin-Kȩplicz, E. Nawarecki (Eds.), From Theory to Practice in Multi-Agent Systems. Proceedings, 2001. IX, 341 pages. 2002. (Subseries LNAI).

Vol. 2297: R. Backhouse, R. Crole, J. Gibbons (Eds.), Algebraic and Coalgebraic Methods in the Mathematics of Program Construction. Proceedings, 2000. XIV, 387 pages. 2002.

Vol. 2298: I. Wachsmuth, T. Sowa (Eds.), Gesture and Language in Human-Computer Interaction. Proceedings, 2001. XI, 323 pages. 2002. (Subseries LNAI).

Vol. 2299: H. Schmeck, T. Ungerer, L. Wolf (Eds.), Trends in Network and Pervasive Computing – ARCS 2002. Proceedings, 2002. XIV, 287 pages. 2002.

Vol. 2300: W. Brauer, H. Ehrig, J. Karhumäki, A. Salomaa (Eds.), Formal and Natural Computing. XXXVI, 431 pages. 2002.

Vol. 2301: A. Braquelaire, J.-O. Lachaud, A. Vialard (Eds.), Discrete Geometry for Computer Imagery. Proceedings, 2002. XI, 439 pages. 2002.

Vol. 2302: C. Schulte, Programming Constraint Services. XII, 176 pages. 2002. (Subseries LNAI).

Vol. 2303: M. Nielsen, U. Engberg (Eds.), Foundations of Software Science and Computation Structures. Proceedings, 2002. XIII, 435 pages. 2002.

Vol. 2304: R.N. Horspool (Ed.), Compiler Construction. Proceedings, 2002. XI, 343 pages. 2002.

Vol. 2305: D. Le Métayer (Ed.), Programming Languages and Systems. Proceedings, 2002. XII, 331 pages. 2002.

Vol. 2306: R.-D. Kutsche, H. Weber (Eds.), Fundamental Approaches to Software Engineering. Proceedings, 2002. XIII, 341 pages. 2002.

Vol. 2307: C. Zhang, S. Zhang, Association Rule Mining. XII, 238 pages. 2002. (Subseries LNAI).

Vol. 2308: I.P. Vlahavas, C.D. Spyropoulos (Eds.), Methods and Applications of Artificial Intelligence. Proceedings, 2002. XIV, 514 pages. 2002. (Subseries LNAI).

Vol. 2309: A. Armando (Ed.), Frontiers of Combining Systems. Proceedings, 2002. VIII, 255 pages. 2002. (Subseries LNAI).

Vol. 2310: P. Collet, C. Fonlupt, J.-K. Hao, E. Lutton, M. Schoenauer (Eds.), Artificial Evolution. Proceedings, 2001. XI, 375 pages. 2002.

Vol. 2311: D. Bustard, W. Liu, R. Sterritt (Eds.), Soft-Ware 2002: Computing in an Imperfect World. Proceedings, 2002. XI, 359 pages. 2002.

Vol. 2312: T. Arts, M. Mohnen (Eds.), Implementation of Functional Languages. Proceedings, 2001. VII, 187 pages. 2002.

Vol. 2313: C.A. Coello Coello, A. de Albornoz, L.E. Sucar, O.Cairó Battistutti (Eds.), MICAI 2002: Advances in Artificial Intelligence. Proceedings, 2002. XIII, 548 pages. 2002. (Subseries LNAI).

Vol. 2314: S.-K. Chang, Z. Chen, S.-Y. Lee (Eds.), Recent Advances in Visual Information Systems. Proceedings, 2002. XI, 323 pages. 2002.

Vol. 2315: F. Arhab, C. Talcott (Eds.), Coordination Models and Languages. Proceedings, 2002. XI, 406 pages. 2002.

Vol. 2316: J. Domingo-Ferrer (Ed.), Inference Control in Statistical Databases. VIII, 231 pages. 2002.

Vol. 2317: M. Hegarty, B. Meyer, N. Hari Narayanan (Eds.), Diagrammatic Representation and Inference. Proceedings, 2002. XIV, 362 pages. 2002. (Subseries LNAI).

Vol. 2318: D. Bošnački, S. Leue (Eds.), Model Checking Software. Proceedings, 2002. X, 259 pages. 2002.

Vol. 2319: C. Gacek (Ed.), Software Reuse: Methods, Techniques, and Tools. Proceedings, 2002. XI, 353 pages. 2002.

Vol.2320: T. Sander (Ed.), Security and Privacy in Digital Rights Management. Proceedings, 2001. X, 245 pages. 2002.

Vol. 2322: V. Mařík, O. Stěpánková, H. Krautwurmová, M. Luck (Eds.), Multi-Agent Systems and Applications II. Proceedings, 2001. XII, 377 pages. 2002. (Subseries LNAI).

Vol. 2323: À. Frohner (Ed.), Object-Oriented Technology. Proceedings, 2001. IX, 225 pages. 2002.

Vol. 2324: T. Field, P.G. Harrison, J. Bradley, U. Harder (Eds.), Computer Performance Evaluation. Proceedings, 2002. XI, 349 pages. 2002.

Vol 2326: D. Grigoras, A. Nicolau, B. Toursel, B. Folliot (Eds.), Advanced Environments, Tools, and Applications for Cluster Computing. Proceedings, 2001. XIII, 321 pages. 2002.

Vol. 2327: H.P. Zima, K. Joe, M. Sato, Y. Seo, M. Shimasaki (Eds.), High Performance Computing. Proceedings, 2002. XV, 564 pages. 2002.

Vol. 2329: P.M.A. Sloot, C.J.K. Tan, J.J. Dongarra, A.G. Hoekstra (Eds.), Computational Science – ICCS 2002. Proceedings, Part I. XLI, 1095 pages. 2002.

Vol. 2330: P.M.A. Sloot, C.J.K. Tan, J.J. Dongarra, A.G. Hoekstra (Eds.), Computational Science – ICCS 2002. Proceedings, Part II. XLI, 1115 pages. 2002.

Vol. 2331: P.M.A. Sloot, C.J.K. Tan, J.J. Dongarra, A.G. Hoekstra (Eds.), Computational Science – ICCS 2002. Proceedings, Part III. XLI, 1227 pages. 2002.

Vol. 2332: L. Knudsen (Ed.), Advances in Cryptology – EUROCRYPT 2002. Proceedings, 2002. XII, 547 pages. 2002.

Vol. 2334: G. Carle, M. Zitterbart (Eds.), Protocols for High Speed Networks. Proceedings, 2002. X, 267 pages. 2002.

Vol. 2335: M. Butler, L. Petre, K. Sere (Eds.), Integrated Formal Methods. Proceedings, 2002. X, 401 pages. 2002.

Vol. 2336: M.-S. Chen, P.S. Yu, B. Liu (Eds.), Advances in Knowledge Discovery and Data Mining. Proceedings, 2002. XIII, 568 pages. 2002. (Subseries LNAI).

Vol. 2337: W.J. Cook, A.S. Schulz (Eds.), Integer Programming and Combinatorial Optimization. Proceedings, 2002. XI, 487 pages. 2002.

Vol. 2338: R. Cohen, B. Spencer (Eds.), Advances in Artificial Intelligence. Proceedings, 2002. X, 197 pages. 2002. (Subseries LNAI).

Vol. 2340: N. Jonoska, N.C. Seeman (Eds.), DNA Computing. Proceedings, 2001. XI, 392 pages. 2002.

Vol. 2342: I. Horrocks, J. Hendler (Eds.), The Semantic Web – ISCW 2002. Proceedings, 2002. XVI, 476 pages. 2002.

Vol. 2345: E. Gregori, M. Conti, A.T. Campbell, G. Omidyar, M. Zukerman (Eds.), NETWORKING 2002. Proceedings, 2002. XXVI, 1256 pages. 2002.

Vol. 2346: H. Unger, T., Böhme, A. Mikler (Eds.), Innovative Internet Computing Systems. Proceedings, 2002. VIII, 251 pages. 2002.

Vol. 2347: P. De Bra, P. Brusilovsky, R. Conejo (Eds.), Adaptive Hypermedia and Adaptive Web-Based Systems. Proceedings, 2002. XV, 615 pages. 2002.

Vol. 2348: A. Banks Pidduck, J. Mylopoulos, C.C. Woo, M. Tamer Ozsu (Eds.), Advanced Information Systems Engineering. Proceedings, 2002. XIV, 799 pages. 2002.

Vol. 2349: J. Kontio, R. Conradi (Eds.), Software Quality – ECSQ 2002. Proceedings, 2002. XIV, 363 pages. 2002.

Vol. 2350: A. Heyden, G. Sparr, M. Nielsen, P. Johansen (Eds.), Computer Vision – ECCV 2002. Proceedings, Part I. XXVIII, 817 pages. 2002.

Vol. 2351: A. Heyden, G. Sparr, M. Nielsen, P. Johansen (Eds.), Computer Vision – ECCV 2002. Proceedings, Part II. XXVIII, 903 pages. 2002.

Vol. 2352: A. Heyden, G. Sparr, M. Nielsen, P. Johansen (Eds.), Computer Vision – ECCV 2002. Proceedings, Part III. XXVIII, 919 pages. 2002.

Vol. 2353: A. Heyden, G. Sparr, M. Nielsen, P. Johansen (Eds.), Computer Vision – ECCV 2002. Proceedings, Part IV. XXVIII, 841 pages. 2002.

Vol. 2358: T. Hendtlass, M. Ali (Eds.), Developments in Applied Artificial Intelligence. Proceedings, 2002 XIII, 833 pages. 2002. (Subseries LNAI).

Vol. 2359: M. Tistarelli, J. Bigun, A.K. Jain (Eds.), Biometric Authentication. Proceedings, 2002. XII, 373 pages. 2002.

Vol. 2361: J. Blieberger, A. Strohmeier (Eds.), Reliable Software Technologies – Ada-Europe 2002. Proceedings, 2002 XIII, 367 pages. 2002.

Vol. 2363: S.A. Cerri, G. Gouardères, F. Paraguaçu (Eds.), Intelligent Tutoring Systems. Proceedings, 2002. XXVIII, 1016 pages. 2002.

Vol. 2367: J. Fagerholm, J. Haataja, J. Järvinen, M. Lyly. P. Råback, V. Savolainen (Eds.), Applied Parallel Computing. Proceedings, 2002. XIV, 612 pages. 2002.

Vol. 2374: B. Magnusson (Ed.), ECOOP 2002 – Object-Oriented Programming. XI, 637 pages. 2002.